Plant Breeding and Whole-system Crop Physiology

Improving Crop Maturity, Adaptation and Yield

PLANT BREEDING AND WHOLE-SYSTEM CROP PHYSIOLOGY

Improving Crop Maturity, Adaptation and Yield

D.H. Wallace

Emeritus Professor of Plant Breeding and Vegetable Crops
Plant Breeding and Biometry Department
Cornell University, Ithaca, New York, USA

and

Weikai Yan

Department of Crop Science
University of Guelph, Canada

CAB INTERNATIONAL

CAB INTERNATIONAL
Wallingford
Oxon OX10 8DE
UK

Tel: +44 (0)1491 832111
Fax: +44 (0)1491 833508
E-mail: cabi@cabi.org

CAB INTERNATIONAL
198 Madison Avenue
New York, NY 10016–4314
USA

Tel: 1 212 726 6490
Fax: 1 212 686 7993
E-mail: cabi@cabi.org

A catalogue record for this book is available from the British Library, London, UK.

Library of Congress Cataloging-in-Publication Data
Wallace, D. H. (Don H.)
Plant breeding and whole-system crop physiology : improving crop maturity, adaptation, and yield / D.H. Wallace and W. Yan.
p. cm.
Includes bibliographical references (p.) and index.
ISBN 0-85199-265-X (alk. paper)
1. Plant breeding. 2. Crops–Physiology. 3. Crop yields.
I. Yan, W. II. Title.
SB123.W328 1998 98-6441
631.5′2–dc21 CIP

ISBN 0 85199 265 X

Typeset in Times by Columns Design Ltd, Reading
Printed and bound in the UK at the University Press, Cambridge

Contents

Preface

This book's goal is to accelerate breeding for crop maturity, adaptation and yield, by elucidation of the complexity of the holistic plant system derived through whole-system research. The system of every successful crop cultivar consists of five essential components. They are the: (i) physiological components of accumulating yield; (ii) product quality; (iii) days to maturity; (iv) resistances against stresses; and (v) appropriate competitive capability against other plants. Superiority for all is desired but the system requires trade-offs.

The central-most concept for assisting breeding for yield is that all quantitative traits (as exemplified by (i) to (v)) collectively occupy a near constant-capacity plant system. This system's level is set by the environment's resources for and stresses against plant growth. The level is due to the always limited but different support of growth by every environment. This becomes a near constant level of intraplant metabolites for all genotypes of each species. Levels of all quantitative traits vary additionally, with negative or positive correlations among them, because each competes with all the others for the large proportion of the available metabolites every quantitative trait requires for its expression. If genetic instruction enlarges the requirement for one trait (system component), this reduces the part of the system capacity that can give expression to one or more other quantitative traits. The environmental resources and stresses and the genotype constitute the whole system. Superior cultivars require genetically directed and compromised levels among all the quantitative traits that maximize the system functioning by optimally using the environment's resources while countering its stresses.

Direct selection for yield has been possible during only late filial generations, after the segregates from crosses approach homozygosity. The yield can be fully accounted for by (i) the maximum potential yield per plant of each genotype and (ii) the negatively correlated reduction of that yield per plant as plant density is raised to maximize the yield per land area. Of several whole-system viewpoints, this one can facilitate effective selection for yield during the F_2 and other early segregating generations.

Another whole-system viewpoint says yield arises through three near fully integrated and penultimate physiological–genetic components: (i) the net accumulated biomass; (ii) its partitioning as measured by the harvest index; and (iii) the days to harvest maturity. A yet more holistic viewpoint follows. Genetic control over days to flowering and to maturity and the yield is modulated strongly by the temperature and photoperiod. Activities of three gene subsets fully account for the days to flowering and to maturity, and also for the yield. Collectively, the majority of all the genes implements the plant growth, and simultaneously implements an autonomous development of each plant through all its stages. The second subset is any photoperiod genes. The third is any vernalization genes. The Q_{10} effect establishes a lowest temperature that allows any activity of every gene, above which, progressively higher temperatures enlarge that gene's activity and its consequences. The days needed to complete the autonomous development through the majority of all the genes is shortened by each higher than the base temperature, this being the thermal time response. That same progressively higher temperature also enlarges the delay of this development caused by any vernalization photoperiod gene activity and/or by any photoperiod gene activity. The photoperiod and vernalization gene activities control whether the available metabolites are partitioned to growth of the reproductive or other organs that become the yield, thereby controlling also the harvest index and the yield per day. The yield per day becomes the rate of growth of the yield organs, and this rate determines the days to flowering or tuber initiation and the days to harvest maturity. With these three gene subsets controlling both growth and development, all control is through the system as a whole.

Whole-system research must begin with focus on a targeted system output, which is illustrated herein by both yield and days to flowering and their correlations with days to harvest maturity. Whole-system research requires comparison of the level of the system output with each of the smallest conceivable number of near fully integrated effects (processes) and recognition of the subsets of genes that fully account for the system output. Using ongoing yield trials to measure each of this smallest conceivable number of component processes for a range of genotypes grown across a range of environments is the cheapest possible way to enlarge whole-system understanding and thereby facilitate more efficient breeding for higher yield. Required also is statistical quantification of the genotype × environment interaction, followed by physiological–genetic interpretation of how that interaction control arises through the difference in genetic direction of each genotype and through the difference in modulation of that gene activity by each different environment.

Acknowledgements

The whole-systems insights presented in this book owe much to collaborative research conducted between D.H. Wallace at Cornell University, Ithaca, NY, USA and plant scientists and graduate students from or working in other countries of most continents of the world. Funding of support for collaborative international agricultural research is gratefully acknowledged.

Cornell University is acknowledged for its policy of supporting research to benefit international as well as local agricultural concerns. A Rockefeller Foundation and Ford Foundation funded post World War II educational development programme at the College of Agriculture of the University of the Philippines provided the senior author with his first international agricultural experience. This led to recognition of the very different agricultural problems and research approaches needed for different areas and societies of the world. Three subsequent years of funding from the Rockefeller Foundation supported research on the physiological genetics of yield of dry bean (*Phaseolus vulgaris* L.), including joint collaboration with Dr M.W. Adams at Michigan State University, and with the Centro Internacional de Agricultura Tropical (CIAT) in Colombia. The longest duration of supplemental support was through USAID funded Bean/Cowpea Collaborative Research Support (CRSP) Program, managed at Michigan State University. This funded 10 years of collaboration with Dr P.N. Masaya, and 2 more years with R. Rodriguez, the leaders of the Bean Research Program of the Instituto de Ciencias y Tecnologia Agricola in Guatemala. Three years of near daily discussion with Weikai Yan, from the Northwestern Agricultural University in China, currently at the University of Guelph, Canada, resulted in numerous preliminary versions and the finalized text of this book. G.Y. Miao of Shanxi Agricultural University in China graciously allowed usage of his unpulished data about photoperiodism and vernalization of wheat. Reviewers from several countries greatly improved the final text. USDA/OICD funded 2 months of collaboration during each of 3 years at the International Crops Research Institute for the Semi Arid Troics (ICRISAT) in India. Special thanks go to Y.S. Chauhan, C. Johansen and L. Singh of ICRISAT. They allowed their unpublished Yield System Analysis (YSA) and the Additive Main Effects and Multiplicative Interaction effects (AMMI) analysis of the major physiological genetic components of the process of accumulating yield of pigeonpea to be published within this book. This YSA/AMMI analysis of a multilocation/multigenotype pigeonpea yield trial presents the most complete application of all the new concepts within the book. The merit is interpretation of the genotype × environment interaction effects due to the genotype, plant habit, photoperiod, temperature, soil and irrigation on the adaptation, maturity, yield, and all of the major physiological genetic components of the process of accumulating the crop yield. Application of YSA/AMMI to yield trials illustrates a philosophy and methodology of basic whole-system research that, when applied to the yield trials already being conducted, inexpensively advances biological understanding of cultivar adaptation, maturity and crop yield.

List of Acronyms and Abbreviations

Abbreviations used in only a few adjacent paragraphs may not be in this list.

a, b, c, d	Unspecified generic terms for a first, second, third and fourth parameter within an equation.
α	The original sensitivity to LD before any vernalization; it can range from positive for spring genotypes through zero for typical winter genotypes to negative for the SLDP.
Age	The plant age at which vernalization is begun; it may be expressed as either the days after germination, or the number of leaves on the plant.
AMMI	Additive Main effects and Multiplicative Interaction effects analysis: a statistical procedure used to quantify the control due to each genotype and due to each environment through the genotype × environment interaction effect ($G \times E$) on the trait being studied.
ANOVA	An acronym for the analysis of variance statistical procedure.
AOTR	The above optimum temperature response.
β	The vernalization by photoperiod interaction coefficient. The relation between a and b is so defined that $a + b = 1$.
b	The increase in responsiveness to LD per unit increase in extent of the vernalization. This extent is measurable as $[v/V_{req}]$.
B	The potential number of leaves that can be eliminated by vernalization.
B0, B1, B2, B3	Each is a fitted parameter derived by equation 7.4a to quantify the parameters of equation 7.4.
BOTR	The below optimum temperature response.
°C	Degrees centigrade.
c	The modification by the plant age of the vernalization sensitivity.
C3	The C3 pathway.
C4	The C4 pathway.
CO_2	Carbon dioxide.
cv.	Cultivar.

CV	Coefficient of variation.
D_1 and D_2	The density of the plants per unit land area of genotype 1 and genotype 2 grown in a mixture.
D_{ab}	The apparent basic (shortest expressed) duration of the days to flowering under the applied photoperiod; it is expressed at the bottom of the U-shaped curve of the DTF in response to the temperature.
D_b	The basic (shortest expressible) duration of the DTF. It is identifiable and measurable only when there is no delay by either non-optimal photoperiod or non-optimal temperature. The D_b is a constant for a given genotype.
Dne	A photoperiod gene in pea.
D_{opt}	The optimum density of the genotype. That stand which will result in the highest crop yield.
DSI	Duration of spikelet initiation.
dt	The derivative of time.
DTF	Days to flowering.
1/DTF	The inverse or reciprocal of the DTF. It is the rate of the development to flowering.
DTFb	Days to flowering measured from the beginning of the duration of vernalization.
DTFe	Days to flowering measured from the end of the duration of vernalization.
DTFL	Days to flag leaf unfolding.
DTHM	Days to harvest maturity.
d*W*	The derivative of the dry weight.
DW	Dry weight.
E	Environment.
Ea	The efficiency of absorption of the solar energy by the leaves and other organs. Conversion to plant biomass requires either direct absorption of the energy by chlorophyll and consequent photosynthesis or a further transfer of the absorbed energy to the chlorophyll.
Ec	The efficiency of conversion of the incoming solar radiation to aerial biomass, i.e. to the accumulated aerial plant weight.
Ei	The efficiency of intercepting solar energy, i.e. the percentage of this light energy that is intercepted by the leaves instead of passing through the leaf canopy and being intercepted by the ground beneath the canopy.
Er	The percent of the intercepted solar energy reflected back to the sky.
Eu	The efficiency of converting the intercepted solar energy into aerial biomass. Calculation of the *Ec* requires that the biomass be measured as the energy it yields during its combustion.
Eul	The efficiency of converting the intercepted solar energy into aerial biomass – when the light energy reflected back from the leaf canopy has not been considered and accounted for.

EuII	The efficiency of converting the intercepted solar energy into aerial biomass – when the light energy reflected back from the leaf canopy has been measured and subtracted so that it is not incorporated into the calculation.
f	The days to flowering as symbolized in the approximately 300 published papers that have applied the additive linear rate model. In this book the days to flowering is always symbolized as DTF.
$F_2, F_3, \ldots F_8$	The second, third, fourth etc. to eighth filial generation of self-pollination after a cross.
FLN	Final leaf number.
G	Genotype.
GDD	Growing degree day(s).
G × E	Genotype × environment interaction.
GPT	The genotype × photoperiod × temperature interaction control over plant development. The term GPT either occurs as the GPT interaction or within the term GPTmodel, which quantifies the control over plant phenology by the genotype, temperature and the photoperiod.
h	Hour.
HI	The harvest index. It is the yield divided by the accumulated aerial biomass at harvest maturity.
Hr	An enhancer (high response) gene. It implements a higher activity of the photoperiod genes *Sn* and *Dne* of pea and *Ppd* of bean.
IPCA1	Interaction principal component axis 1, the value of which is the contribution by the indicated genotype or environment to the largest pattern of a genotype × environment interaction effect on the trait analysed by an AMMI analysis.
IPCA2	Interaction principal component axis 2, the value of which is the contribution by the indicated genotype or environment to the second largest pattern of a genotype × environment interaction effect on the trait analysed by an AMMI analysis.
K	The extinction coefficient for penetrance of light into the leaf canopy.
L	The number of leaves.
L	Leaf area.
L_1	Leaf area at sampling time 1.
L_2	Leaf area at sampling time 2.
L_0	The genetically established smallest final leaf number of a genotype.
L_{max}	The maximum FLN, which is expressed when no vernalization has been applied. This maximum is quantified by $L_{max} = L_0 + B$.
LAD	Leaf area duration. It is the summation over time of the functional leaf area.
LAI	Leaf area index. It is the ratio of the leaf area divided by the land area beneath the leaves.
LAR	Leaf area ratio. It is the leaf per unit of plant dry weight.
LD	Long daylength, i.e. long photoperiod.
LDP	Long-day plant. A plant which flowers earlier under a longer photoperiod.

LSDP — Long-short-day plant; a plant with dual and tandemly occurring phases of development, such that rapid development to flowering (short DTF) requires long daylength followed by short daylength.

NAR — Net assimilation rate; the dry weight gain per unit leaf area as measured by growth analysis.

N_b — The smallest number of nodes (or leaves) to the first flower, which is expressed only when there is no delay by vernalization- and photoperiod-gene activity.

NCE — Net CO_2 exchange rate.

NO_2 — The nitrate molecule.

NTF — Node to flower; the number of nodes or leaves (plus spikelets, for cereal crops) present at flowering.

P — The applied photoperiod.

Pc and P_c — The critical photoperiod; that daylength at which the photoperiod switches from delaying to allowing flowering. This concept is applied by the additive linear rate model and is symbolized as *Pc*. For the GPTmodel it was originally replaced by assumption of an optimum photoperiod. Use of P_{opt} can be replaced by use of the symbol P_c.

Pce — The ceiling photoperiod; that daylength at which a longer daylength for a short-day plant or a shorter daylength for a long-day plant ceases to reduce the rate of development to flowering, i.e. to enlarge the days to flowering. This concept is applied by the additive linear rate model.

P_{opt} — The genotype-established optimum photoperiod for development. It is an LD for a long-day plant and an SD for a short-day plant.

Ppd — The symbol for the dominant allele of a photoperiod gene. Also, a general symbol for multiple photoperiod genes as *Ppd1* and *Ppd2*.

P–T — Photoperiod–temperature combination(s).

$P \times T$ — The photoperiod × temperature interaction effect on plant development. It results from the activity of the photoperiod genes, the level of which depends on both the deviation of the applied photoperiod from the optimum photoperiod and the further modulation of that activity by the Q_{10} effect of the applied temperature.

Q_{10} — The Q_{10} effect from biochemistry. It is a measure of the further enlargement by each higher temperature of each enzyme implemented (gene-directed) biochemical activity.

QTL — Quantitative trait loci.

r — The daily rate of development towards flowering, or heading, or any other stages that relate to the transition from vegetative growth to the commitment of the apex to reproductive development.

r_0 — The maximum daily rate of development to flowering, i.e. the shortest expressible days to flowering. The r_0 is expressed autonomously, provided the development is not delayed by vernalization-gene activity and/or photoperiod-gene activity and/or by non-optimal temperature. Any delay by temperature other than through its modulations of the vernalization- and photoperiod-gene activities is through the effect on the autonomous development through the integrated activities of the majority of all other genes of the plant. These genes collectively implement growth with an attendant implementation of the autonomous capability to develop. This is the thermal effect on the autonomous development, which has been called the accumulated thermal time effect.

R_0 The maximum rate of leaf appearance (or of spikelet initiation), which occurs when the temperature is optimum.

R^2 The R^2 fit derived from the correlation between two sets of observations such as the model described or predicted versus the observed days to flowering.

RGR Relative growth rate, i.e. dry weight increase per unit of existing dry weight.

RLA The rate of leaf appearance.

RLGR The relative leaf growth rate. It is the increase in leaf area per unit of existing leaf area.

RLPI The rate of leaf primordia initiation.

RMSPD Root Mean Square Prediction Deviation.

RSI The rate of spikelet initiation.

RuDP Ribulose diphosphate carboxylase. The enzyme that catalyses CO_2 fixation in the C3 pathway of photosynthesis. For C3 plants it also catalyses photorespiration, i.e. release of O_2 in the light.

S Standard deviation.

S_{at} Apparent sensitivity of the days to flower to deviation of the applied temperature (T) from the optimum temperature (T_{opt}). The S_{at} is quantified by the steepness of the side(s) of the U-shaped response to temperature of the days to flowering.

SD Short day, i.e. short daylength or photoperiod.

SDP Short-day plant, a plant for which a SD (short photoperiod) promotes rapid development to the reproductive stage.

sla Specific leaf area, i.e. leaf area per unit of leaf weight.

SLDP Short-long-day plant; a plant with dual and tandemly occurring phases of development, such that rapid development to flowering (short DTF) is promoted by short daylength followed by long daylength.

Sn The symbol for a photoperiod gene of pea.

SN Spikelet number.

S_p Sensitivity of the genotype to the deviation of the applied photoperiod from the optimal photoperiod (P_{opt}). This sensitivity is established by the photoperiod gene(s).

SR The single-ridge stage.

S_t Sensitivity of the plant to deviation of the applied temperature from the optimum temperature (T_{opt}). The S_t is controlled collectively by all the majority of all the genes if the plant other than the photoperiod- and vernalization-genes.

S_v The vernalization-gene established sensitivity to vernalization. Quantitatively, it is the enlargement (delay) of the node or days to flower due to each rise of the temperature above the base (lowest) temperature that allows vernalization gene activity.

t_1 The initial sampling time 1.

t_2 The second sampling time 2.

T The applied temperature, i.e. the average temperature in which the plants are grown while developing to a specified developmental stage.

T_{aopt} The apparent optimum temperature. It is that applied temperature at which the days to flowering is smallest under the given applied photoperiod. It is a photoperiod-dependent parameter, i.e. it changes with each different photoperiod.

Tb The base temperature below which no developmental advance occurs but above which advance does occur.

T_{bp} The base (lowest) temperature that allows activity of the photoperiod gene(s).

T_{bv} The base (lowest) temperature that allows activity of the vernalization gene(s).

To The optimum temperature obtained by determining the smallest DTF across a range of temperatures but a single photoperiod. It is the calculated optimum temperature using the additive linear rate model, but it changes with each different applied *P*, and also with each different sensitivity of the genotype to the photoperiod. Consequently, unless the applied *P* is the P_{opt}, the observed *To* is only an apparent optimum temperature for development. The term *To* is never used in conjunction with the GPTmodel. This model's genotype-established true optimum temperature for development is symbolized as the T_{opt}.

T_{opt} The optimum temperature. It is that temperature which causes the smallest expressible days to flowering of the genotype. Identifying the T_{opt} requires that simultaneously the associated applied photoperiod is the optimal photoperiod (P_{opt}). The T_{opt} is collectively controlled by the majority of all the genes of the plant other than the photoperiod- and vernalization-genes.

TSI The terminal spikelet initiation stage.

v The duration in days of the low temperature applied to promote rapid plant development through the vernalization response. When $v \geq$ the vernalization requirement (V_{req}) the plant has been fully vernalized.

VIA The 'vernalization-insensitive age' of a plant that has a vernalization requirement. It may be measured as either the leaf age or as the number of days from the beginning of seed germination until the plant development is not accelerated by further low-temperature vernalization.

VLDP A vernalization-requiring long-day plant.

v/V_{req} The applied vernalization duration divided by the vernalization requirement. It is the percentage of the vernalization requirement that has been met.

V_{req} The duration in days of a low temperature required to cause expression of the smallest number of nodes and/or smallest number days to flowering. The V_{req} is established by the vernalization gene(s). It is exactly the shortest duration of vernalization needed to result in the full LD response of the given genotype.

vrn The symbol for a vernalization gene.

vrn1vrn2	The recessive alleles of vernalization genes 1 and 2.
Vrn1Vrn2	The dominant alleles of vernalization genes 1 and 2.
VSDP	Vernalization-short-day plant. A plant with dual and tandemly occurring phases of development. Its shortest expressible days to flowering requires low temperature vernalization followed by short daylength.
W	The accumulated aerial plant weight.
W_1	The accumulated aerial plant weight at sampling time 1.
W_2	The accumulated aerial plant weight at the next sampling time 2.
Y_1	The average yield per plant of genotype 1, genotype 2 and genotype 3 grown in a mixture.
Y_{mix}	The yield per plant of a genotype in a mixed stand of multiple genotypes.
Y_{pure}	The yield per plant of a single genotype in a pure stand, i.e. of a monoculture.
YSA	Yield system analysis. It is a procedure which measures the days to flowering, days to maturity, accumulated aerial biomass and yield of every genotype grown in a yield trial. Calculated from these measures are the days of seedfill, yield per day of seedfill, yield per day to harvest maturity, biomass per day to harvest maturity and the harvest index.
YSA/AMMI	A procedure that applies yield system analysis (YSA) to ongoing multiple location yield trials followed by application of AMMI analysis to each trait already quantified by the YSA. The AMMI analysis quantifies the effect through the genotype × environment interaction due to each genotype and due to each environment. The correlations between the levels of the traits measured by YSA quantify the system-established interconnections between the traits.
Z	The crop yield potential, i.e. the maximum crop yield of a given genotype under a given environment, when the density is optimum [$D_{opt} = (a_1 + b_1)/(2b_1)$].
Z_{max}	The maximum crop yield potential.
Σ	The summation of that which follows.

1 Shifts from Current Paradigms Suggested by Near-whole and Whole-system Research

The major conclusions and concepts within this book were derived through either near-whole-system, or whole-system research. The latter is defined in shift 25 of the 32 paradigm shifts presented below. Most conclusions of this book are summarized within this Chapter 1. Each conclusion is presented as a shift from a current paradigm. The shifts are a consequence of the whole-system research approach. These paradigm shifts introduce the conclusions as you begin reading this book. At the end of each paradigm shift the chapter(s) are listed within which that shift is derived and/or is reinforced. The introductory paragraph of each successive chapter indicates the paradigm shifts derived within that chapter. Each chapter begins with the knowledge base and background as the research it reports began. The research methods are then presented, followed by the results and conclusions. The conclusions often are seemingly or actually contradictory to the commonly accepted paradigms, because the near-whole-system or whole-system research has revealed insights that reductive research has failed to elucidate and/or to emphasize.

A concise summary of this book consists of reading all 32 paradigm shifts presented in this chapter. These paradigm shifts are all interconnected by the biological system that accumulates the plant yield. Each shift presents a slight to strongly different perspective about the interconnections between the parts of this yield system.

Before you continue to read Chapters 2 to 13, it is suggested you read the all inclusive book summary. It consists of reading Chapter 1, followed by reading the introductory paragraph and then the summary at the end of each successive chapter. The 32 paradigm shifts of Chapter 1 present evidence that whole-system research reveals insights that have not arisen through prior applied and/or reductive research. Many of the shifts will seem to be contrary to what you know to be the commonly accepted biological explanation. Reading first the successive chapter introductions and their summaries presents the successively derived shifts as simple conclusions, without dealing in full with the holistic network of interconnected tandem, parallel, feedback, feedforward and circular interconnections among the parts of the system. These complexities are dealt with within the chapters.

After reading the above described inclusive summary, continued readings of the successive entire chapters will bring out these complex interconnections that constitute the yield system. The understanding of these system-established interconnections will elucidate why whole-system research reveals insights that reductive research does not.

The reader should recognize that the network of tandem, parallel, feedback, feedforward and circular interconnections among parts of the system repeatedly returns our writing and your reading to components of the system already discussed. It is difficult to decide when enough is enough about these returns to previously discussed concepts. For those concepts you struggle with as paradigm shifts away from the well known 'facts', you will benefit from the repetition and altered viewpoint. We have tried to reduce excessive redundancy but undoubtedly have not fully succeeded.

1.1 Whole-system Research Will Complement Reductive Research

Current paradigm 1. *Reductive research is preferred and deserves more funding and support than whole-plant research because reductive research more rapidly provides definitive yes-versus-no answers to research hypotheses.* Reductive research has been preferred since shortly after World War II. It has become synonymous with the term basic research.

Paradigm 1 should be shifted to: *Basic whole-system plant research complements reductive research, by providing insights about complexities of the holistic plant system that reductive research cannot provide; therefore, whole-system research deserves more societal and financial support than it currently receives.* Need for this paradigm shift is argued in Chapter 2. Its complementation of reductive research is demonstrated by the results presented in all subsequent chapters.

1.2 Partitioning of the Photosynthate is Strongly Controlled by Photoperiod-gene Activity

Paradigms 2 to 13 all deal with multiple effects due to partitioning of the available photosynthate between the organs that become the yield versus the other plant organs.

Current paradigm 2. *Photoperiod directly controls the time of initiation of flower buds and thereby controls the time to flowering.*

Paradigm 2 should be expanded to: *A strong direct control over both the time of initiation of flower buds and the time to flowering arises through photoperiod-gene established negative control over partitioning of the net accumulated photosynthate toward the reproductive organs that become the yield versus toward the vegetative organs of the plant.* This is a control over competing relative sink strengths.

Depending on the genotype and environment, the photoperiod may control any or all stages of reproductive development (yield accumulation), including the development of both sexual organs and storage organs such as tubers, bulbs and roots. The need for this paradigm shift toward greater recognition of the consequences of partitioning is derived in Chapters 4 and 5. It is reinforced by the research results of Chapters 6, 7 and 9.

This paradigm shift has not occurred earlier for the following reasons. First, basic research on photoperiodism has been conducted with genotypes (plant species) deliberately selected to be extremely photoperiod sensitive, i.e. to have a qualitative photoperiod response. The focus has been on the biochemistry of photoperiodism. These most highly sensitive species (genotypes) were selected to avoid the variability of the quantitative response to photoperiod expressed by most crop species. Consequently, a far larger literature base exists for the qualitative photoperiod response (Salisbury, 1963; Vince-Prue, 1975; Salisbury and Ross, 1981; Bernier, 1988) than for the quantitatively variable photoperiod response many plant species express. This larger research base consists predominantly of reductive research directed toward understanding the molecular, and intracellular events of the earliest stages (until flower bud appearance) of the response to the photoperiod. Therefore, books such as those cited below result in reading perhaps 395 pages for which the criterion for response to photoperiod is the days to flower bud initiation. In contrast, just ten pages in Vince-Prue's (1975) final chapter, entitled 'Miscellaneous effects of daylength', are directed toward whole-system-research of photoperiodism. A consequence of the reductive research approach was recognized by Vince-Prue (1983).

> The emphasis of interest in the switch from vegetative to reproductive growth, often using 'model' systems which require only a single photoinductive cycle to effect the switch, has tended to obscure the fact that the primordia formed after one or a few inductive cycles frequently do not complete their development into fertile flowers. When induction is minimal (with only a few inductive cycles or daylengths near the critical), flowers often abort at an early stage or they may develop various degrees of leafiness of the flower.

Current paradigm 3. *Competition occurs between yield accumulation and continued vegetative growth.*

Paradigm 3 should be shifted to: *Depending on the species and the environment, competition between reproductive growth (yield accumulation) and concurrent vegetative growth can occur at any and/or all stage(s) of sexual and/or vegetative reproductive development.* Competition for photosynthate may begin before visibly evident flower bud initiation (or tuber or corm initiation). Only if prior competition has occurred, will the photoperiod gene activity control the time of initiation of the flower buds. The need for this paradigm shift toward greater recognition of the consequences of partitioning is derived in Chapters 4 and 5, and is supported by the modelling results of Chapters 6, 7 and 9.

Current paradigm 4. *The days to harvest maturity (DTHM) of a cultivar establishes its adaptation to the duration of the growing season at each site, and thereby strongly controls yield of the cultivar.*

Paradigm 4 is valid. But, it should be expanded to incorporate the control

by partitioning over the DTHM in addition to over the yield: *The proportion of the available photosynthate partitioned to the sexual (or vegetative) organs that become the yield controls the rate of growth of these organs. This partitioning and the leaf area (capacity for photosynthesis) established before flowering (Damisch and Wiberg, 1991) establish the rate of growth of these organs, which simultaneously is the rate of accumulation of the yield. A faster rate of growth of the organs that become the yield results in an earlier DTHM, thereby establishing the adaptation or maladaptation of each cultivar to the growing season duration of the site-season, and strongly controlling the yield of the cultivar.* The need for this paradigm shift toward greater understanding of the role and consequences of partitioning is derived in Chapter 4, detailed in Chapter 5, and is supported by the research results of Chapters 6, 7 and 9.

Current paradigm 5. *The days to flowering (DTF) and DTHM have no interconnection with the harvest index (HI), the latter being the endpoint measure of the partitioning.* This paradigm is invoked whenever selection is independently applied for the DTF and/or DTHM and the HI, and when this is intended to be indirect selection for higher yield.

Paradigm 5 is invalid since: *Both the DTF and DTHM are interconnected (i.e. negatively correlated) with the HI; these negative correlations arise because any photoperiod-gene activity strongly controls the partitioning, with consequent controls over the HI, DTF and the DTHM.* The need for this paradigm shift toward greater understanding of the role and consequences of partitioning is derived in Chapter 4, detailed in Chapter 5, and is supported by the research results of Chapters 6, 7 and 9.

Current paradigm 6. Control over competition for photosynthate between continued vegetative growth versus continued reproductive growth is independent of the genetic control over the DTF and DTHM.

Paradigm 6 is invalidated by the paradigm shifts 2, 3, 4 and 5. The need for this paradigm shift toward greater understanding of the role and consequences of partitioning is derived in Chapter 4, detailed in Chapter 5, and is supported by the research results of Chapters 6, 7 and 9.

Current paradigm 7. *Partitioning of photosynthate becomes important only subsequent to flowering or after tuber initiation, i.e. during actual accumulation of the yield.* This is a commonly used but usually unstated assumption.

Paradigm shift 7: *Paradigm 7 is invalidated by paradigm shift 3, which indicates that differential partitioning between the reproductive versus vegetative growth controls the time of the initiation (visible appearance) of flower buds.* That paradigm 7 is invalid is also supported by demonstration (Beveridge *et al.*, 1992) for sweet pea (a long-day species) that daylength alters the partitioning even before flower bud initiation. The need for this paradigm shift toward greater understanding of the role and consequences of partitioning is derived in Chapter 4, detailed in Chapter 5, and is reinforced by the research results of Chapters 6, 7 and 9.

Current paradigm 8. *The appropriate daylength induces flowering.*

Paradigm 8 should be shifted to: *Activity by a sensitive photoperiod-gene allele negatively controls and thereby inhibits or delays the autonomous capability possessed by every plant to develop to all its stages.* Expression of the autonomous capability to develop to flowering is demonstrated by: (i) lack of need by photoperiod insensitive genotypes for any specific photoperiod to cause their development to flowering; (ii) lack of control by photoperiod over the time of initiation of flower buds of photoperiod sensitive genotypes of bean in spite of control over time to flowering, and; (iii) lack of any photoperiod control over the DTF of peanut genotypes, when for peanut as for bean, (iv) the post-flowering photoperiod effect by an LD is a negative control over partitioning of photosynthate toward the reproductive organs (yield accumulation). The latter is indicated by lowering of the HI, enlargement of the DTHM and biomass, and reduction of the yield per DTHM (Witzenburger *et al.*, 1988). The need for this paradigm shift toward greater understanding of the role and consequences of partitioning is derived in Chapter 4, detailed in Chapter 5, and is reinforced by the research results of Chapters 6, 7 and 9.

1.3 Interacting Modulations of Development by Photoperiod and Temperature

Current paradigm 9. *The photoperiod and temperature are the two major environmental controls over the DTF.*

Paradigm 9 should be enlarged to: *Photoperiod and/or temperature are the major environmental modulators within each given environment of the cultivar's DTF, DTHM, HI, adaptation and yield of all crops.* Paradigm 9 exists because the 'qualitative response' to photoperiod and reductive research of photoperiodism have a much larger research and literature base (Salisbury, 1963; Vince-Prue, 1975; Salisbury and Ross, 1981; Bernier, 1988) than research directed toward understanding of the system-established interconnections (correlations) between (i) the photoperiod response and the crop yield and (ii) the yield and the thermal response to temperature measured as the accumulated growing degree days. The thermal-time response has been studied independently of the photoperiod response, and vice versa. Richards (1991) stated: 'Crop phenology is the most important single factor influencing yield and adaptation'. The need for this paradigm shift toward greater understanding of the role and consequences of partitioning is derived in Chapter 4, detailed in Chapter 5, and is supported by the research results of Chapters 6, 7 and 9.

Current paradigm 10. *The control by photoperiod over development can be studied independently of the controls by temperature.* This assumption is invoked every time the DTF in response to a range of photoperiods is compared at a single temperature, and whenever the thermal-time response is measured without reference to any simultaneous response to the photoperiod.

Paradigm 10 should be shifted to: *Any photoperiod-gene activity will always be enlarged by each higher temperature that remains below the optimal temperature for that gene activity.* This shift is merely application of the concept that every enzyme activity (and therefore genetically-directed process) is enlarged by the Q_{10} effect of each higher temperature that (i) does not degrade the functional protein of the enzyme by being higher than is optimal for that activity; but (ii) is higher than the lowest (base) temperature that will allow any of the photoperiod gene activity. The need for this paradigm shift is hinted at in Chapters 4 and 5 and detailed in Chapters 6 and 7.

Paradigm 11. *The control by temperature over the DTF is measurable using the growing degree days to flowering, while ignoring any photoperiod response.* This is applied by the theory for the growing degree day (thermal-time) concept.

Paradigm shift 11 is: *Accurate quantification of the control by temperature over the DTF requires quantification of the Q_{10} effect on the photoperiod-gene activity, and quantification of the simultaneous Q_{10} effect known as the thermal-time response.* The need for this paradigm shift is hinted at in Chapters 4 and Chapter 5, detailed in Chapters 6 and 7, and is reinforced in Chapter 9.

Current paradigm 12. *The DTF is controlled by the photoperiod genes and a few other maturity genes such as vernalization genes.*

Paradigm shift 12 is: *The gene action that implements expression of the DTF and all stages of development includes (i) the integrated activity of the majority of all the genes of the plant, and (ii) a negative control over this integrated activity by either any photoperiod and/or by any vernalization gene activity.* The majority of all the genes collectively implement (i) the plant growth, and (ii) the attendant autonomous capability to develop mentioned in paradigm shift 8 above. The control over the DTF by the photoperiod genes (or by the vernalization genes) becomes a strong and obvious control – because this activity due to one or a few photoperiod or vernalization gene(s) so strongly inhibits the autonomous development. The need for this paradigm shift is derived in Chapters 6 and 7. In Chapter 7 this shift is validated by an often 80% to 99% effectiveness of the genotype × photoperiod × temperature interaction (equation 7.4 for the GPTmodel) to predict the DTF.

Paradigm 13. *There is no base (lowest) temperature for photoperiod control over the DTF comparable to the base temperature needed to properly quantify the accumulated growing degree days for prediction of the DTF.*

Paradigm shift 13 is: *The photoperiod-gene activity is prevented completely by a sufficiently low temperature.* As for paradigm shifts 10 and 11, this shift is based on application of the Q_{10} concept that every enzyme activity (and therefore every genetically-directed process) is enlarged by the Q_{10} of a higher temperature. The Q_{10} concept also indicates: each lower temperature will reduce the photoperiod-gene activity (and all other gene activities) until it is totally prevented by a temperature below the base (lowest) temperature that allows any activity of the gene. The need for this paradigm shift is derived in Chapters 6 and 7. That the photoperiod-gene activity is reduced by low temperature is reinforced for two crops in Chapter 9.

1.4 Shifts of Paradigm Based on the Vernalization Response to Temperature

Current paradigm 14. *Vernalization-gene activity occurs at low temperature.* This paradigm usually is not stated specifically. But, it is implied by statements that low temperature applied to imbibed seeds and/or seedlings results in earlier flowering due to the genotype of the vernalization gene(s) and due to fulfilment of part or all of the vernalization requirement.

Paradigm 14 should be shifted to: *Low temperature reduces vernalization-gene activity, and if sufficiently low totally prevents all the activity. Prevention of the* *vernalization-gene activity eliminates the delay of the autonomous development that will occur and will be enlarged by each progressively higher than the lowest (base) temperature that allows any vernalization-gene activity.* This paradigm shift aligns description of vernalization-gene activity with the concept that all enzyme-implemented (gene-activity directed) processes are enlarged by the Q_{10} effect of a higher temperature that (i) is above the lowest (base) temperature that allows any of the gene activity, but (ii) does not degrade the functional protein of the enzyme by being higher than optimal for the gene activity. The need for this paradigm shift is derived in Chapter 6 and is detailed in Chapter 8.

Current paradigm 15: *Control by vernalization over the DTF should be measured from the end of the low-temperature treatment.*

Paradigm 15 should be shifted to: *Control by vernalization over the DTF should be measured from the beginning of the low-temperature treatment.* This will facilitate observation of (i) the first observable evidence of the thermal response to low temperature. That response is an always functioning Q_{10} caused delay of the expressed DTF in response to low temperatures like those used for vernalization. This first evidence is any delay of the DTF with further continuation of the low-temperature for vernalization. This first observable delay of the DTF by the low temperature indicates that (ii) the vernalization requirement has been completed. The vernalization response (change toward earlier DTF) becomes gradually smaller with each initial extension of the low-temperature vernalization. Eventually, further continuation of the low temperature delays the DTF through the thermal response, rather than continuing to accelerate it through the vernalization response. The time duration of the low temperature at which the thermal-response-caused delay of the DTF becomes evident because it is larger than the acceleration of the DTF by the vernalization response indicates the completion of the required vernalization duration. This endpoint of the vernalization requirement can be approximated but never accurately determined when, as is the common practice, the DTF is measured from the end of the low-temperature vernalization. The need for this paradigm shift is derived and detailed in Chapter 8.

Current paradigm 16. *Any reduction by low temperature of the DTF that can be replaced (substituted for) by a long or short photoperiod is a vernalization response.* This paradigm has given rise to the terms short-day vernalization and long-day vernalization.

Paradigm shift 16 is: *Any change toward earlier DTF due to low temperature that can be replaced by a photoperiod is due to a photoperiod-gene activity that the low-temperature reduces to totally prevents; it is not due to vernalization-gene activity because the activity of these genes is modulated by only the temperature.* The need for this paradigm shift is derived in Chapter 6 and detailed in Chapter 8.

Current paradigm 17. *The vernalization response does not interact directly with the photoperiod response.* This becomes an implied assumption whenever either the vernalization control or any post-vernalization control over the DTF is quantified without reference to the other.

Paradigm 17 should be shifted to: *Any vernalization response, the always-functioning thermal-time response, and any photoperiod response always jointly and interactively control the expressed DTF.* This conclusion is forced by understanding, from paradigm shift 12, that all the genes of the plant collectively control the DTF. The need for this paradigm shift is derived in Chapter 6 and is detailed in Chapter 8.

Current paradigm 18. *The original hypothesis of phasic development is invalid.*

Paradigm 18 should be shifted to: *The original hypothesis of phasic development needs to be updated to recognize the following. For wheat genotypes having a vernalization requirement, and an initial short-day response followed by a subsequent long-day response, the vernalization-gene activity mediates a gradual change from the initial SD photoperiod response to expression of the LD photoperiod response.* This paradigm shift is derived in Chapter 6 and detailed in Chapter 8.

1.5 Shifts of Paradigm Based on Yield System Analysis and Control by the Genotype × Environment Interaction

A yield system analysis (YSA) consists of measuring in yield trials the levels of traits 1 to 4 below. From their levels, the levels of traits 5 to 9 are calculated.

1. Days to flowering	Time used for development to flowering	
2. Days to harvest maturity	Time used to develop to harvest maturity (*major component* 3)	
3. Aerial biomass	The overall photosynthetic efficiency (*major component* 1)	
4. Yield	The economically important output (the system output)	
5. Days of seed fill	Time used for actual yield accumulation	
6. Yield day^{-1} to maturity	Efficiency of yield accumulation	The rate of partitioning
7. Yield day^{-1} of seedfill	Efficiency of yield accumulation	
8. Biomass day^{-1}	Efficiency of photosynthesis	
9. Harvest index	Endpoint efficiency of partitioning to yield (*major component* 2).	

Application of the YSA procedure fulfils the criterion for whole-system research presented in paradigm shift 25, because traits 2, 3 and 9 constitute the smallest number of component processes the accumulation of crop yield can be divided into, while accounting in full for the yield which is the system output. This makes traits 2, 3 and 9 the three major components of the holistic process of accumulating the crop yield. The other traits measured by the yield system analysis are the yield *per se* and some subcomponents of the component traits 2, 3 and 9.

Current paradigm 19. *Phenology provides the major control over adaptation and yield.* This is a message of all the paradigm shifts 2 to 18 and of most that follow. But, except for paradigm shift 17, which states that the ever-present thermal response and any photoperiod response and/or vernalization response interactively and jointly control the adaptation and yield, each prior shift has tended to emphasize either: (i) only the combined photoperiod and temperature modulations of the photoperiod-gene activity, or (ii) only this photoperiod response and its control over the DTF, HI and yield, or (iii) only the controls by the photoperiod response interacting with the thermal response, or (iv) only the vernalization response interacting with the thermal response.

The shift toward the enlarged physiological genetic interpretation needed for paradigm 19 is that: *The holistic control over phenology and then yield by the photoperiod and the temperature involves three distinct genotype × environment (G × E) interaction sources. Each is due to one of three distinct subsets of genes. The three G × E interactions are: (i) any genotype × photoperiod × temperature (GPT) interaction caused delay of the autonomous development due to photoperiod-gene activity; (ii) the always-functioning genotype × temperature interaction of the thermal time control over the autonomous advance of development, which arises through the integrated activity of the majority of all the genes; and (iii) any genotype × temperature interaction due to the vernalization genes and their implementation at high temperatures of a delay of the autonomous development.* These three *G* × *E* interactions jointly establish the holistic *G* × *E* interaction control over the DTF, adaptation and yield. These three simultaneous controls over the DTF should receive far greater whole-system research and attention than they have received previously. This paradigm shift toward the necessity to consider all three subsets of gene action and control was indicated first in Chapters 6. It is derived in theoretical terms in Chapter 8.

Current paradigm 20. *The photoperiod response and thermal-time responses can only be quantified separately and independently.* (This is another viewpoint about Paradigm 11.)

Paradigm 20 calls for shift to: *Every expressed DTF arises through the holistic genotype × photoperiod × temperature (GPT) interaction control over the DTF.* The control by temperature is predominant to that by the photoperiod because the temperature modulates the level of activity of every gene. But, this control by the temperature is less obvious than the control by the photoperiod, because one or a few photoperiod genes direct all the response to the photoperiod, including the level of the modulation of this photoperiod-gene control by the temperature. The predominant control by the temperature – interacting with strongly obvious control also by the photoperiod – was verified by measurement of the DTF during yield system analysis (Chapter 9). This was followed by quantification of the genetic control over the *G* × *E* interaction effect on the DTF and the genetic control over the *G* × *E* interaction effect on the growing degree days (GDDs) of both pigeonpea and bean. These two species have no vernalization genes, and therefore no control over their DTF by a vernalization requirement. The largest genetically-directed *G* × *E* interaction control over the DTF and the largest-genetically directed *G* × *E* interaction control over the GDDs due to the 30 genotypes of pigeonpea were quantified across 10 photoperiod-temperature combinations by an Additive Main effects and Multiplicative Interaction effects (AMMI) analysis. The directions by the genotypes over the largest *G* × *E* interaction control over the DTF and those over the GDDs had correlation with each other of 0.98 and correlations no smaller with both the DTF and the GDDs to flowering. These correlations were across both the photoperiod insensitive and the photoperiod sensitive cultivars. These same correlations were not smaller than 0.97 across ten bean cultivars of both photoperiod insensitive and sensitive genotype grown in five temperatures combined with one photoperiod. This paradigm shift is partially derived in Chapters 4 and 5, further derived in Chapters 6, 7 and 8, and is shown most thoroughly in Chapter 9.

Current paradigm 21. *Use of the accumulated GDDs (i.e. of thermal-time) to indicate the expected days from planting to flowering (or other stages) is valid only when each higher temperature causes linear change toward earlier DTF.*
Paradigm 21 requires a shift to: *The accumulated GDDs (thermal-time) will indicate the DTF of all genotypes, regardless of whether each higher temperature causes change toward earlier DTF as occurs for the photoperiod insensitive genotypes, or causes change toward later DTF as occurs at the higher temperatures for the photoperiod sensitive genotypes.* This paradigm shift is required by the near perfect correlations presented in paradigm shift 20. This shift is derived in Chapter 9.
Current paradigm 22. *It is too time consuming and expensive to measure the aerial biomass in ongoing yield trials.*
Paradigm 22 requires a shift to: *Measuring the DTF, DTHM, aerial biomass and yield at harvest maturity of the genotypes compared in multilocation yield trials that are already being conducted is the cheapest possible way to advance understanding through whole-system research about the interacting genetic and environmental controls over cultivar adaptation, the accumulation of crop yields, and the control by the holistic G × E interactions.* Selection for higher biomass *per se* will result in longer DTHM than optimal for the environment. To avoid the indirect selection toward a longer and maladapted DTHM, the selection should be for biomass per day. The advances that can be gained in understanding of the yield system; the interconnections among its components; their genetic direction, and; the modulations of the levels of implementation of this direction by the environment are derived in Chapters 4 and 5. These advances are most fully shown in Chapter 9.

1.6 Shifts of Paradigm Based on a Hypothesized Constant-capacity System

Current paradigm 23. *The resources of the environment strongly modulate and determine the level of all quantitatively variable traits, those traits that are controlled by many genes.*
Paradigm shift 23 expands both the environmental and genetic viewpoints about quantitative traits as follows: *First, the resources available from the environment versus the stresses of each environment establish a potential for small, medium to large plant growth. Second, this potential becomes a continuum of a small to large level of the available intraplant metabolites and this level constitutes a near-constant-capacity system for all the genotypes of the species. Third, whatever the available quantity of the intraplant metabolites, the expression of every quantitative trait requires a large proportion of these metabolites. Fourth, this forced sharing within the plant system of the available metabolites accounts for much of the control by many genes over each of all the quantitative traits. Fifth, therefore, the interconnections indicated by the recognized biochemical pathways of each physiological process account for only a previously most obvious part of the control by many genes over the levels of the quantitative traits. Sixth, necessity to share the available metabolites causes the genetically-directed level of every quantitatively variable trait to competitively and simultaneously partially control the level of all the other quantitatively variable traits. Seventh, the consequences are system-established interconnections measurable as a negative or else positive correlation between the expressed levels of most quantitatively variable*

traits. The bases for this paradigm shift are derived most thoroughly in Chapter 11, but existence of the system-established interconnections is derived and illustrated within every chapter.

1.7 Shifts of Paradigm Relative to Whole-system Versus Reductive Research

Current paradigm 24. *Whole-system research should be minimized, because of the longer time it requires to obtain statistically-significant and definitive yes-versus-no answers to research hypotheses.* This is an alternative way of stating paradigm 1.

Paradigm 24 should be shifted to: *Whole-system research may require longer time to get definitive answers to research hypotheses, but it reveals insights about the interconnections between components of the system that will not be revealed by reductive research. Demonstrated throughout this book is that whole-system research and reductive research are complementary; each reveals insight that the other cannot.* This paradigm shift is derived in Chapters 2 to 11. The shift is detailed most thoroughly in Chapters 9, 10 and 11.

Current paradigm 25. *The best way to achieve full understanding of the functioning of a complex biological system is to conduct reductive research on each of all its many components, followed by fitting all the resulting findings together to explain the whole system.* The current great predominance of reductive research applies this paradigm.

Paradigm 25 should be complemented by a shift to: *Continued advance of understanding of the functioning of a complex biological system must also involve whole system research, which is defined as follows. Begin with focus on the system output. Then, divide the holistic biological process of achieving that system output into the smallest conceivable number of near-fully integrated genetically-directed processes it can be divided among, while accounting in full for that system output. Next, each of this smallest number of near-fully integrated processes should be quantified across multiple genotypes and also multiple environments. Then, the negative or positive correlations between the levels of all pairs among these components, and between the level of each and the level of the system output should be quantified across (i) all the compared genotypes and also across (ii) all the environments. Respectively, (i) and (ii) will quantify the system-established interconnections among the system's components and traits that are due to variation in the genetic instruction and those due to variation in the environmental modulations of that instruction. Finally, the G × E interaction control over each component and the system output should be quantified.* This is the definition for whole-system research presented in Chapter 2. This systems thinking and definition for whole-system research originated from the research results presented in Chapters 3, 4 and 5. Application of this definition is illustrated in all subsequent chapters.

The system output focused on most within this book is the accumulation of yield, and its three major components are: (i) the net accumulated biomass, (ii) the partitioning of this biomass as measured by the HI, and (iii) the time used (needed) to accumulate the biomass and to partition it, which is the DTHM. The second most focused on system output is the DTF. Its major components are its control by just three subsets of genes: (i) the majority of all the genes of the plant and their implementation of the autonomous capability to develop to all stages; (ii) any photoperiod gene(s) and their inhibition at sufficiently high temperature of this autonomous development; and (iii) any vernalization gene(s) and their inhibition also at sufficiently high temperature of the autonomous development.

Current paradigm 26. *Statistical analysis of control over a trait as being due to only (i) the holistic genetic instructions of the genotype, and (ii) the holistic environment does not specify the controlling factors precisely enough for meaningful interpretation of the control by the G × E interaction.*

Paradigm 26 should be shifted to: *Statistical analysis of control over a trait as being due to only (i) the holistic genotype and (ii) the holistic environment will implement the requirement for whole-system research given in paradigm shift 25 that the holistic process of achieving the system output be divided into the smallest number of near-fully integrated component processes that account for the system output.* Such statistical analysis facilitates superior interpretation of the genetic instruction and meaningful interpretation of the *G* × *E* interaction control – **if the researcher plots on to a biplot graph the statistically quantified**: (i) additive control due to each distinct genotype against (ii) the also quantified multiplicative control due to the same genotype over the *G* × *E* interaction, and (iii) the additive control due to each different environment against (iv) the multiplicative control due to the same environment over the *G* × *E* interaction. **Then, the researcher must also:** (v) label each point on the graph that represents a holistic genotype for a gene or allele, or for an expected level of the genetic instruction that the researcher hypothesizes to direct the expressed phenotype (system output). And, the researcher also (vi) must label each point representing a holistic environment for its level of the factor(s) expected to modulate the level of the gene action that directs and implements the phenotypic level. Finally, (vii) the researcher must interpret whether his/her hypothesized genetic and environmental controls do or do not account for the statistically quantified *G* × *E* interaction. Gain in understanding of the *G* × *E* interaction derived by this procedure is detailed most thoroughly in Chapter 9.

Current paradigm 27. *As commonly presented in the literature, the term genotype × environment (G × E) interaction is a statistically defined terminology. And, statistically quantified G × E is always attended by the two main (additive) components and the two interaction (the multipicative) components like those numbered (i) to (iv) in the previous paragraph.*

Paradigm 27 should be accepted but the physiological-genetic viewpoint should be recognized also. It is that: (i) *Biologically, every expressed phenotype results solely (100%) from a genotype × environment (G × E) interaction.* (ii) *It is impossible for any gene action to occur unless the environment provides all the resources needed for that gene activity to occur.* (iii) *Every environmental influence is implemented by one to many gene activities.* These shifts are needed because (from Fig. 1.1 and section 9.3), all phenotypes result from a biological system that consists of (i) a subset of the genetic directions and (ii) the level of the implementation of the genetic direction as modulated (allowed) by a subset of the factors of the environment. From this biological viewpoint, the statistical main (additive) effect due to the genotype and main (additive) effect due to the environment listed in paradigm shift 26 are artefacts of the statistical analysis procedure used

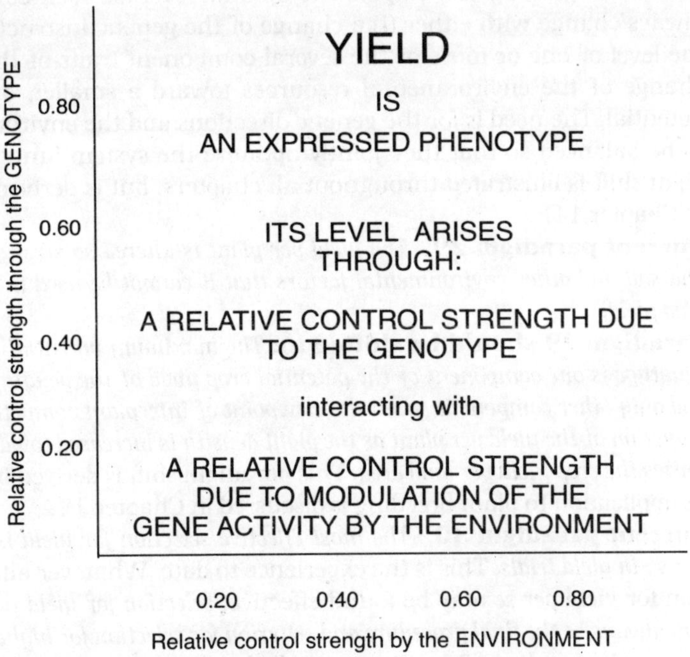

Fig. 1.1. The genotype and environment constitute the smallest number of components into which the physiological genetics of yield or of any other system output can be subdivided. There is a tendency to think two components are too few to be informative, because both are too broad and non-specific to give precise understanding of the biology through which the system output is achieved. The results throughout this book, and best described in section 9.4, show support of whole-system research through assignment for statistical analysis of all control to just these two major components of the holistic control over any phenotype.

to quantify the $G \times E$. The holistic physiological-genetic basis for the expression of every trait and of its level is 100% $G \times E$ interaction. Section 9.4 details how the joint applications of paradigm shifts 26 and 27 apply to a full extent to whole-system research as it is defined in paradigm shift 25.

1.8 Shifts of Paradigm Relative to Breeding for Adaptation and Yield

Current paradigm 28. *Select for a higher level of the component trait, when that selection is expected to result in indirect selection toward a higher level of a system output such as yield.*

Paradigm 28 requires a shift to: *Due to the necessity for all quantitatively variable traits to share the available metabolites (paradigm shift 23), selection for a physiological-genetic component trait intended to be indirect selection toward a higher level of a system output must be toward that (equally likely) higher or lower level of the component trait which will optimize the system functioning within the given environment. If the indirect selection is toward either lower or higher than the optimum level of the component trait, the selection will reduce the system functioning and lower rather than* *raise the targeted system output.* The optimal level for each component trait will always change with either (i) a change of the genetic instruction (genotype) for the level of one or more of the several component traits of the system, or (ii) a change of the environmental resources toward a smaller, or a larger growth potential. The need is for the genetic directions and the environmental resources to be balanced so that they jointly optimize the system functioning. This paradigm shift is illustrated throughout all chapters, but is derived most thoroughly in Chapter 11.

Current paradigm 29. *The yield per plant is altered so strongly by plant density and soil and other environmental factors that it cannot be used in breeding for higher crop yield.*

Paradigm 29 should be shifted to: *The maximum potential yield per plant of a genotype is one component of the potential crop yield of the genotype. And, the second and only other component (from the viewpoint of interplant competition) is the relative reduction of the yield per plant as the plant density is increased toward that which maximizes the crop yield per land area.* This paradigm shift is derived in Chapter 10, and its application to plant breeding is discussed in Chapter 12.

Current paradigm 30. *The most effective selection for yield is selection for yield* per se *in yield trials.* This is the experience to date. Whatever alternatives to selection for yield *per se* may be found effective, *selection for yield* per se *in yield trials will always be the final procedure and criterion for selection for higher yield.*

Current paradigm 31. *The way to find and use genetic diversity is to incorporate exotic germplasm.*

Paradigm shift 31 is: *Another way to find and rapidly use genetic diversity is to recurrently intercross parents (even related ones) that collectively span the range of the observed levels for each of that smallest number of (major) component processes that account in full for the expressed levels of a system output. Yield is the exemplary system output.*

1.9 Shifts of Paradigm Relative to the Research Agenda

Current paradigm 32. The individual scientists in collaboration with the scientific community are fully capable of planning their basic research agenda.

Paradigm shift 32 is: The societal and cultural community as a whole will not continue to allow the scientist, scientific community and agricultural community to independently and completely establish their research agendas. This is a conclusion of Chapter 13.

Chapter 1 Summary

This chapter lists 32 original paradigms, and the shifts from them suggested by the interpretations and conclusions from the near-whole to whole-system research results detailed in Chapters 2 to 13. Each paradigm shift is followed by designation of the chapter(s) within which that shift is derived, detailed and reinforced.

The Beginning of Systems Thinking about Breeding for Yield

2

Within this chapter, the origin of paradigm 1 is described. It is a shift from applied research on whole plants to reductive research. It occurred after World War II. Reductive research more easily achieves yes-versus-no conclusions from research hypotheses applied to less than whole systems. Recent calls by plant scientists for shift back to (i) more whole-system research and (ii) application of optimization theory are cited. Most of these calls have not suggested a philosophy or methodology for how to conduct the needed whole-system research on plants. A definition for whole-system research is proposed in this chapter. It was summarized in Chapter 1 as paradigm shift 25. This definition arose from the recognition derived in Chapters 3, 4 and 5 of system-established interconnections measurable as the negative or else positive correlation between the levels of all of the major component processes of accumulating crop yield

2.1 Crop Yield Trials Have Focused on the Economic Goal

Maximization of yield is a goal of all who grow crops, whether the grower's intent is food or fibre for direct use by their family, or is to sell the crop yield to obtain economic income to purchase foods and fibres grown by others, or is to buy prepared foods or manufactured consumer products. Any of these intents is an economic goal. During the 20th century, yield trials and statistical comparison of the yields of the available cultivars of each crop species have become the primary tool used to identify the highest yielding cultivars available for each geographical region. Cultivars that give the highest yields are recommended to farmers. They are often but not always accepted and grown. Farmers, processors and consumers reject the highest yielding cultivars if unacceptable taste, colour, shape, protein content, milling or other defects of quality lower the economic merit of the yield *per se*, or if for them superior economic merit exists for some part of the plant other than the seeds or tubers.

Yield trials conducted with the sole intent of identifying the highest yielding cultivars and recommending them for use by farmers represent applied research. The focus is economic, even though it is recognized the yield arises through a biological system. Every yield trial with this sole focus ignores opportunity simultaneously to use the yield trial for basic whole-system research directed toward advancing understanding of plant biology. Every yield trial, in addition to fulfilling its economic goal, can be used to quantify the variations of the major components of the complex biological system that accumulates the different yield of each tested cultivar, and/or to quantify the variations of these components that result in different yields for the same cultivar grown in the environments of different locations, years or growing seasons.

2.2 Recent Research on Crop Yield has Bypassed Whole-system Research

Because of the large variation in crop yield due to the genotype × environment ($G \times E$) interactions, and because the biological system as a whole is extremely complex, beginning after World War II, most investigations directed toward elucidating the processes that accumulate crop yield have adopted reductive research. The term reductive indicates research about less than the whole system. It is research on one or a few components of the complex biological system. Accumulation of crop yield requires functioning of the holistic plant system. This biology involves partial control over the system's output which is the yield, by each of most – if not all – of the thousands of genes of the plant. In general, each gene or subset of a few genes codes for a specific enzyme. Multiple enzymatic steps (gene actions) are integrated into a physiological process such as photosynthesis or dark respiration, etc. These physiological processes are further integrated toward accumulating the crop yield. For example, dark respiration fulfils a role that is essential to plant growth, yet is virtually the reverse of the process of photosynthesis. Other subsets of genes implement another reverse process called photorespiration. Yet additional gene actions implement nitrogen uptake and/or its reduction or synthesis into amino acids, proteins and RNA or DNA, or mineral uptake and the transport thereof. The level or intensity of each gene action is modulated by the variation of some to many environmental factors. Specific minerals are required for given enzymatic activities to occur. A higher temperature accelerates the activity of most if not all genes. The duration of the light, i.e. the photoperiod, modulates the activity of a relatively few 'photoperiod genes'. The light intensity, i.e. radiant energy level, and/or the light quality also modulate the activity of some genes.

Reductive research has been adopted, because complexity of the whole system is considered too great to allow informative results from direct study of it. The reductive research is to bypass the quantitatively variable, complex, and therefore difficult to interpret results that arise from integration of the many as yet not-understood genetic instructions within the system, and from the modulations of the levels of all these gene activities by the environmental factors. Any change of genetic instruction from one cultivar to another alters the level of the system output (the yield) through gene × gene interaction effects. And, any change of either the genotype or the environment alters the system output through differential environmental modulations of the overall gene activity and the resultant $G \times E$ interaction modulations of the yield.

2.3 Currently, Reductive Research is Equated with the Term Basic Research

Greater repeatability of conclusions results because the reduction isolates the specific component of the system under study from the always interacting (Fig. 2.1, compare Fig. 1.1 and Fig. 11.1) influences by other components of the holistic system. It reduces to eliminate the quantitatively variable results that will arise through whole-system research, because of change of genotype between every cultivar, and unwanted or unrecognized change of some environmental factor(s). Reductive research facilitates more precise control of the relevant environmental and genetic factors that influence the small part of the system being studied. The problem is: difficulty for researchers to integrate their reductively derived interpretations toward understanding the system as a whole. Reductive research equates with what now is called basic research.

Levels of reductive research range from near holistic to highly reductive. Research is near holistic if the traits measured result from all but one or two of all the processes integral to expression of traits at the output end of the system, such as yield. Research is highly reductive, if it studies traits that lie deep within the system. They probably are integral to the output from the system. But, the research method isolates the study of the trait(s) from influence by other genetic variabilities of the system as a whole, and/or from other environmental influences upon the trait. Studies of intracellular processes such as individual enzyme functions are highly reductive research.

2.4 More Basic Whole-system Research is Needed to Complement Basic Reductive Research

Recent thinking suggests need for whole-system research to complement the predominantly reductive research on the biology of plants. Gain in understanding of the yield system through reductive research is not questioned. What is questioned is continued reliance solely on search for yet more individual and possibly limiting components of the system. Need for elucidation of the system as a whole is being recognized. A system is described by Watt (1966) as:

> An interlocking complex of processes characterized by many reciprocal cause-effect pathways. . . . Everything affects everything else, and the complexity of the system of interlocking cause-effect pathways confronts us with a superficially baffling problem in scientific analysis.

The current calls for more holistic research are exemplified by a recent statement by Loomis (1993).

> An important question, is whether simple yield trials are not only powerful but also

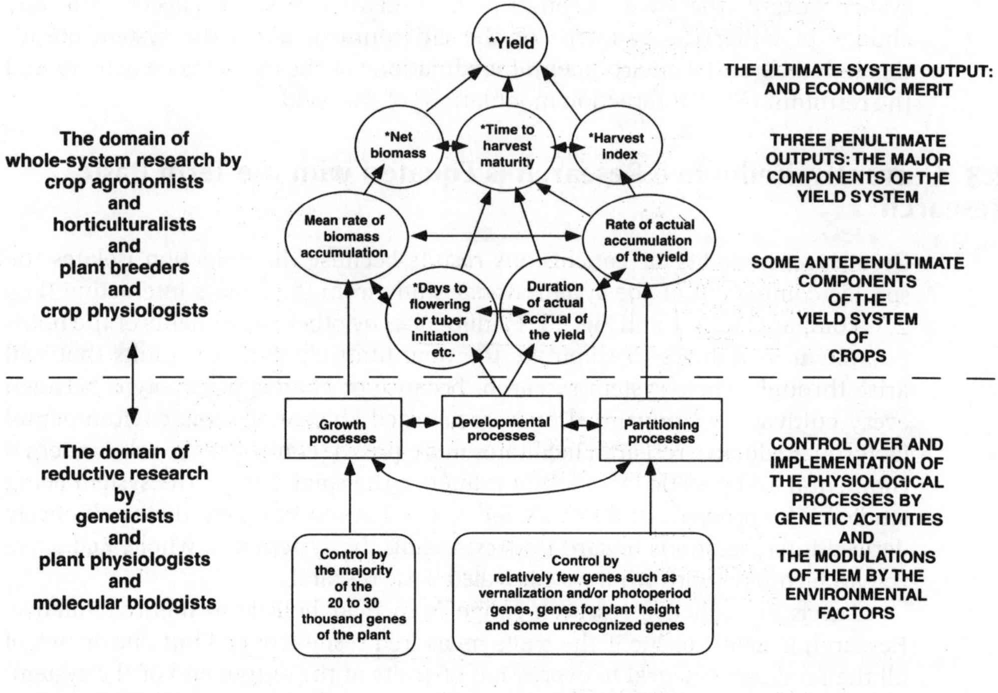

Fig. 2.1. An overview of the system-established interconnections among the processes of growth and development, and the ultimate, penultimate and antepenultimate outputs from the crop yield system.

The penultimate and antepenultimate components of the process of accumulating crop yield and the system-established interconnections among them can be quantified by using yield system analysis as an adjunct procedure applied to every yield trial.

Single headed arrows indicate the cause to effect direction, some of which are interconnected circularly. Double headed arrows indicate reciprocal interdependency, where the level of one or more traits must decrease whenever the level of another increases and vice versa.

The intended messages of this diagram are to indicate that: (i) yield trials can be used to enhance basic biological understanding of the crop yield system; (ii) there is need for more philosophical encouragement and funding of whole-system research; and, (iii) whole-system research will complement the reductive (basic) research that currently is more encouraged and better funded.

* Yield system analysis measures these four traits within the yield trial *per se*, while the other traits must be calculated from them.

sufficient for yield advancement. Further yield advance for most species is probably remote from individual genes, i.e., we already know about most genes that have enough influence to be found in this way. Molecular biology in fact offers new tyrannies – having found genes or moved them into new combinations, one is faced with a myriad of experiments to see what they do. If the genotype is new, such trials are not complete unless modifications of the cropping system (and other environmental factors) are also considered.

> Current frustrations with crop physiology as a source of 'bandwagons' and for failing to identify yield-enhancing traits are not without merit. Among the reasons: too many poorly conceived fads, such as proline for osmotic adjustment and NO_3 reductase for enhanced N assimilation, over emphasis of reductive research relative to integration, too little progress with bad models, and failure to turn physiological knowledge into simple screens. Sadly, many of these failures find roots in present systems of funding and rewarding research. . . . This problem is complemented by plant breeding books that omit competition, density, nutrients, lumping such matters vaguely under 'genotype × environment interaction'. . . . Times have changed because the easy yield advances have been achieved. My thesis is that rewards for careful attention to mechanistic insights offered by optimization will increase.

It is shown repeatedly in this book that the level expressed of one quantitatively variable component within the yield system influences negatively or positively the expressed level of other quantitatively variable components. Such system-established interconnections between the components verify the need Loomis (1993) indicates to select for the level of quantitatively variable traits that will optimize system functioning under each given environment. Selection for any quantitatively variable component trait is intended to indirectly select (i) a genotype that establishes a level of gene actions which (ii) allows for the environmental modulations of these genetic directions to thereby maximize the system functioning. The system includes the given genotype and the given environment. Often, selection for the highest possible level of individual component traits of the system causes negative rather than positive change in level of the system output. With minor modifications to fit the current text, Wallace and Zobel, (1995) stated:

> Almost all yield trials that compare genotypes are directed solely toward the applied economic goal of determining which cultivar yields the most. Then, that cultivar is recommended for planting by farmers. This is essential but purely-applied research. This is inefficient use of the yield trial data, because every yield trial can advance understanding about the biology of how crop yield is achieved.

All that is needed for yield trials to fulfil the goals of basic whole-system research, in addition to their economic goal, is to use each trial to answer basic questions such as the following. Why do some of the compared genotypes yield more and some less? How does the environment cause another cultivar to yield the most for another year at the same location? Why are some cultivars best adapted to the photoperiods and/or temperatures of the current and/or other environments? Improved understanding is needed of both of these but two holistic components (the genotype and the environment, Fig. 1.1) of the system-functioning, and of their relative control strengths over the component processes of accumulating the yield. The systems thinking, inexpensive yield system analysis and modelling procedures described in this book can derive data from every yield trial, to fulfil both the applied and the basic whole-system research objectives.

Feedforward, feedback, circular, reciprocal, tandem and parallel interlocking of component processes occur within whole biological systems. They arise through: (i) multiple (pleiotropic) effects by the same gene(s), (ii) competitive sharing of metabolites with other genetically directed components of the system

(Figs 2.1, 5.1 and 9.1; Tables 4.3 and 4.4), and/or (iii) due to modulations of the gene activity levels by the variations of the environmental factors.

The yield system *per se* is but a complex component of the plant system as a whole. More of the reductive research should be planned with a preconceived concept as to how it will contribute toward enhanced understanding of the system as a whole. Improved understanding of the crop yield system is required to continue to grow sufficient food for the ever-increasing population of the world. Miller and Kay (1993) present the following arguments.

> Most of us, from the beginning of our careers have followed the disciplinary, reductionist approach with considerable success. It is understandable that we continue to train future agricultural scientists in this mode. Our institutional organization and reward systems encourage continuation of this approach. But, this must change if we are to successfully meet the challenges of ecosystem research. This need for change in our undergraduate teaching programs has been emphasized by Boyer (1990). He suggests the need to enlarge the perspective of scholarship to include the 'scholarship of integration'.

Miller and Kay cite Van Doręn as saying in 1959: 'The connectedness of things is what the educator contemplates to the limits of his capacity.' Miller and Kay continue:

> We need to consider whether, in the longer term, we need to develop, within agriculture, specialists in systems science with a broad understanding of system components, . . . to act as integrators in future multidisciplinary teams. . . . The transition to a more holistic approach must begin with senior scientists. Senior scientists are much less likely to suffer in professional development and they have a greater influence than younger scientists on administrators and on both the research and training strategies.

Tollenaar *et al.* (1994) indicate:

> Different approaches can be taken to obtain an insight into the physiological basis of genetic improvement of . . . yield. In the first approach, major functional components of yield formation, such as seasonal dry matter accumulation and dry matter distribution, are conducted before proceeding to an examination of lower levels of organization. In the second approach, basic processes are studied at the molecular or biochemical level before proceeding to the cell, organ, plant, and crop levels. The second [reductive] approach is difficult to follow to completion (i.e., to the crop level), because of the large number of processes involved and the existence of multiple compensatory mechanisms when moving to higher levels in the hierarchical organization of plant growth and development. In the first [holistic] approach, physiological traits associated with genetic improvement are studied at the highest level of organization [i.e. agronomic yield] and, subsequently, at lower levels of organization [i.e. crop canopy, leaf, cell, and molecular levels], once an elementary understanding of the physiological impact on grain yield improvement at a given higher level makes studies at a lower level of organization relevant.

Cram (1980) made the following statement:

> Historically the understanding of biological systems has proceeded from the whole organism to the component processes and to cellular, biochemical and biophysical levels of organization. The success of the last since 1950 is clearly evident. However,

> . . . we appear to understand the cellular processes better than we do the whole plant. At the same time, most writers, when they have the opportunity, present their work as a contribution to understanding the whole plant. . . . While one must accept as valid an alternative claim that understanding a biological principle at a molecular level is a legitimate objective, I suggest that the profit and interest to be gained from such studies will, nevertheless, lie in understanding the whole organism.

2.5 Definition of Basic Whole-system Research

Watt (1966) stated:

> Clearly, if we are to study systems as whole systems and not just collections of fragments of systems, we must use a strategy of research which at every step is designed in terms of the problem of fitting all the fragments together correctly at the end of the research program. Otherwise, we get to the end and find out that the fragmental results obtained from various components cannot be fitted together to make a meaningful 'big picture'. In essence, the means of avoiding this horrendous outcome is to design the whole research program in terms of a conceptual model into which submodels growing out of the various parts of the whole research program can be fitted as components. . . . The first step in studying a complex system is to develop a comprehensive list of the variables and causal pathways that seem of potential importance in determining the function of the system.

As suggested by Watt in the preceding paragraph, Wallace (1973, 1976, 1980) and Wallace and Zobel (1982) developed listings of 30 to 40 traits that are obvious components that control the accumulation of crop yield. These listings were followed by attempts to describe the myriads of interconnected feedback, and tandem and parallel causal-pathways. The tandem and circular interconnections between these many components made it difficult to describe succinctly the integrated effect on yield due to these relatively few of the tens of thousands of gene-directed processes within the system. Detailing the component functions of complex systems within which everything affects everything else always requires circular returns to previous starting or intermediately discussed processes.

After attempting such descriptions for 30 or so components, guidance by peer reviewers, and noting failures of reductive research to provide effective assistance to breeding for higher yield (Wallace *et al.*, 1993a, b, c), Wallace and Zobel (1995) concluded: 'We suggest that basic and truly whole-system research requires measurement of the ultimate output from the system, plus at least all of the penultimate outputs from the system under study, and there must be measurements across both multiple genotypes and across multiple environments.' The penultimate outputs from a system consist of the smallest number of near-fully integrated-processes (component traits) into which the whole system can be conceptually subdivided, while fully accounting for the system output. Each of the smallest set of these major components that the accumulation of yield can be divided among must – through their further integration – be directly causal of and collectively essential to the ultimate system output. If any component's level is either zero or lower than optimal there will be no or low system output. For high yield, each component's level must be optimally balanced against the level of the other components. When not balanced against the other components – whether a component's level is smaller or larger than optimal – its deviation from optimal will lower the yield. Essential also to whole-system research on yield is comparison – across multiple genotypes grown in multiple environments – of the yield level with the level of each of the penultimate physiological-genetic components of the holistic process of accumulating the yield.

A succinct summary of the prior paragraph follows. *Basic whole-system research of the process of accumulating crop yield requires measurement of the ultimate output from the system, which is the yield, plus at least all of the major (penultimate) outputs from the crop yield system. These measurements must encompass a sufficient range of the available genotypes of the species, and a sufficient range of the environments in which the crop will be grown, that a near complete range of expressible levels of yield and of its major physiological genetic components will be observed and compared.* Ultimately, the comparison of multiple genotypes should be extended to include comparison across multiple species. Research reports focused on a given species are commendable. However, a holistic plant-science approach requires at least a literature review to ask if the phenomenon observed in the species studied is expressed in the same, a similar or a different way in other plant species.

The correlations among the levels of all pairs of these penultimate system components will quantify the system-established interconnections between them. The correlation between each component and the expressed yield, measured separately across the genotypes and across the environments, will quantify the relative control strength for each major component that is due to the genetic direction as compared with being due to the environmental modulation. Such focus at the output end of the system will also facilitate a quantification of the $G \times E$ interaction control over each component and the yield *per se*.

2.6 All Modulations by Environmental Factors are Implemented by Gene Activity

During their biological training all plant scientists are taught that every expressed plant phenotype (trait) is a consequence of control by both the genotype and the environment (Fig. 1.1). Nevertheless, with the current predominance of reductive and disciplinary approaches to research and scholarly learning, rather than whole-system approaches, plant breeders and geneticists tend to focus their research on the genetic controls over the expressed phenotype, while standardizing the environment to limit variation due to the environmental modulations of these gene activities. Contrariwise, crop and plant physiologists tend to vary the environmental factors, while standardizing the genotype, and not comparing multiple genotypes, thereby eliminating the variation of the phenotype due to variation of the genetic direction. Both approaches are reductive. Both lack focus on the extreme interconnectedness among all the parts of the system (Figs 2.1, 11.1). Whole-system thinking demands recognition that all alterations of the phenotype by different levels of environmental factors are implemented by the activity of one or more genes. The specificity of each gene activity, and of its modulation by environmental factor(s), emphasizes that every expressed phenotype is the result of a specific $G \times E$ interaction. $G \times E$ specificity is evidenced throughout this book. $G \times E$ accounts in full for every phenotype as illustrated in Fig. 1.1 and detailed best in section 9.3.3.3.

Chapter 2 Summary

A shift from applied whole-plant research to reductive research about each of the many components of plant systems occurred about the middle of the 20th century. As the 20th century ends, plant scientists are calling for more basic, whole-system research on the plant system that accumulates crop yield. There are two approaches toward improved understanding of the yield system. The reductive one is to conduct research about all of the many components, and then fit them together to explain functioning of the system as a whole. The whole-system approach begins with focus on the output from a system, as represented by the yield. Then, the process of accumulating this yield or expressing any other output is to be conceptually divided among the smallest number of near-fully integrated components (the penultimate system outputs) the system can be divided among while accounting in full for the yield. The reductive and whole-system approaches each reveal details at different organizational levels of the system that cannot be revealed by the other. Each complements the other. There should be increased funding and support for basic whole-system research.

3 Biomass Accumulation: The First Major Physiological-Genetic Component of Yield

None of the 32 paradigm shifts are fully derived within this chapter. The research results presented illustrate the focus after World War II that reductive research is preferable to whole-plant research. This is paradigm 1 of Chapter 1. The research repeatedly shows that the holistic system establishes interconnections among its components. These interconnections cause concentrated focus on individual components of the system to usually be ineffective for indirect selection toward higher yield. The research reported in this chapter led to the more nearly whole-system research results of Chapters 4 and 5.

The plant biomass is an obvious and first near-fully integrated major component of the process of accumulating the yield of a crop. It is called the net accumulated biomass, to recognize that respiration and its efflux of CO_2 are as essential to the accumulation of this biomass as the CO_2 influx through photosynthesis. The influx of every essential mineral is equally essential. An influx and an efflux of both water and oxygen are likewise equally essential. Each is required for occurrence of all the others. Each is a component of the biological system. For crop yield, the biological system is the holistic plant system. This chapter summarizes the history of research during the last half of the 20th century about how to raise crop yield indirectly, by selection for efficiencies of photosynthesis, light interception, and its absorption for use as energy to drive accumulation of the net biomass. Selection for rate of respiration as a means of indirect selection for yield is also summarized.

3.1 The Earliest Indirect Selection for Yield

Crop scientists consider yield to be about the least heritable of all plant traits, due to control by almost all the genes of the plant, and to simultaneous control by the many variable factors of the environment. Low heritability means a segregate selected for high biomass or yield in one generation or growing season does not express that same high level in the subsequent generations or years. To explore the potential for more effective selection for yield, plant breeders have attempted indirect selection. The assumption has been that one of the components of the yield is the most limiting factor for the yield; therefore, selection for that component should be indirect selection for yield *per se*.

Watson (1952) indicated the first attempted indirect selection for yield was selection toward enlargement of the morphological components of yield. For cereal crops these are the heads land area^{-1}, grains head^{-1} and weight seed^{-1}. Selection showed these component trait levels to be correlated; they are interconnected by and within the yield system. Enlargement of one component was compensated by reduction of the level of one or both of the others. It was interpreted that these compensations arise because they share the available photosynthate. They also share the time of the growing season duration. Measuring them describes morphological differences among the genotypes, but provides little understanding of the underlying physiological-genetic controls over accumulating the yield.

Watson (1952) indicated the following. Yield arises through system-established interconnections between a series of processes involving environmental influences and the plant genetics. More effort had been applied toward study of the environmental factors that limit yield than toward understanding plant traits that are limiting. The goal of agronomic research should be to understand all linkages between all the processes. Understanding the physiological bases for variation in crop yield requires observations on crops growing in field conditions, in addition to laboratory research. The field studies should measure simultaneous changes with time across the growing season of as many growth attributes as possible, especially of traits susceptible to simple physiological interpretation. Only simple measurements, easily and quickly made, are practicable for field studies. Data from individual leaves and canopies and plots within enclosed chambers or unrealistic field environments should be extrapolated with caution.

3.2 Description of Growth Analysis Methodology

3.2.1 The growth analysis method

Watson (1952) concluded study of yield should be based on changes in the net accumulated aerial biomass that occur during growth. He considered biomass accumulation to arise through a process analogous to continuous compound interest. The incremental gain in biomass during any interval will add to the 'capital' for subsequent growth. The rate of interest, or relative growth rate is:

$$\mathrm{RGR} = \frac{1}{W} \cdot \frac{\mathrm{d}W}{\mathrm{d}t}$$

where W is the aerial dry weight (biomass) of the plant at any time. The RGR is the efficiency of the plant as a producer of additional biomass. Gain in biomass will depend on (i) the initial capital, i.e. the seed weight, (ii) relative growth rate (RGR), and (iii) the growth duration (time). Variations in the net accumulated biomass can be analysed based on these three quantities. Watson indicated not all aerial biomass is productive capital. Much of it is skeletal material. Biomass gain arises through photosynthesis. Therefore, the productive capital (growing material) of the plant can be considered to be the leaf area. The net biomass gain per unit leaf area $(1/L)(\mathrm{d}W/\mathrm{d}t)$, where L is the total leaf area of the plant, and W is a measure of the excess of the rate of photosynthesis over the rate of biomass used for respiration. This is called the net assimilation rate (NAR). The RGR is the product of the NAR and the ratio of leaf area to the total weight. This ratio (L: W), called the leaf area ratio (LAR), indicates the amount of growing material per unit biomass.

Watson (1952) found both RGR and L: W to be complex and difficult to interpret. An analysis without using them can be based on the fact that the product of NAR and total leaf area gives the absolute growth rate in dry weight ($\mathrm{d}W/\mathrm{d}t$). The total biomass accumulation across any time interval is the integral of this product. Therefore, the biomass accumulation and its end point yield at harvest can be completely described in terms of the NAR and the leaf area. Measuring NAR and leaf area at intervals, usually of a week or multiples of a week, facilitates growth analysis that determines the mean RGR and NAR for successive intervals. If W_1 and W_2 are the total dry weight at times t_1 and t_2, respectively, whatever the form of the growth curve, for the time interval t_1 and t_2, the mean RGR is

$$\frac{\log_e W_2 - \log_e W_1}{t_2 - t_1},$$

and the mean NAR is

$$\frac{(W_2 - W_1)(\log_e L_2 - \log_e L_1)}{(t_2 - t_1)(L_2 - L_1)}$$

3.2.2 Interpretations from the earliest applications of growth analysis

Growth analysis of wheat, barley, potato and sugar beet (Watson, 1952) showed variation in nutrient supply over a wider than normal range to cause little change in NAR. NAR was high when leaf area index (LAI) was low. The LAI is the leaf area of the crop divided by the land area beneath it. Increasing LAI decreased NAR by reducing the average light intensity of the leaf surfaces, due to mutual shading of the leaves. Watson's (1952) conclusion was: 'As the possibility of increasing NAR appreciably by plant breeding or cultural means appears to be small, improvement of yield must be sought mainly through control of leaf area'. He concluded that most growth analysis research has been directed toward understanding variation in the NAR. Therefore, there is need for more study of the physiology of leaf growth, which will require experiments on plants grown in controlled environments.

3.2.3 A plant breeder's experience with growth analysis

D.H. Wallace enrolled as a graduate student in the Department of Plant Breeding and Biometry at Cornell University in 1953. Guided by Dr H.M. Munger, he accepted a PhD research project of determining why six dry bean cultivars differed in yield. The six included three commercial classes of dry bean, with a high and lower yielding cultivar within each. Munger stated he did not want information about how the yield was packaged, i.e. about numbers of pods plant^{-1}, seeds pod^{-1} and weight seed^{-1}. It was known that selection to enlarge one of these morphological components of yield always reduces one or both of the others, to thereby prevent yield gain. The goal was to identify the physiological-genetic bases causal of the yield differences. Dr Munger instructed Wallace to search the literature and identify an approach to this objective. After reviewing Watson (1952), Wallace suggested the growth analysis procedure.

A summary from Wallace and Munger (1965) of the growth analysis results for bean follows. Within two of the three bean classes the higher yielding cultivar had both the larger leaf area and larger LAR. But, because of initially smaller seed size, both cultivars of the third (pea bean) class initially had both smaller leaf areas and smaller aerial biomass than the other cultivars. One of the two (cv. Michelite) had the smallest seed size of all six cultivars and, therefore, the smallest leaf area and aerial biomass. Nevertheless, before harvest maturity, it accumulated the largest of all leaf areas and biomass. Therefore, (i) highest yield was not associated consistently with the largest leaf area, and (ii) Michelite had the highest relative leaf growth rate (RLGR) and highest RGR. The higher RGR and RLGR of both pea bean class cultivars, and highest of all for Michelite, were accompanied by a much higher LAR. The higher LAR added more leaf area for each unit gain of biomass. This suggested the higher LAR may be the chief factor responsible for the higher RGR.

The growth analysis results (Wallace and Munger, 1965) for four of the six bean cultivars agreed with the results Watson (1952) reported for four crops. *Leaf area was more closely associated with the yield than was the NAR. Leaf area seemed to be the more promising trait for indirect selection for yield.* Leaf area was measured directly, and was less variable across the sampled growth stages than the calculated NAR. Greater variability of the NAR suggested it is more strongly influenced by environmental factors such as moisture availability and/or incident light energy. The largest interaction effect was due to the sampling-time for the NAR, but smaller stage-of-development × plant-age interaction occurred also for leaf area and total aerial biomass. *These plant-age × genotype interactions indicated differences in the time used and needed by the cultivars to develop to the same stage. The interactions resulted in no trait being consistently associated with higher yield.*

One peer reviewer of Wallace and Munger (1965) asked the following questions. Why did you not compare just two cultivars within one of the bean classes, rather than comparing six cultivars and involving two additional commercial classes of bean? Would that not give more conclusive answers? These questions illustrate lack of appreciation for whole-system research, and the bias toward reductive research. Although not obvious at that time, greater gain in knowledge has resulted through the more holistic research with six rather than just two cultivars. Within two of the three bean classes, the hypothesis that larger leaf area is generally associated with the larger yield would have been accepted. Chapters 4 and 5 show this to be an invalid conclusion.

3.2.4 Deficiencies of growth analysis

The growth analysis results have not considered two aspects of the holistic process of accumulating crop yield. One is that measures of plant biomass, leaf area and yield result from harvesting the available radiant energy from the sun. Leaf area was measured by the growth analysis. However, differences among the genotypes (cultivars) in leaf thickness, and angle of display of the leaves relative to the energy source (the sun) alter the efficiency with which light energy is absorbed. Understanding the basis for yield differences among genotypes requires accounting for each cultivar of the energy intercepted by its leaves; how much is absorbed rather than reflected, and; how efficiently the absorbed energy is transformed by the photosynthesis and respiration into the net accumulated plant biomass.

A second physiological genetic component of crop yield never mentioned in conjunction with the growth analysis procedure is the harvest index (HI), which is the percentage of the net accumulated biomass partitioned to the yield. The HI can be measured during growth analysis, provided the last sampling time is at harvest maturity.

3.2.5 Interconnectedness among traits measured by growth analysis

The results from growth analysis demonstrated the NAR, to be negatively correlated with the total leaf area (*L*) and the LAI. All three change with the duration of growth (plant age and stage of development). This indicates a plant-age × genotype interaction effect on whether biomass gain is achieved predominantly through the LAI or NAR. The total accumulated biomass is positively correlated with the duration of plant growth, and with the summation over time of the functional leaf area, the latter being called the leaf area duration (LAD). The RLGR will influence the leaf area and the LAD, and it is negatively correlated with the LAR. These system-established interconnections illustrate a constant conclusion. *The level of every quantitative trait within the plant and yield systems is influenced by the level of virtually all the other quantitatively-variable components.* The biological basis is examined most thoroughly in Chapter 11.

3.2.6 Applicability of growth analysis to breeding for higher yield

Growth analysis improved understanding about how leaf area and NAR interactively determine the accumulated plant biomass and crop yield. These gains in understanding influence positively the judgments made thereafter by plant breeders, as they select among segregates from crosses for higher yield. However, for use as a routine tool to assist selection toward higher yield, the sequential samplings of the aerial biomass and leaf area, negative correlation between leaf area and the NAR, and changes of their relative effects on growth across the growing season make complete growth analysis impractical for direct assistance to plant breeding. Plant breeders can afford to and will benefit most from applying the growth analysis procedure to just the harvest ready stage of plant development (Chapters 4, 5, 9).

3.3 Light Energy Interception, Absorption and Reflection

3.3.1 Concepts of capturing and transforming radiant energy into plant biomass

A paper in Russian (Nichiporovich, 1967) was translated into English by the Israel Program for Scientific Translations. Nichoporovich's illustration of the complex interconnections among components of the yield system is presented (Fig. 3.1). The following all occur within the space of the aerial atmosphere occupied by a plant canopy. Some of the radiant energy is intercepted, part of which is absorbed by the plant organs. Another part is reflected back to the atmosphere. Some absorbed energy results in heat that energizes transpiration of H_2O from the leaves. Often, heat is excessive during mid-day when incident energy is highest, which causes excessive transpiration that limits photosynthesis. Some of the absorbed energy drives photosynthesis to fix CO_2 and release O_2. Some photosynthate is used by dark respiration to drive the plant metabolism, which releases CO_2 back to the atmosphere while taking up O_2. If not absorbed by the leaves, light energy is either absorbed by the soil, or reflected from the leaves or soil back to the sky. Nichiporovich considered photosynthesis to be the fundamental process of plant nutrition. Other plant nutritional processes, such as uptake of minerals and water from the soil, he considered to help increase crop

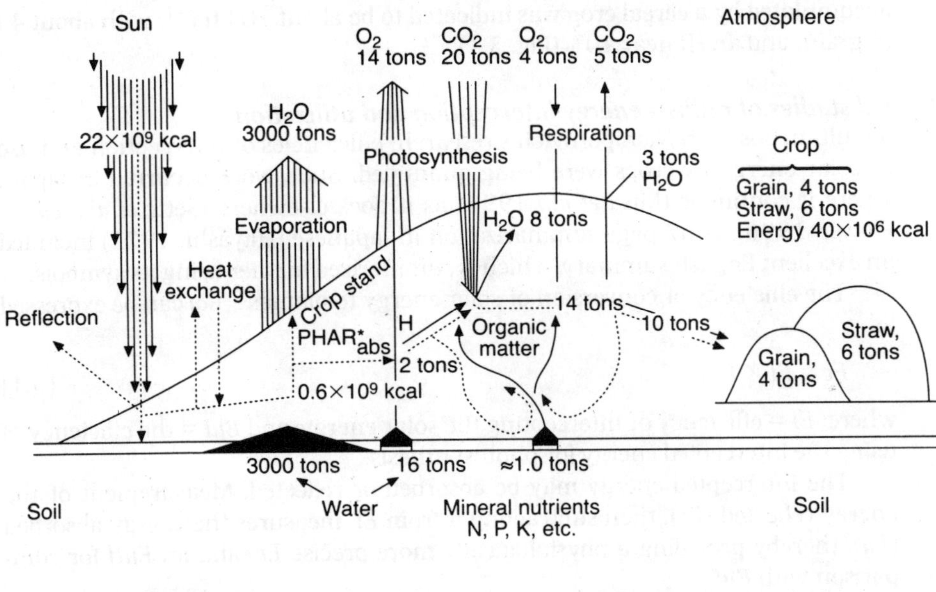

Fig. 3.1. Balance of energy, mineral nutrients, carbon dioxide and water involved in the formation of a wheat crop of 4 (metric) tonnes of grain ha^{-1}(0.32 cwt $acre^{-1}$). (Reprinted from Nichiporovich, 1967.)
*PHAR is photosynthetically active radiation from 0.4 μ to 0.75 μ.

yields, but only by enhancing photosynthesis and facilitating the best use of its products for growth and development.

Nichiporovich discussed how efficiently the available radiant energy is captured by the plant and transformed into net accumulated biomass. He found optical density of the leaves to be an important control, with leaves of most plants having 3–4 mg of chlorophyll per square decimeter of leaf area. He concluded this chlorophyll content will allow a single continuous layer of leaves to absorb 85–90% of the photosynthetically active radiation. This energy would not be used efficiently, however; especially at mid-day its highest energy causes excessive heat. Nichiporovich suggested crop stands can achieve greater photosynthesis, particularly tall plants such as maize, because their leaves are arranged at varied angles, occupy large aerial space, and may form 4 m^2 to 7 m^2 of leaf area per 1 m^2 of field area. This will facilitate absorption of 85–90% or 95% of the incoming photosynthetically active radiation, similar to the single horizontal leaf layer. But, the absorbed energy will be distributed over four to seven times more leaf area. The weaker energy level per unit leaf area will involve less heat, and enable higher photosynthetic efficiency. The greater efficiency will be due to vertical orientation of the leaves and their less crowded occupation of vertical and horizontal space. Nichiporovich believed the major need to be improvement of the canopy structure, and other properties of the photosynthetic system. Learning how to straighten the daily photosynthetic light curves by preventing low mid-day rates of photosynthesis was an important goal. With adequate water and mineral nutrients, the targets should be 4% or 5% efficiency, rising to 12% or 15% during periods of most rapid growth in crop stands with an optimum canopy structure. The expected net photosynthate accumulated by a cereal crop was indicated to be about 10 t ha^{-1}, with about 4 t of grain, and an HI near 40% (Fig. 3.1).

3.3.2 Field studies of radiant energy interception and utilization

Simultaneous with Nichiporovich's research, efficiencies of interception and use of solar energy by crops were being quantified. Such work occurred in Japan, where it continues (Misa *et al.*, 1994) as it does elsewhere (Settr *et al.*, 1995; Flower, 1996). A 67-page summarization in Japanese (Hayashi, 1972) included an excellent English summary, which is summarized further using its symbols.

The efficiency of conversion of solar energy to biomass (*Ec*) can be expressed as

$$Ec = Ei \times EuI \qquad [3.1]$$

where: *Ei* = efficiency of intercepting the solar energy, and *EuI* = the efficiency of using the intercepted energy for photosynthesis.

The intercepted energy may be absorbed or reflected. Measurement of the energy reflected (*Er*), then subtracting it from *Ei*, measures the energy absorbed (*Ea*), thereby providing a physiologically more precise *Ec*, and an *EuII* for comparison with *EuI*.

$$Ec = Ea \times EuII \qquad [3.2]$$

In a first paper Hayashi (1966) measured only *Ei*, and calculated *EuI*. Later, Hayashi (1972) extended this to report measurement of *Er* and of *EuII*.

To identify genetic traits that increase *Ec*, growth analysis and equations (1) and/or (2) were applied to rice. Variables included three planting densities, an early and normal planting date, and five cultivars. The results follow.

1. Each higher plant density increased *Ei* of all five cultivars (Fig. 3.2). *EuI* increased up to the densest planting for two of the five, but decreased at above intermediate density for three.
2. Both *Ei* and *EuI* were higher for the normal planting time than for the earlier planting for all five cultivars. Consequently, *Ec* was higher for normal than early planting.
3. Over the growing season, the reflected energy (*Er*) was 17–20%, transmission through the canopy to the soil was 30–47%, *Ea* was 35–51%, *EuII* was 2.8–3.98%, and *Ec* was 1.12–1.71%. The rice cultivars did not differ significantly for these traits.
4. There was high positive correlation between *Ea* and the product of [LAI] × [the light extinction coefficient (*K*)] with depth into the canopy. The value of *K* is larger as the percentage of interception (*Ei*) of the light is larger at any compared depth down into the leaf canopy. There was high negative correlation between *EuII* and the product of [*K* × the specific leaf area (*sla*)], the *sla* being the leaf area per unit of leaf weight)].
5. The *Ec* was correlated with each of the LAI, *K*, and *sla*. Cultivars for which *Ec* was highest had three characteristics: higher LAI, smaller *K* and larger *sla*. High *Ec* depended most on a higher LAI and leaf display to advantageously distribute the *Ei*, i.e. on having a smaller *K*. These and a larger *sla* (thinner leaves) are the genetic traits that gave high *Ec*. The *Ec* depended minimally on NAR of the leaves. High NAR was negatively correlated with LAI or a smaller *K*.
6. Changes of *K* attended changes of morphological characters such as angle and thickness of the leaves. Cultivars with thinner and/or erect leaves had smaller *K*, i.e. superior light penetration down into the canopy.
7. The smaller the *sla*, i.e. the thicker the leaves, the higher the NAR under intense light, and the larger the regression of NAR on the light intensity.
8. The LAI, *K*, and *sla* were all highly repeatable across the stages of growth. After mid growth, the repeatability was more than 90% for all three traits. High repeatability indicates strong genetic control.
9. Compared with older rice cultivars, newer ones tended to have higher *Ec* after heading during actual accumulation of the yield. They also tended to have higher LAI, smaller *K* and larger *sla*.
10. The *Ei* ranged from 45% to 66%, *Eu* from 1.5% to 2.1% and *Ec* from 0.7% to 1.4%. These ranges (Fig. 3.2) corresponded to 714–1429 g of plant dry weight (DW) per square metre of land. Norin 22 had relatively low *Ei*, and relatively high *Eu*. These gave Norin 22 nearly the same DW as three cultivars but less than Kinmaze, which achieved its highest *Ei* and also *Eu* at the highest planting density to thereby give the largest DW.

The highest association was between *Eu* and the DW day^{-1} (Table 3.1). This correlation was low during early growth but increased progressively until harvest, being nearly linear during ripening. The DW increases were highly correlated with *Ei* during early growth, and this correlation decreased through the season.

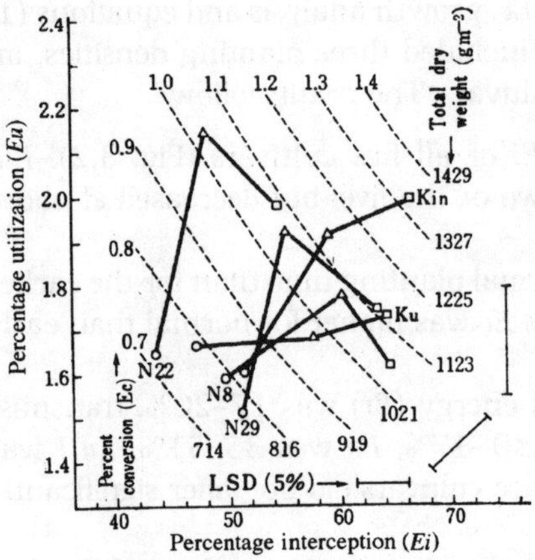

Fig. 3.2. Total dry matter production in relation to *Ei*, *Eu* and *Ec*, of Norin 8 (N8), Norin 22 (N22), Norin 29 (N29), Kinmaze (Kin) and Kusabue (Ku). (Reprinted from Hayashi, 1966.)

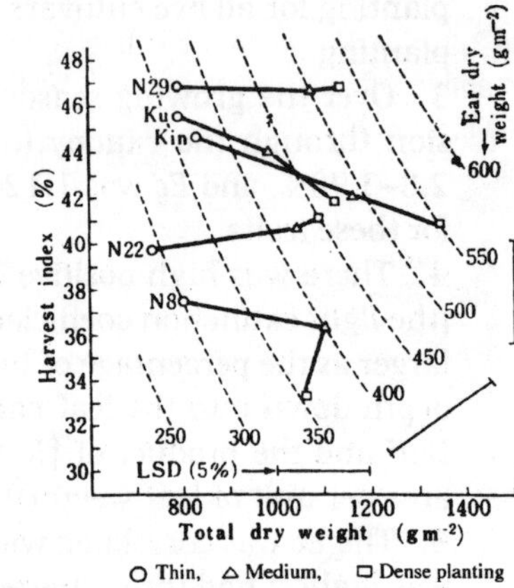

Fig. 3.3. Ear dry weight in relation to total dry weight and HI of Norin 8 (N8), Norin 22 (N22), Norin 29 (N29), Kinmaze (Kin) and Kusabue (Ku). (Reprinted from Hayashi, 1966.)

The DW increases were most poorly related to *Eu* during the mid-season (Table 3.2).

Cultivar performances were compared in relation to their total DW and grain yield. Norin 29 and Norin 22 both had large increase of DW and *Eu* with change from least to intermediate plant density (Fig. 3.2), but decrease of *Eu* at the highest plant density. This differentiated one group among the five cultivars. Norin 8, the oldest cultivar, responded the same way to these densities, but its *Eu* was always low. For both Kinmaze and Kusabue, increase from intermediate to the highest density resulted in their highest Eu.

For yield one more measure of efficiency (HI) has to be compared. The HI of Norin 29 and Norin 22 decreased little if any across the three densities, whereas HI of Kinmaze, Kusabue and Norin 8 decreased with each greater density (Fig. 3.3). An unchanged HI stabilizes yield for a given DW. Reduction of HI forces need for greater increase of DW in order to raise yield. Because Kinmaze gave the largest DW of all cultivars at the highest density, combined with a relatively high HI, it gave the highest of all yields. Because Norin 29 maintained the highest HI at all densities, at the intermediate density it gave yield about equal to Kinmaze.

Hayashi's (1972) summary statement follows:

> During the earlier stage of growth the DW of rice varieties growing in the field was determined mainly by the *Ei*, whereas during the middle and later stages it was determined mainly by *Eu*. The average efficiencies over the whole growing season ranged from 45 to 66% for *Ei*, 1.5 to 2.1% for *EuI*, and the corresponding efficiency of converting the total incident energy to dry matter (*Ec*), the product of *Ei* and *Eu*,

Table 3.1. Correlation coefficients between the efficiency of utilizing the radient energy (*Eu*) and the photosynthetic efficiency (dry weight day^{-1}) measured across six successive sampling dates (*t1*–*t6*). (Reprinted from Hayashi, 1966.)

Sample	*t1*	*t2*	*t3*	*t4*	*t5*	*t6*
Eu: DW/day^{-1}	0.288	0.292	0.723**	0.862**	0.893**	

** = significant at 1% level.

Table 3.2. Correlation coefficients between the dry weight increase (DW) and the efficiency of intercepting energy (*Ei*) determined across six successive sampling dates (*t1*–*t6*). (Reprinted from Hayashi, 1966.)

Sample	*t1*	*t2*	*t3*	*t4*	*t5*	*t6*
DW : *Ei*	0.946**	0.888**	0.321	0.426	0.288	
DW : *Eu*	0.808**	0.474	0.558**	0.983	0.959***	

** = significant at 1% level
*** = significant at 0.1% level.

ranged from 0.7 to 1.4%. Two responses to the increase in planting density were distinguished among the cultivars. One compensated for the decrease in *Eu* by a relatively stable HI (Norin 29-type), and the second compensated for the decrease in HI by a consistent increase in *Eu* (Kinmaze-type).

A simple subtraction of the energy transmitted through the canopy to the ground from the total incident energy does not give the real energy effective for photosynthesis, since the transmitted energy is rich in green which is less effective for photosynthesis. All reflected light included in *Ei* is another unmeasured loss of efficiency.

3.3.3 *Interconnectedness among traits of light interception and usage*

Hayashi's light interception results advanced understanding of the prior evidence from growth analysis that the NAR, is negatively correlated with the LAI. The efficiency of interception of radiant energy (*Ei*) increased as the LAI became larger, i.e. with increasing age, growth and leaf area of the plant. But, this *Ei* is negatively correlated with the efficiency of utilization (*Eu*) of the energy. Additionally, the *Eu* is positively correlated with greater penetration of light into the canopy; it is enlarged by the more equal distribution of the light across all the leaf area, and is reduced by shading of the leaves. Equality of light distribution is increased by both wider vertical and horizontal distances between the leaves. Both penetration of light into the canopy and *Eu* are negatively correlated with the thickness of the leaves. There are genetic and environmental influences on all these traits. *Eu* is reduced also, when excessive light on directly exposed leaves raises their temperature to excessive levels.

Hayashi's (1966, 1972) measurements of light energy interception and efficiency of its utilization were combined with growth analysis of multiple genotypes grown in six environments that represented early versus late planting and three plant densities. He measured also the aerial biomass

(Fig. 3.2) and its partitioning between the yield and the net biomass to give the HI (Fig. 3.3). The results redemonstrated the following. *Within the plant system, the level of a quantitatively variable trait is negatively or positively interconnected with the level of most if not all other quantitative traits.* Biological bases for the positive or negative correlations among the components of the yield system are detailed in Chapters 3, 4 and 9. From the whole-plant viewpoint, they are detailed best in Chapter 11.

3.3.4 *Was growth analysis combined with measures of light interception whole-system research?*

No one would consider the combined light interception and growth analysis results above to be highly reductive research. Few would suggest it is not whole-system research. We will test this against the definition for whole-system research presented in paradigm shift 25 of Chapter 1 and section 2.5. Required to decide if Hayashi's results represent truly whole-system research is knowing the penultimate output components from the yield system. There are three (Figs 2.1, 5.1, 9.1), of which Hayashi's results fail to consider one. Figure 3.2 quantifies the net accumulated biomass. Figure 3.3 shows the partitioning of the net biomass between the yield and other plant organs as measured by the HI. Hayashi's results (section 3.3.3) failed to present only the variation in the days used by each genotype to develop to harvest maturity (the DTHM). Time usage is component 3 of the process of accumulating crop yield. A biological basis for the system established interconnections between the time from planting to harvest maturity and at least 25 traits will be detailed in Chapter 4 (Tables 4.3, 4.4).

3.3.5 *Applicability of measuring light interception to breeding for higher yield*

Like for complete growth analysis, measurement of efficiencies of light interception and utilization provided gain in understanding about how leaf area and its interception of light energy jointly control the accumulated plant biomass and associated crop yield. The enhanced understanding now influences positively judgments made by plant breeders as they select for higher yielding segregates from crosses. Breeders previously paid little attention but now consider leaf orientation and equality of distribution of the light energy across all the leaves. However, the multiple sampling times required for repeated measures of aerial biomass, leaf area and the associated light interception and efficiency of utilization render these procedures impractical for routine direct assistance to selection by plant breeders.

3.4 Insights Provided by Research on CO_2 Exchange

3.4.1 *The infrared gas analyser*

During the 1950s, the infrared gas analyser accelerated research about the relationships of photosynthesis and respiration with crop yield. It replaced previous necessity to determine the weight of freshly excised leaves, or of discs cut from leaves, then to expose them to light, followed by weighing them again to determine the gain in dry weight due to photosynthesis, or the loss due to respiration. The infrared gas analyser instantaneously measured the level of CO_2 in an air stream entering and then exiting a sealed chamber. Within this chamber a leaf still attached to the plant (or a canopy of leaves) could be exposed to light of a measured intensity. Measured air flow at entry, and of reduction of CO_2 content at its exit, facilitated calculation of the net CO_2 exchange rate (NCE), i.e. the rate of photosynthesis per unit leaf area and of time.

Measures of the NCE for leaves and leaf canopies with an infrared analyser do not fulfil the requirement (sections 3.3.3 and 2.5) for whole-system research to measure a penultimate and therefore major component of the system output. Gas analysis measurement of photosynthesis is reductive research because its instantaneous measures of the net photosynthesis are across only a short time. Photosynthesis is a component of the holistic process of growth and biomass accumulation, but it accounts for only about 95% of the gain in dry weight. While accounting for less than 1% of the biomass, an insufficient uptake of any essential mineral will reduce to completely eliminate the rate of photosynthesis. Appropriate temperature and humidity in a leaf or canopy chamber cannot be maintained for long times. This prevents results that duplicate those of natural unenclosed environments.

3.4.2 A symposium on CO_2 and crop productivity

Availability of the infrared gas analyser established an outburst of optimistic predictions similar to current predictions for conceived advantages from the techniques of biotechnology (Toenniessen, 1993). The following statement by Ogren (1976), at a symposium on CO_2 and crop productivity, demonstrated early predictions and cautions based on the then new technology:

> Photosynthetic CO_2 fixation and photorespiratory CO_2 evolution are two primary determinants of net photosynthesis in soybean and other C3 species. Beneficial modification of the reductive pentose phosphate pathway in these species will consist of changes in the biochemical characteristics of the pathway so that the rate of CO_2 fixation is increased, the rate of photorespiration is decreased, or there is a combination of both. One approach to modifying the biochemistry of photosynthesis is to identify rate-limiting or regulatory sites, characterize the nature of the regulation, and then proceed to alter the regulatory sites genetically or chemically to achieve the desired goal of increased photosynthetic productivity. . . .
>
> Beneficial modification of the Calvin cycle, with respect to agricultural productivity, is a plausible but unproved objective. With the recent recognition that RuDP carboxylase is a primary regulator of both photosynthesis and photorespiration, one potential solution to the problem of increasing net photosynthesis has become apparent – the enzyme must be altered so that CO_2 fixation is increased, O_2 fixation is decreased, or both. Biochemical activation of the enzyme has been achieved *in vitro*, but it is not known whether this activation represents a potential for increasing activity *in situ*, or whether the enzyme is already in its most active form in the illuminated leaf. Variation in carboxylase specific activity within and between species has been demonstrated, but it is not known whether these differences are expressed in the leaf or are artifacts of isolation. The relative rates of carboxylation and oxygenation can be altered by changing leaf temperature, and pH can be permanently incorporated into the enzyme structure. Nonetheless, considering the importance of RuDP carboxylase in limiting CO_2 fixation and regulating photorespiration, it is apparent that substantial increases in photosynthetic productivity will require modification of this enzyme.

Moss (1976) gave the following statement at the same symposium on CO_2 and crop productivity. He indicates hope for progress but lack of confidence that selection for higher rates of photosynthesis will assist breeding for higher yield. Moss identified multiple other components of the process of accumulating the yield that interact with the rate of photosynthesis.

> Higher yield is always a goal for the plant breeder. Obviously, yield can be no greater than the net photosynthesis from which it is derived. Thus, it seems logical that one way to increase yield is to build into crop varieties the capacity for greater photosynthesis. Yet *it is difficult to document that economic yield and photosynthesis are indeed related in any direct way.*
>
> Some positive evidence of this is found in the literature; yields of many crops are depressed in shaded environments. Data for 10 corn varieties grown in tobacco shade tents show that when light was reduced by one-third, yields averaged 29% lower than the same varieties grown in the open. Presumably the primary effect of shade is to reduce photosynthesis, but in this case the difference in yield was due partially to failure of silking. In fact, varieties which tolerate dense planting averaged only a 19% yield reduction in the shade while those varieties which do not tolerate dense planting were reduced in yield by an average of 43%.
>
> CO_2 feeding experiments may give more convincing evidence of the importance of photosynthesis to yield. The world-wide practice of enhancing CO_2 concentrations in greenhouses needs no documentation. This yield enhancement occurs despite the fact that light intensity often is low during greenhouse cropping seasons. . . .
>
> The conventional breeding procedure for enhancing photosynthetic capacity of species is to select among genotypes within a species for ability to fix CO_2. Many workers have reported varietal differences in the rate of CO_2 uptake in soybeans, barley, bean, maize, alfalfa, ryegrass, and cotton. Unfortunately, there is a large genotype × environment interaction for photosynthetic rate because the ranking of the genotypes changes with the environment. Furthermore, there is great variability for measurements within a genotype at a given time, so multiple replicates are required to get reliable rankings. A system more reliable than measuring CO_2 uptake is required (or a way to control stomatal apertures, the source of much of the variability is needed) before progress can be made in breeding for photosynthetic enhancement. Thus, research is needed on methodology to measure photosynthetic rates and to rank genotypes. . . .
>
> One of the serious handicaps in breeding crop plants with enhanced photosynthetic capacity is that the internal control mechanisms of the process in plants are not understood. There is no doubt genetic variability for photosynthetic capacity exists. However, sometimes the environmental effect is great enough to mask the genetic component. Perhaps the area of greatest ignorance about photosynthesis is the lack of understanding of the mechanisms by which a plant senses the environment and translates that message into physiological responses controlling photosynthesis.

3.4.3 Restraints to breeding by low heritability

Wallace was the only plant breeder invited to present a paper at the symposium of CO_2 and crop productivity. The basis for this invitation was Izhar and Wallace (1967), which presented evidence from plant breeders for significant cultivar differences in net CO_2 exchange rate. The paper presented at the symposium (Wallace *et al.*, 1976) began with the original first paragraph of the growth analysis data discussed briefly in section 3.2.3 above from Wallace and Munger (1965).

For the yelloweye and marrow pairs, the higher-yielding variety consistently had through the growing season about 10% higher leaf area than the low-yielding one, a finding in agreement with Watson's (1952) conclusion that leaf area was the factor most correlated with varietal differences in yield. When all six varieties are compared, however, yield is not highly correlated with leaf area. At emergence, pea bean variety Michelite had slightly less leaf area than the other pea bean variety, Monroe, but by mid and late season, Michelite's area was often 30 to 60% greater. This variety × developmental stage interaction for leaf area was more obvious when comparing pea beans with the marrow and yelloweye types. Pea beans had 30 to 50% less leaf area shortly after emergence, but as much as 75% more area near maturity. Similarly, at emergence, plant dry weights of pea beans were about 50% of those of the other varieties, but ultimately they accumulated as much as 20% more, indicating that pea beans had the highest RGR. Also, through the entire growing season they had a leaf area ratio (leaf area/plant dry weight) about 10% higher. Net assimilation rate changed irregularly with time among the varieties, i.e. there was a definite, but irregular, variety × stage of development interaction with no consistent varietal difference in net assimilation rate (Wallace and Munger, 1966). It was also observed that marrow variety Cornell 7–16 and yelloweye variety Charlottetown had HI of 64% and 66%, respectively, compared with 55% for pea bean and red kidney types, and 60% for the other marrow and yelloweye varieties (Wallace and Munger, 1966).

Since the RGR equals the NAR × the leaf area ratio (LAR) (RGR = NAR × LAR), the higher LAR of the pea beans might account for their higher RGR. To test the possibility that NAR might also contribute, infrared CO_2 analysis was done with greenhouse-grown plants. Under light intensity varying from 5 to 48 klux, it was demonstrated that individual leaves of pea bean variety Michelite-62 had net CO_2 exchange (NCE) rates about 25% higher than those of Perry Marrow, Red Kidney, and Steuben Yelloweye (Izhar and Wallace, 1967). In this and subsequent studies with growth-chamber grown plants, care was taken to see that all tested leaves had just achieved full expansion. Growth-chamber grown plants were slightly more variable, but Michelite-62 again averaged 25 to 30% higher NCE than Red Kidney, which suggested that Michelite-62 had superior genetic potential for rate of photosynthesis, and supported the possibility that high NAR contributes to the higher RGR of Michelite. But this finding was inconsistent with the growth-analysis observation that pea beans had the lowest NAR near the end of growth, when they also had the largest leaf area.

The genetics controlling NCE, and the possible use of NCE rate as an effective selection criterion in segregating progenies, were studied with the crosses Michelite-62 × Red Kidney and Michelite-62 × Redkote (Izhar and Wallace, 1967; Wallace *et al.*, 1972). F_1 plants generally had NCE intermediate between parental rates. Segregation in F_2 and F_3 was extensive and indicated quantitative genetic control (by many genes). Some segregates had lower NCE than Red Kidney, and some were higher than Michelite-62. From four observations per plant (two primary leaves and two trifoliates), statistically significant variation was demonstrated among F_2 plants (0.05) and highly significant (0.01 level) variation among F_3 progeny means. The infrared CO_2 gas analyzer system clearly had sufficient sensitivity to identify genetic variation in segregating progenies, but the process was time consuming.

The merit of selecting for NCE was tested with two statistical-genetic parameters, broad and narrow sense heritabilities (Wallace *et al.*, 1972). Broad sense heritability is estimated as the percentage of the total variance among the F_2 or F_3 (segregating) plants that results from genetic control, the remainder being caused by environmental influence. It averaged 60%, a sufficient degree of genetic determination to encourage selection. In contrast to broad sense heritability, narrow sense heritability indicates effectiveness of selection. It was measured as the regression of 14 F_3 progeny means on the respective F_2 parent plant mean. The regression was 0.07, indicating a disappointingly low 7% effectiveness of selection for NCE. Thus, in spite of the genetic determination being 60%, selection in F_2 was virtually non effective. This apparently occurs because photosynthetic rate is controlled by many genes, as seems obvious when one considers the one-gene-one-enzyme concept and the many enzymes functioning in the photosynthetic pathway or otherwise affecting it. This low effectiveness of selection (low narrow sense heritability) is undoubtedly due to segregation of numerous gene-enzyme systems, each with attendant but varied small, medium, or large plus or minus effects on NCE.

The genetic and heritability studies of net CO_2 exchange rate were accompanied by similar studies of dark respiration rate (Wallace et al., 1972). The F_1 was again intermediate between the parents. The F_2 means differed at the 0.01 level, as did F_3 progeny means and plant means within some F_3 progenies. Broad sense heritability was a low 26%, and narrow sense heritability a very low 3%.

The above message from a plant breeder to crop physiologists exploring the role of CO_2 exchange in crop yield was a blunt statement. Measure of CO_2 exchange won't assist plant breeding. It is not a viable approach to more efficiently breeding for higher yield.

3.4.4 A crop physiologist's response to a plant breeder's conclusion

In answer to the blunt statement summarized in the previous paragraph, the following assessment was made by Schrader (1976) at the summary session of the symposium on CO_2 and crop productivity.

Wallace concludes that there is not much promise in selecting for net carbon exchange alone, because too many genes are involved. I concur with his assessment that no one process will always be limiting. However, we must continue to seek those physiological and biochemical factors that are limiting productivity, determine whether they are heritable traits, identify genetic variability within lines that are available, and then classify or catalog the germplasm that is available. As lines that are high in the desired traits are identified, they can be used in a breeding program to obtain progeny with a higher level of the desired trait(s). As limiting factors are eliminated, the potential for higher productivity should improve. We cannot afford to be discouraged from using a physiological or biochemical approach to plant breeding. Much of plant breeding in the past has been empirical. Better selection tools are certainly needed. Cooperative research with scientists from several disciplines (e.g., plant breeders, geneticists, biochemists, and plant physiologists) should be intensified.

Other areas that may offer promise for increasing productivity include attempts to initiate flowering earlier, so that the reproductive phase can be lengthened. This means that senescence must be prolonged, but offers the possibility of producing less vegetative biomass in non-forage species, so that more assimilates can be transported directly to the seed or other economically important part of the plant.

This physiologist's response to a breeder's statement indicated views of the crop yield system from different perspectives. The physiologist's assessment accepts the breeder's conclusion of complexity because many genes are controlling different parts of the system. His response to the breeder's negative conclusion did not, however, incorporate the breeder's interpretation that the activity of each gene interacts with the activity of other genes. Consider just the system then under discussion, the process called photosynthesis. After any cross for a trait controlled by many genes, many new genotypes will arise due to the segregations and recombinations among those many genes. This will include a few crossover events and occasional mutations. Virtually every segregate will have a genotype that differs by many genes from the genotypes of both parents, and from most other segregates. Thus, most segregates have a differently functioning system than all other genotypes. The physiologist's assessment did not adequately incorporate the consequent low level of heritability, due to segregation of hundreds and likely thousands of genes that segregate after a cross. As compared with the superior parent, for example, segregation of some genes results in negative effects, while segregation of others causes positive effects on any trait controlled by many genes. Neither the positive nor negative magnitude of change in the level of a trait controlled by the segregation of the many genes can be known in advance. Further, the effect on the trait due to the gene × gene interaction and/or to the different $G \times E$ interaction effects on the phenotype of the new segregated genotypes will also be unknown. The consequence is the very low heritability the breeder described above in section 3.2.3.

Rather than comprehending and accepting these assessments based on the breeder's experience and interpretation, the crop physiologist concluded it may not be photosynthesis that is the limiting factor. He did not indicate how the limiting factor can ultimately be identified, but did suggest successive improvements of one limiting factor followed by the next, and next and next, etc. He suggested successive eliminations of one limiting factor after another will lead to an ultimate maximization of all components and a high genetic potential for yield.

In 1975, this breeder also accepted need to identify the first and then second, etc. limiting factors. *Both crop physiologists and plant breeders were unaware of the need for whole-systems thinking and especially for compensatory optimization of the quantitatively variable components.* Neither recognized that the level of every quantitatively variable component within a system affects the level of every other quantitatively variable component trait. Neither sufficiently recognized that change of genetic factor(s) establishes need for different levels of the environmental factors, and vice versa. *We were not aware of need for the selection for every component within a system to be toward that lower or higher level that will optimize functioning of the holistic plant system of which the environment is half of the system.* We did not consider that an above-optimum level of a quantitatively variable trait will force other quantitatively variable traits to function at suboptimal levels, and vice versa.

The following statement by Watt (1966) also was not fully recognized. *Within a system: 'Everything affects everything else, and the complexity of the system of interlocking cause-effect pathways confronts us with a superficially baffling problem in scientific analysis.'* If everything affects everything else, the many changes of genotype, consequent gene × gene interaction effects, and $G \times E$ interaction effects due to either genetic change and/or change of the environmental factors will all alter the limiting factor(s) within the system. The limiting factor will change with day-to-day changes of weather and with advance of the developmental stage of the plant.

A statement from Nichoporovich's (1967) description that crop yield is based on light interception and its utilization was not applied. *His statement does incorporate Watt's concept of interlocking cause-effect pathways.* Nichiporovich stated:

> These concepts, frequently formulated, provide a starting point for numerous theoretical and experimental investigations. These investigations were concerned with the individual processes and parameters of the photosynthetic activity of plant communities, their relationships, their dependence upon environmental conditions and the type and condition of plants. The relative significance of these factors in the formation of crops has been studied and the conditions for maximum productivity have been determined theoretically.
>
> *Thus, the formation of very productive crops calls for an optimum combination of conflicting requirements, trends and conditions.* On the one hand, it is necessary to obtain stands with the largest possible leaf area growing as rapidly as possible, and reaching an optimum size as soon as possible. On the other hand, it is necessary to avoid the effects, possibly unfavorable, of too rapid growth, leading to the excessive and untimely shading of organs responsible for crop development, as well as the rapid desiccation of soil by very dense stands. Thus, it is important not only that the leaves should reach an optimum area but also that their expansion should proceed at an optimum rate specific for different cultures.

Functioning of an optimum combination of conflicting requirements, trends and conditions is demonstrated within every part of the crop system. *The inappropriateness of aiming to maximize the expressed level of every trait that contributes to the economically important system output is shown repeatedly.*

3.4.5 Conclusions of crop physiologists after one and two decades

Twelve years after the symposium on CO_2 and crop productivity a crop physiologist (Gifford, 1987) made the following statement, 'Selection for rate of photosynthesis will not lead to higher yield until we understand feedback regulations arising from environmental limitations in the field'. As described by Wallace *et al.* (1991; 1993a, b, c; 1995a, b), Wallace and Zobel (1995) and Mmopi *et al.* (1995), and throughout this book, these limitations are the environmental control and the $G \times E$ interaction control over the yield. The quantified smallest control strength (Fig. 1.1) over the yield is by the genotype, control is intermediate through the $G \times E$ interaction, but control by the environment is largest as shown in sections 9.2 and 9.3.

Two decades after Schrader's summarization of the potential from studies of photosynthesis to assist breeding for yield, Medrano and Vadell (1993) stated:

> Attempts to breed crop varieties with higher rates of photosynthesis have met with no success although considerable genetic variation in photosynthesis rates exist in several crop species and a positive correlation between leaf photosynthesis and productivity is reported in a number of experiments. *An efficient partitioning of assimilated carbon seems to be more critical in determining plant productivity.* Selection for low dark respiration or enhanced light interception have successfully increased net carbon gain by the plant. Selection by survival under low CO_2 atmosphere in

> tobacco haploids and on *Lolium multiflorum* L. (the Italian ryegrass cultivar RvP) showed a significant increase in plant productivity even when the leaf photosynthesis rate was not high. For Italian ryegrass RvP population, the surviving plants under low CO_2 had significant increase in initial dry weight but the difference disappeared in the second regrowth. No single character of ryegrass or tobacco genotypes could account for their survival under low CO_2. Even though photosynthesis and plant production are closely related, large environmental and ontogenic-induced variations in leaf photosynthesis rate make it difficult to achieve a good estimate of its contribution to the entire plant carbon economy. . . . Nevertheless, the effort devoted to photosynthesis has been compensated with a significant progress in the knowledge of relationships between photosynthesis and plant growth. Photosynthesis and plant growth are obviously related, but it is difficult to establish the relationship because of the large variability in photosynthetic rate induced by leaf age, plant development, source sink ratio, and environmental factors.

None of the statements cited above from Schrader (1976), Gifford (1987) or Medrano and Vadell (1993) adequately emphasize the interpretation by Watt (1966) that in a system: 'Everything affects everything else'. All three consider environmental variability to be the major part of the difficulty, without indicating it as a full half of the biological system – with genetic instruction being the other half (Fig. 1.1).

3.4.6 *Toward a view of the crop yield system that is applicable to plant breeding*

Most of this section is recited from a paper presented by Wallace (1980) at the Fourteenth International Congress of Genetics in Russia in 1979. It follows.

> The low narrow sense heritability of photosynthetic rate and poor correlation between rate and yield [section 3.4.3] led us to consider progressively more holistic models [Fig. 3.4, 3.5] of the biological system that gives yield expression [Fig. 3.4] is based on the often used argument that photosynthesis is the most important physiological component of yield expression. This argument arises because net photosynthesis (net carbon assimilation) accounts for 90–95% of the total accumulated plant weight [section 3.4.1]. The proposed system was both detailed and complex [Fig. 3.4], yet it was not sufficiently holistic. It recognized that net CO_2 assimilation resulted from the balance of CO_2 influx and CO_2 efflux, the influx being gross photosynthesis and the efflux being dark plus photorespiration. However, this model failed to consider that for carbon assimilation (photosynthesis) to occur, many additional net-assimilation processes must proceed simultaneously. Three such additional processes are: net H_2O assimilation, net mineral assimilation and net nitrogen assimilation [Fig. 3.5]. All four of these net-assimilation processes must proceed simultaneously for any one of them to occur, fully negating the argument, from the viewpoint of a functional biological model, that photosynthesis is the most important process. Negation is also indicated by the fresh-weight proportions of C, H_2O, minerals and N. H_2O accounts for 90% and thus water assimilation could, as photosynthesis was for dry weights, be designated as the most important process. However, biologically significant net assimilation of H_2O cannot occur without functioning of all of the other three net-assimilations.

A review of the physiological genetics of crop yield by Wallace *et al.* (1972) was criticized for excessively emphasizing photosynthesis and failing to mention

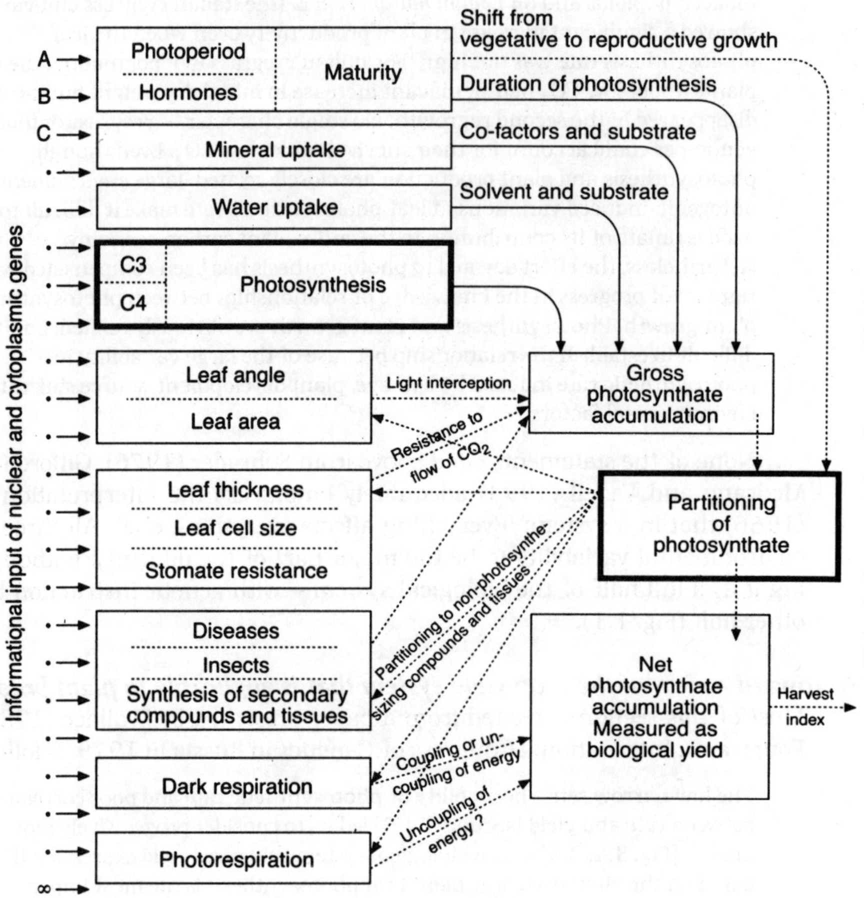

Fig. 3.4. A systems analysis of phenotypic expression of yield that is based on the assumption that photosynthesis is the most important process. (Reprinted from Wallace *et al.*, 1976.)

any role for nitrogen fixation, uptake and utilization. Figure 3.5 included the role of nitrogen, but Fig. 3.4 did not.

Each of the four net-assimilation processes may have at least two components: one or more for influx of the assimilated substance, and one or more for its efflux from the plant. This multiple-component nature is obvious for mineral nutrition. Each of the essential mineral elements is assimilated by its own distinct mechanism(s) (Pitman, 1977). Lack of simultaneous, i.e. parallel, and balanced assimilation of any one of the numerous essential mineral elements will negate the effectiveness of achieved assimilations of all the other minerals. It will also reduce or, if extreme, stop the net assimilations of CO_2, H_2O, and N. A multiple-component nature for the four major net-assimilation processes is further illustrated by the many sequential enzyme-catalyzed biochemical steps that constitute the pathways of photosynthesis,

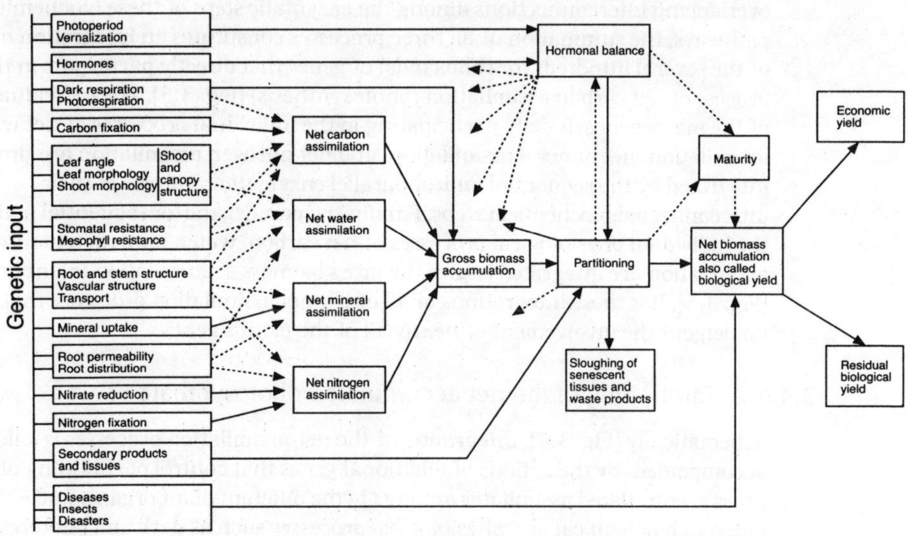

Fig. 3.5. An approach to systems analysis of phenotypic expression of yield. The analysis begins with informational input from the nuclear and cytoplasmic genes. It considers implementation of these genetic instructions via enzyme catalysis, integration of parallel and sequential enzyme reactions to give expression of individual physiological processes, and integration of parallel and sequential physiological processes to give net biomass accumulation and phenotypic expression of economic yields. Net biomass accumulation, and economic and residual yields are all quantitative outputs of the system. The days to harvest maturity provide a measure of the rate of output. Direct pathways or pathways that modulate the major physiological process of gross biomass accumulation and its partitioning are shown with solid lines, indirect pathways with dashed lines, and feedback pathways with dotted lines. (Reprinted from Wallace, 1980.)

dark respiration, photorespiration, translocation, water and mineral uptake, nitrogen uptake and utilization, etc.

3.4.6.1 Most genes are involved in the expression of yield

The model [Fig. 3.5] suggests that essentially every gene of the plant has either a major or a minor effect on crop yield. Relying on a recent measure of the quantity of DNA in plants, and allowing for replicated DNA sequences, about 27,000 genes function in a plant leaf (Goldberg *et al.*, 1978). These all-inclusive gene actions include both all nuclear genes located on the chromosomes, and all non-nuclear genes like those of the DNA in chloroplasts and mitochondria. It includes all structural genes that code for enzymes or other proteins and also regulatory genes that modify the quantity of enzyme transcribed and translated under instruction from the structural genes. . . .

The numerous sequential and/or parallel enzymatic steps of each of the biochemical pathways of photosynthesis, dark respiration and photorespiration constitute integrations of the respective distinct yet interconnected and overlapping sets of participating genes (enzymes). Because of the sequential, parallel and overlapping interconnections among the enzymatic steps of these biochemical pathways, the summation of all three processes constitutes an integration of actions of the several hundreds (or thousands) of genes that directly participate in the process of net carbon assimilation (photosynthesis) [Fig. 3.5]. The individual actions of the many genes directly participating in the individual processes of net water assimilation, net mineral assimilation, and net nitrogen assimilation are similarly integrated by the sequential and/or parallel enzymatic catalyses of their interconnected biochemical steps. Finally, the parallel and/or sequential actions of the individual physiological processes of net carbon, water, mineral and nitrogen assimilation are integrated to give the gross biomass accumulation (plant growth) [Fig. 3.5]. It is these integrations of all of the net assimilation processes that implement the involvement of nearly all of the plant's genes.

3.4.6.2 Partitioning of the net accumulated photosynthate

Schematically [Fig. 3.5], integration of the net-assimilation processes is followed (or accompanied) by the actions of additional genes that control partitioning of the gross accumulated assimilates among (1) the different plant organs, and (2) the different biochemical and physiological processes such as dark and photorespiration and growth. Directly, or indirectly through effects on hormone balance, partitioning provides major feedback signals that modulate rates of the four net-assimilation processes and of their component processes. Because some assimilates are partitioned to processes such as respiration that effect an efflux of assimilates from the plant, partitioning leads directly to the net biomass accumulation, which is commonly called the biological yield [Fig. 3.5].

Partitioning has importance equal to that of each of the assimilation processes. Unless the accumulated assimilates are properly partitioned, and then translocated to the actively metabolizing tissues, all of the component biological processes of plant growth will cease. Functioning of each of the four assimilation processes plus partitioning is dependent upon simultaneous, or near simultaneous, functioning of each of the other of these processes.

In a macrosense, partitioning is measured as the ratio of the weight of plant parts having economic value, the economic yield, to the weight of the biological yield. This ratio is the HI. In a more detailed macrosense, partitioning refers to division of the assimilates among roots, stems, leaves, fruits, seeds, etc. In a microsense, partitioning includes the addition (or deletion) of any radical such as C, CH_3, or NH_2 to (or from) another molecule. Thus, biochemically, partitioning is fully interconnected and integrated with the assimilation processes.

The consequences of partitioning are detailed most thoroughly in Chapters 4 and 11.

3.4.6.3 Linking gene action(s) to modulations by environmental factors

C, H, O, N and the essential minerals are all assimilates of the processes of plant growth. Therefore, plant growth is linked to availability of such essential assimilates. Shortage of any one assimilate constitutes a factor that limits gene action and consequent growth.

Presence vs. absence of light, varied light intensity and varied light quality can each turn specific genes on or off or alter their rate of action. Also, the rate of each biochemical and physiological component of plant growth and tissue and organ development is controlled by temperature. This temperature control is achieved through an alteration of the rate of action of each individual gene, this being implemented by the Q_{10} effect on the rate of the gene-associated enzyme catalyses.

The pigment phytochrome links the plant to variations in light quality and also to variation in duration of exposure to this light, e.g. to daylength (Vince-Prue, 1975). Genetic studies with numerous plant species indicate that one or a few genes implement this responsiveness of plant growth to daylength and light quality (Murfet, 1977a, b). Genotypes with alternative alleles of these genes may be less photoperiod sensitive, or may be photoperiod insensitive. The enzyme-catalyzed components of the photoperiod-sensitivity response are biochemical, so rate of these processes is influenced by the Q_{10} effect of temperature. On the other hand, the component constituting the light effect on phytochrome is a physical process that is insensitive to temperature. Therefore, plant responses to photoperiod have a large daylength × temperature interaction effect. An increase of daylength for beans, which are short day plants, or of temperature, and especially an increase of both, will curvilinearly intensify the photoperiod responses [Figs 4.1, 4.2, 5.3, 6.1, 6.2 and 6.7]. These responses include delay of flowering and consequent later maturity, which results in longer duration of active photosynthesis and growth [Fig. 5.1]. These responses also include additional branching and consequent additional leaf area, and an attendant decrease in the proportion of the net biomass assimilate that is partitioned to the seed, i.e. to the economic yield. Thus, a photoperiod (daylength–temperature controlled) response gives altered levels of many of the physiological and also morphological components of the net biomass accumulation and its partitioning. As already stated, the strong and simultaneous control over these numerous components of yield expression is for numerous photoperiod sensitive plant species implemented by the daylength–temperature altered action of one or a few photoperiod genes. The resultant variations in these physiological components of yield expression such as maturity or duration of active plant growth, have usually been considered to be controlled by quantitative genetics, with attendant large influence by environmental factors. The point being emphasized here is the large and simultaneous effect on the actions of numerous other (quantitative) genes of the yield system which can be exerted by one or a few photoperiod-temperature sensitive genes.

That one or a few genes simultaneously alters so many physiological and morphological components of plant development and growth suggests that the numerous gene actions that are controlled by variations in daylength and temperature are implemented through environmental control of hormonal balance with consequent control of partitioning. Hormonal balance and partitioning are the major components of feedback control within the biological system of yield [Fig. 3.5].

3.4.6.4. Support of plant breeding by statistical- versus physiological-genetics

Studies of single-gene controlled enzyme actions and of the consequent regulation of molecular transformations leading to chlorophyll synthesis and chloroplast assembly, plus similar studies of single-gene directed biochemical steps within the component pathways of photosynthesis, constitute one extreme of the possible approaches to genetic analysis of photosynthesis and crop productivity. Each single-gene study provides precise detail about the action of that one gene from among the thousands that function nearly simultaneously in implementing photosynthesis and crop productivity. That within one species hundreds of different genes have been identified that have activities related to chlorophyll biosynthesis and photosynthesis emphasizes that a very small part of the plant biological system is elucidated by each study of a single gene. A further limitation is that little knowledge is gained about any changes of action by these single genes as one, ten, hundreds or thousands of the other genes of the system change their actions to varied extents as the segregational and recombinational genetic events occur.

The most effective progress toward thorough understanding of the physiological-genetic system of yield expression will come from the efforts of multi-disciplinary teams. These efforts should combine single-gene biochemical-genetic and physiological-genetic studies. There should also be interactive contributions from the disciplines of biophysics, plant breeding, plant pathology, entomology, agronomy and horticulture.

The previously cited paper (Wallace *et al.*, 1976, from the symposium on CO_2 and crop productivity) indicates that the conductor of yield trials is a key member of scientific teams attempting to acquire genetic understanding of photosynthesis and crop productivity. Readily available to conductors of variety-yield trials, much more so than to other scientists, are *the following three [Fig. 3.5] endpoint outputs from the yield system:* 1) the economic yield; 2) the biological yield, and; 3) the days to maturity.

The citation from Wallace (1980) spanning sections 3.4.6.1 to 3.4.6.4 summarized his thinking about crop yield physiology and its potential assistance to plant breeding through the 1960s and 1970s. Listing within the preceding paragraph of three endpoint outputs from the yield system failed again to focus on all of the near-fully integrated penultimate system outputs of the process of accumulating the crop yield. The three endpoint outputs listed above did not specifically include partitioning of the biomass (the biological yield) between the yield and the other organs of the plant, although both Fig. 3.4 and 3.5 do. *The thinking by 1980 has moved toward focus on the output end of the yield system. But, it has not yet recognized how doing so provides the philosophical basis and methodology for whole-system research.*

3.4.7 Dark and photorespiration

Discussion of dark and photorespiration begins with a citation from the summary chapter of Amthor (1989). The citation discusses only dark respiration. The italics for emphasis is by the authors of this book rather than by Amthor.

In view of the functional model and biochemistry and physiology of respiration . . ., three means of increasing crop productivity through manipulation of respiration are implicated.

The first is by *decreasing maintenance respiration rate*, assuming that this can take place without detriment to the crop; that is, if some respiration not directly associated with growth can be eliminated, then growth should increase due to an effective increase in the supply of substrate for growth.

The second is by *increasing the rate of growth respiration*. If growth rate and growth respiration are indeed coupled, then an increase in growth respiration should result in an increased rate of structural phytomass accumulation. Source-sink studies show that an increase in substrate demand, which in this case would be due to increased growth respiration, often results in an increase in substrate production. Because of this, an increase in growth respiration need not result in simply more rapid depletion of existing substrate, but could lead to more rapid growth coupled to more rapid substrate production in photosynthesis. An additional factor also implies that an increase in growth respiration will lead to an increase in photosynthesis.

Increased growth rate will increase the rate of leaf expansion, and this will in turn increase the leaf surface area and therefore increase the amount of photosynthetically active radiation intercepted.

The third is by an *increase in respiratory efficiency* and, in particular, a decrease in (any) activity of the alternative pathway. . . . Examples of increases in growth or productivity by crop species due to each of these three mechanisms have been demonstrated, although the paper by Obenland *et al.* (1988) casts some doubt on the third. These results should stimulate research efforts to increase crop growth and productivity through an increase in respiratory efficiency.

In his last chapter, Amthor (1989) presented and briefly answered four questions regarding respiration and crop productivity.

1. Is respiratory metabolism of crop plants growing in the field operating efficiently? . . . Future research focused on an application of these principles may be able to ascertain the efficiency of respiration of field-grown crops.
2. How much do crops and crop plants respire? . . . A simple answer to the question at hand, therefore, is *Crops respire a lot.*
3. Is it likely that increased productivity can be gained by modifying respiration rate? . . . Information contained in Chapter 8 [of Amthor] indicates the answer to this question is 'Yes'. It must be noted, however, that *both increases and decreases in respiration could lead to increased productivity.*
4. Do we now know enough to make wise breeding and management decisions from the viewpoint of respiration, or must we look only at other physiological and morphological characteristics of a crop? [Amthor's opinion was:] *Yes, we do know enough to more seriously consider respiration in breeding programs, but there is an enormous amount of fundamental research that yet needs to be conducted and interpreted.*

3.4.8 Potential of respiration for indirect selection for higher yield

As interpreted by the authors of this book, the citations above from Amthor (1989) say: Yes, at some later time, after enough additional understanding has been achieved about dark respiration, we can select for it and achieve successful indirect selection and breeding for higher yield. Amthor (1989) indicates current insufficiencies of understanding of multiple biochemical pathways of dark respiration, of what functions they fulfil for the plant, and how to measure the efficiency of these functions. He indicates many difficulties to overcome for acquiring the necessary understanding, and that much basic research, at the reductive level, is required before the hoped for effective indirect selection toward higher yield can be realized.

Based on necessity of knowing whether to select toward lower or higher rate of respiration in order to indirectly select for higher yield, we argue for requirement to meet our definition for whole-system research. With yield as the system output, this definition requires measurement of the net accumulated biomass, its partitioning measured as the HI, and the time used for their completion. These penultimate system outputs will include all consequences due to all the combined CO_2 fixation and CO_2 efflux pathways (Fig. 3.6, and compare Figs 2.1, 5.1, 9.1). Photosynthesis definitely contributes positively to the net accumulated biomass and yield, but contribution by dark respiration is ambiguous. In spite of its definite negative contribution to the net accumulation of biomass and yield, dark respiration is essential to all of them. Part of this CO_2 efflux is required to use the photosynthetically fixed CO_2 for growth. Efflux of 1–4% of the already existing biomass is essential (Fig. 3.6) to maintaining the biological viability (living status) of the plant system (Chapter 11), including its major subcomponents that constitute the yield system. Photosynthesis, dark respiration, the alternative pathway of respiration which results in heat rather than growth, and photorespiration are each a part of compensatory and conflicting requirements. Each needs its compromised integrated-optimum level in order to maximize yield. Simultaneous quantifications of all three respiration pathways is too time consuming to assist selection across an adequate number of segregates, among which subsequently there must also be selection for the expected higher yield *per se*.

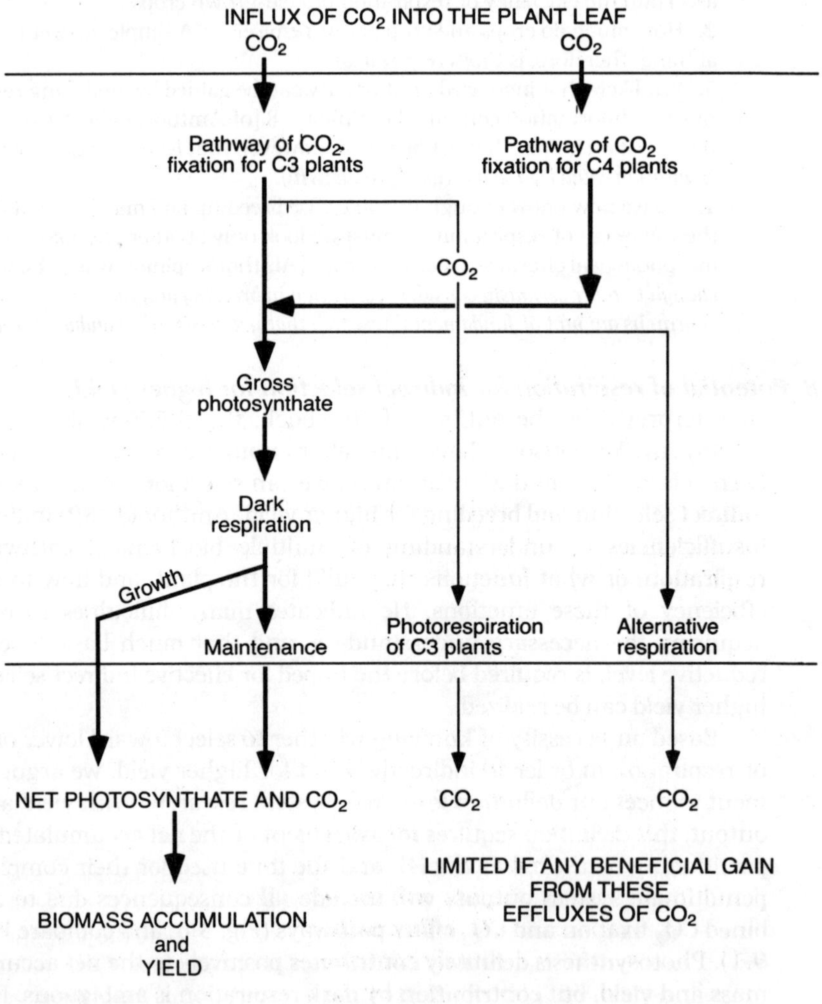

Fig. 3.6. A schematic illustration of the possible pathways in plant tissues of CO_2 fixation by photosynthesis versus efflux of CO_2 by respiration.

The requirements for CO_2 influx (photosynthesis) and CO_2 efflux (respiration) are each dependent on the level of the other. For example, the rate of dark respiration (efflux) required for growth depends on the current rate of photosynthesis, and on the efficiency with which the respiration implements growth. Growth respiration, maintenance respiration and any alternative respiration must share and therefore compete for the available photosynthate. The maintenance respiration depends upon the total biomass accumulated before the recent net accumulation of photosynthate, while the other respiration pathways do not.

Amthor (1989), as cited above, indicated *it may be an increase or a decrease of the dark respiration rate that is required for increased crop yield.* This fails entirely to indicate to the breeder a specific direction to apply selection pressure in order to use the dark respiration rate for indirect selection for higher yield. Every gene-action and physiological-process within the plant and yield systems interacts with all the other processes because all share the available metabolites. Therefore, comparison among one subset of a few genotypes may show the higher yield to be associated with a higher rate of dark respiration whereas, for comparison of that same subset in other environments, or of other genotypes within the same environment, the higher yield can be expected to require the reverse selection for lower rather than higher dark respiration rate.

A breeder interprets reversal of associations (correlations) between rate of dark respiration and yield to result from either or both of: (i) differences for (usually many) genes between every pair of cultivars and/or (ii) the differential control over the expressed phenotype that different environments may modulate through the same genetic instructions. Virtually every gene action interacts to some extent with virtually every other gene action, and interacts differentially after different environmental modulations of those gene actions. Thus, everything does affect everything else. Consequently, low heritability is an expected consequence of recombinations among the many genes and their effects on the multiple biochemical steps within each of the several complementing biochemical pathways that collectively control the overall rate of respiration (CO_2 efflux).

Chapter 3 Summary

In this chapter the growth analysis method of weekly or biweekly measuring the biomass and leaf area was evaluated. Some cultivar comparisons suggested selection for leaf area would be indirect selection for higher yield, but others invalidated this. Selection for higher rates of CO_2 exchange as a way to raise biomass and yield showed lower-than 3% effectiveness of selection for dark respiration, and below 7% for selection for the rate of photosynthesis. This and far smaller expected effectiveness for the intended indirect selection for yield led to the following conclusion. A breeder cannot expect effective indirect selection for yield by selecting for system traits expressed before the penultimate and antepenultimate outputs from the holistic yield system. It is concluded failure to give indirect selection results from (i) segregation of the many genes controlling quantitative traits being causal of the low heritability, and (ii) interpretation that photosynthesis and respiration are traits expressed so early within the

system that their levels per unit leaf area measured across short time spans do not carry through to strong control over the yield.

This research during the 1950s and 1960s led to recognition that the net accumulated biomass varies with efficiency of the light interception, absorption and reflection, which in turn each vary with the leaf area per land area, and mutual shading among the leaves versus equality of distribution of the light across all the leaves. Another control over yield that began to receive attention was partitioning of the biomass as measured by the harvest index. The concepts of growth and maintenance respiration, of dark and photorespiration and of an alternative respiration that is not linked to growth currently render it impossible to easily determine whether yield will be enhanced by selection for a higher or for a lower rate of respiration.

The results reviewed indicate the following. Within a system the level of every quantitatively variable trait tends to be connected (correlated) with the level of all other quantitatively variable traits. A consequence is that the control is by the system as a whole.

4

Partitioning of Photosynthate: The Second Major Physiological-Genetic Component of Yield

This chapter derives paradigm shifts 2 to 13. Shifts 3 to 13 are all consequences of the recognition summarized succinctly in paradigm shift 2. It is that photoperiod-gene activity controls partitioning of the available photosynthate between the cells and tissues that become the reproductive organs versus competitive partitioning to continued growth of the vegetative organs. An obvious consequence of partitioning a larger proportion of this photosynthate to the reproductive organs is enlargement of the harvest index (HI). This is described in paradigm shift 5. This partitioning as measured by HI is the focus of this chapter. Photoperiod-gene control over the partitioning will be shown to simultaneously control all three of that smallest number of physiological-genetic components that the process of accumulating crop yield can be divided into, while remaining inclusive of the yield as a system output. This causes the level of every major component of the yield system to have system-established interconnections (correlations) with and influence on the level of the other quantitatively variable components of the system.

For forage crops, and some edible crops consumed by humans (such as spinach), the expressed yield is the aerial biomass. This net accumulated biomass is the first major control over the yield of all crops (Chapter 3 and Figs 2.1, 5.1, 9.1 and 11.4).

4.1 Empirical Selection for Yield Raises the Harvest Index

For most crops, partitioning of the net accumulated biomass between the organs that become the yield and the other plant organs is the second major control over the yield. Discussion of this partitioning begins with recognition that it is measurable as the HI. Discussion continues with a statement about the effectiveness of empirical selection for yield by Evans (1993).

For many crop plants the main source of improved yield potential so far has been a rise in the proportion of the biomass that is partitioned to the harvested organs, i.e. in the HI. Increase in biomass has, in many cases, been slight, so the rise in HI must come from reduced investment in the non-harvested organs, most notably the stems in recent years. Indeed for most of the cereals *there is a close relation between the reduction in stem height and the rise in HI*. . . .

Two conclusions can be drawn, one of them uncomfortable for crop physiologists, from the rise in HI. The first is that, by and large, this rise has been made possible by the improvement of agronomic support for the crop, and is not an example of our improving on natural selection. *Irrigation and fertilizers promote rapid growth, so that its earlier flowering, after fewer leaves, is possible*. More effective weed control allows stem height to be radically reduced. Better control of pests, diseases and other stresses could allow the level of reserves to be reduced, although it is not clear that this has happened so far. Similarly, improved supplies of water and nutrients could allow the root system to be reduced, but again it is not clear that this has happened, possibly because more prolonged leaf photosynthesis and growth require more prolonged root activity.

The second lesson is that, although our understanding of the physiology and genetics determining HI is slight, this has not hampered plant breeders in selecting for greater yield potential, *which has brought higher HI along with it, once again, emphasizing the power of empirical selection for yield*.

The HI of spaced plants can correlate well with the crop yield, as in wheat (Fischer and Kertesz, 1976), but although deliberate selection for high HI raised it in oat (Takeda and Frey, 1985) and wheat (Sharma and Smith, 1986), grain yield was not increased because growth [*biomass accumulation*] was reduced, while in soyabean yields also fell (Kenworthy and Brim, 1979). As so often happens with yield potential, undue attention to any one feature in selection, other than yield itself, can be counter productive.

Within the above citation we inserted the words *biomass accumulation*, to emphasize that growth represents enlargement of the net accumulated biomass. The four italicized phrases within this citation from Evans (1993) collectively summarize much of the message of Chapters 4 and 5. The first italicized phrase indicates selection for reduction of plant height indirectly enlarged the proportion of the net accumulated aerial biomass partitioned to the reproductive organs that become the yield. Trade-off between the biomass partitioned toward plant stem height versus toward yield is the mechanism through which higher yield of cereal crops has resulted from a larger HI.

The second of the four italicized phrases states Evan's thinking that earlier flowering results from rapid growth. His implication that high yield can arise through earlier flowering is one of the major messages of this and Chapter 5, and it carries through to all subsequent chapters. This message is that rapid growth of the yield organs arises through partitioning to them of a large proportion of the net accumulated biomass. After flowering, continued partitioning toward rapid growth of the reproductive organs, or of tubers or bulbs or other organs, that become the yield results in: (i) early harvest maturity, (ii) high HI, and (iii) high rate of yield per day to harvest maturity. Derived in this chapter, but not mentioned in the citation by Evans is that the high HI and associated early flowering and maturity can result because: (i) the genotype is photoperiod-insensitive so there is no photoperiod-gene activity; or (ii) progressively less delaying photoperiod–temperature (*P–T*) combinations quantitatively reduce the photoperiod-gene activity of the photoperiod-sensitive genotypes; and (iii) any photoperiod-gene activity reduces the proportion of the photosynthate partitioned to the reproductive or other organs that become the yield. Points (i), (ii) and (iii) are the main message of the remainder of this chapter.

Whether altered by genetic direction or by environmental modulations of their implementation, quantitative reduction to elimination of all photoperiod-gene activity quantitatively increases the proportion of the available photosynthate partitioned to the sexual reproductive organs that become the yield. On the contrary, progressively greater photoperiod-gene activity causes the available photosynthate to be partitioned competitively, and quantitatively more predominantly, toward continued growth of more vegetative organs. This increases shoot elongation and initiation of more shoots as described by Murfet (1985) and Beveridge *et al.* (1992). In turn, the latter results in more branches, leaves and leaf area, later flowering, more days to harvest maturity (DTHM), and a lower HI. Control over partitioning by *P–T* combinations results through their joint modulations of the level of any photoperiod-gene activity. Such control occurs for crop species having potential for growth of the reproductive (yield) organs to compete with any continued growth of more vegetative shoots, branches, leaves and leaf area. For some crops, in contrast, including the monocotyledonous cereals, limited or no further growth of additional vegetative organs can occur after initiation of the reproductive organs that terminate the plant stems. Nevertheless, for these crops, any photoperiod-gene control over the number of nodes before initiation of the reproductive organs is a control over vegetative versus continued reproductive growth.

Similar control over partitioning occurs for vegetative organs that become tubers and yield of potato (Kooman and Rabbinage, 1996), bulbs that become the yield of onion (Steer, 1980), the roots for the yield of radish (Keiler and Smith, 1989), and fruits for yield of strawberry (Baumann *et al.*, 1993). For gladiolus long photoperiod (*P*) causes predominant partitioning toward the large flowers, and short *P* results in predominant partitioning toward the vegetative reproductive corms (Halevy, 1985).

The first italicized phrase within the above citation from Evans (1993) emphasizes agreement within this book that empirical selection for yield raises the HI of self-pollinated crops. The genetic explanation is that both HI and DTHM have far higher heritability than both the biomass and the yield, because of the strong control over them by relatively few photoperiod or other maturity genes. Biomass and yield have extremely low heritability, because their control is due to virtually all the genes of the plant. When there is strong control by a single gene, one of every four segregates after a cross will breed true for the trait level of the selected plant. For control by two genes only one plant of every 16 will breed true. With this exponential change per additional gene, for control by just 10 genes, only one of 4,194,304 segregates will breed true for the trait level of the selected plant (Table 4.1). With control by virtually all the genes, selection cannot be as effective as for strong control by one or a few genes. Delaying selection from the F_2, to F_3 and finally to the F_8 generation reduces the one of 4,194,304 that will breed true to one out of 1024. This explains why selection for yield is

Table 4.1. The probabilities that a randomly selected F_2 or F_8 plant is homozygous for the preferred allele of each of one to ten genes.

Number of genes	Generation of first selection						
	F_2	F_3	F_4	F_5	F_6	F_7	F_8
1	1/4		Progressively intermediate				1/1
2	1/16		Progressively intermediate				1/4
3	1/64		Progressively intermediate				1/8
4	1/256		Progressively intermediate				1/16
5	1/1,024		Progressively intermediate				1/32
6	1/4,096		Progressively intermediate				1/64
7	1/65,536		Progressively intermediate				1/128
8	1/262,144		Progressively intermediate				1/256
9	1/1,048,576		Progressively intermediate				1/512
10	1/4,194,304		Progressively intermediate				1/1,024

Table 4.2. The percentage of the plants within the successive filial generations after a cross for which segregation for each single gene difference between the two parents will give the homozygous genotype of each parent.

Filial generation	Segregants homozygous like parent 1	Segregants heterozygous like the hybrid	Segregants homozygous like parent 2	Total percentage homozygous progeny
F_1	0.00	100.00	0.00	0.00
F_2	25.00	50.00	25.00	50.00
F_3	37.50	25.00	37.50	75.00
F_4	43.75	12.13	43.75	87.50
F_5	46.88	6.06	46.88	93.75
F_6	48.44	3.03	48.44	96.88
F_7	49.22	1.52	49.22	98.44
F_8	49.66	0.76	49.66	99.22

usually delayed until at least the F_5 or F_6 generation. Table 4.2 shows the proportion of the segregates that will breed true for each single gene to be 50% for the F_2 generation. This proportion increases by a further 50% of the non-true-breeding segregates for each succeeding generation. A 99% probability of breeding true occurs at the F_8 generation.

The statement cited from Evans suggested that early flowering causes an effect toward higher yield, and can arise through the greater growth resulting from irrigation and added fertilizer nutrients. This and also Chapters 5 and 9 describe the system-established trade-off effects for high yield that are associated with early versus late maturity.

4.2 Deriving Explanation of the Interconnections among the Major Physiological-Genetic Components of Accumulating Crop Yield

It has not been intuitive that selection for either higher HI or larger aerial biomass will lower the yield. It is intuitive that selection for higher yield will lead to strong indirect selection toward an optimum DTHM and a consequent adaptation to the growing season duration. The yield will be reduced below the potential of the genotype, by any genetic and/or environmental control causal of either a shorter or longer than optimal DTHM. The level of each of the three major components of accumulating the crop yield is in part a response to the level of the other two components (Figs 2.1, 5.1, 9.1).

While conducting the growth analysis research summarized in section 3.2.3, Wallace began measuring for each cultivar compared in yield trials the net accumulated biomass at the DTHM. He called measurement of the biomass at harvest only, instead of at intervals through the entire growing season, a simplified growth analysis (Wallace, 1973; Bravo, 1975). It was learned that even though bean leaves abscise following the onset of senescence and before harvest maturity, the compared genotypes (cultivars) maintained nearly the same rankings at maturity for the plantings of different years at the same location for their (i) net accumulated biomass, (ii) HI, and (iii) DTHM. Each genotype established a relatively repeatable, and therefore heritable (yet quantitatively variable) level of these three major components of the process of accumulating yield. Even based on just air dry weights of the stems, branches, pod walls and seeds at maturity, the aerial biomass and HI of the genotypes, with minor exceptions, ranked in the same order as when the weight of the leaves collectable at and just before maturity was included and all tissues were oven dried. To reduce variability as much as possible, leaves that remain on the plants at maturity of some cultivars were removed and discarded. Quantification and interpretation of the measurements of the partitioning of the aerial biomass by calculation of the HI is summarized in the following quote from Wallace and Munger (1966).

> The extent of variation in distribution of accumulated dry matter among aerial organs at maturity was compared for 11 varieties of *Phaseolus vulgaris* L. representing marrow, red kidney, yelloweye, and pea types of dry bean. The lowest percentages of aerial dry weight present as fruit (seed and pod wall) were exhibited by the red kidney and pea bean varieties. 'Charlottetown' yelloweye exhibited the highest percentage as fruit and the highest HI (percentage of aerial dry weight as seed), but it had the lowest seed yield because of early maturity and accompanying low dry matter accumulation. The second highest HI was exhibited by 'Cornell 7–16' which regularly out-yielded all other varieties. The HI of the varieties varied from 50% to 67%. The high seed yield and HI of 7–16 were accompanied by the highest ratio for seed weight divided by pod wall weight. HI was not correlated with time to maturity nor with determinate versus indeterminate growth habits.

In retrospect, the interpreted lack of correlation between the DTHM and the HI is probably due to comparison across too narrow a range of genotypes and too narrow a range of environments to fulfil the criteria for whole-system rather than reductive research that are described in section 2.5 and paradigm shift 25 of Chapter 1.

Wallace was told by plant physiologists that the average rates of biomass and yield accumulation were calculated from air dry tissue weights. The average rates do not display all the changes in rate across time, including immediate effects by current or past environments. Plant breeders argued that they already have too many traits to measure and consider. Measuring the aerial biomass would involve too much time and effort, and subtract from rather than accelerate breeding progress.

Because of near-consistent ranking of the relative aerial biomass, HI and DTHM for cultivar comparisons across the different environments of yield trials across years, Wallace drew the following conclusions. The aerial biomass, HI and DTHM are each one of but three genetically-directed near-fully-integrated physiological-genetic components of the process of accumulating crop yield. Each is a penultimate output from the yield system. Therefore, as measured using simplified growth analysis applied only at harvest maturity of yield trials, a breeder should let each cultivar indicate the comparative contributions to its yield through its: aerial biomass, HI and duration of growth (DTHM).

4.2.1 The empirical power of selection for near-fully-integrated as compared with reductive traits

In section 4.1, the citation from Evans (1993) described the power of empirical selection for yield. Yield has been raised by selection for yield *per se*, in spite of limited understanding of the system that accumulates the yield. The power of empirical selection for higher yield results because yield is the output from the integrated levels of all the gene activities that occur within the yield system.

Evans cited evidence from Kenworthy and Brim (1979) that repeated (i.e. recurrent) selection for HI of soybean indirectly lowered the yield. It also lowered the aerial biomass and the DTHM. Evans suggested the reduced yield was due to excessive attention to a single feature (yield system component). This is a trade-off due to the power of empirical selection for the HI. The reductions in yield, rather than the breeder's hoped for indirect selection for higher yield, indicate conceptual failure to recognize the system-established interconnectedness among the three major components of the physiological process of accumulating yield. These system-established interconnections are succinctly illustrated in Figs 2.1, 5.1 and 9.1.

When they were first observed, a physiological explanation was not apparent for the system-established interconnections among the major components of accumulating yield. Wallace proposed measurement of just the (i) yield, (ii) HI and (iii) DTHM, because this is a simple enough procedure a plant breeder can incorporate it into his/her ongoing yield trials. Wallace and Zobel (1982) argued that measuring them on each yield trial plot will indicate the integrated consequences on the yield due to all of each genotype's many gene actions. These integrated consequences for all three major components incorporate also all modulations of all gene action levels by the environment of the yield trial. The expressed levels of each of the (i) aerial biomass, (ii) HI, (iii) DTHM, (iv) average rate of biomass and (v) rate of yield accumulation per day represents an integrated consequence within the system functioning. In turn, the expressed level of the yield represents the completed integration of all five traits. Evans

(1994) emphasized such integration by noting trade-offs in the levels of the component traits, and that balance and coordination among them is essential. He indicated it is unlikely there is just one optimum genotype. There can be different physiological pathways to the same high yield.

4.2.2 Harvest index is interconnected and negatively correlated with both the aerial biomass and days to harvest maturity

The marrow type bean line 7–16 and the yelloweye cultivar Charlottetown were both shown (section 4.2) by simplified growth analysis to have a high HI. Therefore, both were crossed with the red kidney class of beans. Serendipitously, the cross to Charlottetown provided the first indication of pleiotropic genetic control over Charlottetown's high HI and early maturity. However, emphasis on the cross to Charlottetown, rather than the cross to the higher yielding Cornell 7–16, began because of a many-fold higher proportion of the segregates having a commercially desired product-quality trait. These segregates had the seed colour of the red kidney class of bean.

The early, yelloweye class, dry bean cultivar Charlottetown was crossed with the late, red kidney cultivar Redkote in 1967. The reasoned goal was the 50% gain in yield that would result if the higher HI of this early parent was combined with the larger aerial biomass of the late parent. The red-kidney cultivar Redkloud was selected from among the segregates. Redkloud inherited the early DTF, DTHM and high HI of Charlottetown. It accumulated the same yield in 85 days as Redkote in 105 days. Equal yield in 23% less time arose because, as measured by the HI, Redkloud partitioned a higher proportion (Tables 4.3, 4.4) of its smaller biomass to the reproductive (yield) organs (Bravo, 1975; Kueneman, 1978; Kueneman *et al.*, 1979; Scully *et al.*, 1991; Wallace *et al.*, 1993b).

The early days to flowering (DTF) and DTHM of Redkloud suggested the possibility its genotype is photoperiod (*P*) insensitive. The first experiments showed Redkloud to be photoperiod insensitive and made earlier by each higher temperature (*T*), whereas the DTF and DTHM of its parent Redkote are delayed by both long *P* and higher *T* (Masaya, 1978; Wallace and Enriquez, 1980; Muhammad, 1983; Gniffke, 1985; Wallace *et al.*, 1991). These responses versus no response to *P*, and opposite changes of the DTF by higher *T* through the photoperiod response versus the thermal response, were shown for the insensitive and photoperiod sensitive genotype grown in 32 controlled *P*–*T* combinations (Fig. 4.1). Each higher *T* caused the photoperiod insensitive genotype to flower earlier under both short and long *P*. Short *P* prevented the photoperiod-gene activity of the photoperiod sensitive genotype, so under 9 h it responded to higher *T* with earlier DTF just as the insensitive genotype did under all *P* (Fig. 4.1). On the contrary, under 14 h and 16 h, the sensitive genotype responded to every higher *T* with delay of the DTF. Under 12 h, the sensitive genotype expressed progressively shorter DTF with rises from the lowest to an intermediate *T*, but its DTF was delayed by yet higher *T*. Figure 4.2 reinforces evidence, using field environments, that high *T* interacts with a photoperiod sensitive genotype to synergistically enlarge the delay of the DTF of bean caused by a long *P*.

As we began to study inheritance of the insensitive versus sensitive responses to the *P*, we were unaware a narrow range of expressed DTF is inherent across plants of the insensitive genotype, in contrast to a fivefold larger

Table 4.3. A listing of traits with levels pleiotropically controlled by the photoperiod × temperature interaction caused modulation of activity of the photoperiod-gene alleles *Ppd* and *ppd*. (Modified from Wallace *et al.*, 1993b.)

	Trait with differential levels	Function or closest relationships to the three major component(s) of yield
1	Sensitivity to photoperiod	*The genetic control over all the traits*
2	Days to flowering	Related to days to maturity
3	**Days to maturity**	**Major component No. 3**
4	**Aerial biomass**	**Major component No. 1**
5	Number of branches	Related to aerial biomass
6	Number of nodes on shoots	Related to aerial biomass
7	Number of leaves	Related to aerial biomass
8	Leaf area	Related to aerial biomass
9	Rate of yield accumulation per day of seedfill	*A control of harvest index and days to maturity* (viewed from a physiological perspective)
10	Rate of yield accumulation per day of growth	*A control of harvest index and days to maturity* (viewed from an economic perspective)
11	**Harvest index**	**Major component No. 2**
12	Variability of days to flower	Related to gene effect on nodal position
13	Variability of nodal position	Related to gene effect on nodal position
14	Bud number at main stem apex	Related to harvest index and days to maturity
15	Bud size at main stem apex	Related to harvest index and days to maturity
16	Preflowering partitioning rate	*A control of days to bud initiation and flowering*
17	Flush of 1st flowers versus only 1 or 2	Related to harvest index and days to maturity
18	Postflowering partitioning rate	*A control of harvest index and days to maturity*
19	Rate of biomass accumulation	Related to aerial biomass
20	Rate of vegetative development	Related to aerial biomass and days to maturity
21	Seedfill duration	Related to harvest index and days to maturity
22	Rate of development to flowering	= 1/days to flowering
23	Rate of development to maturity	= 1/days to 1/maturity
24	Node of the first flower	Morphological equivalent of the days to flowering
25	Reproductive versus vegetative sink strengths	*Physiological cause of the alternative partitioning and thereby of the variable levels of all the traits*
26	Yield	The economic output of plant's yield subsystem system

Traits and functions in bold are major components, i.e. the three functions constitute the smallest conceivable list of near-fully integrated processes that account in full for accumulated yield level.
Functions in italics are all subcomponents of one to all three of the major components and thereby are a control over their level.

range of first DTF being inherent across plants of the homozygous photoperiod sensitive genotype grown under long days. Because of its greater variability, reviewers thought the late flowering genotype must be heterozygous. Its homozygosity was verified, by comparing F_8 progenies derived by single seed descent from random individual F_2 plants. The F_8 generation was used, because 99% of all the segregates are expected to be homozygous for one or the other

Table 4.4. Pleiotropic effects on nine developmental traits by photoperiod-sensitive genotype *Ppd* versus insensitive genotype *ppd* under 16 h daylength in the growth chamber. (Reprinted from Wallace *et al.*, 1993b.)

Output trait from the yield system	Photoperiod genotype		Least (99%) significant difference	Association with days to first flower
	Sensitive	Insensitive		
Days to 1st flower	39.0	29.0	2.4	
Bud no. at apex of main stem	2.0	8.4	1.5	Negative
No. primary branches	8.9	7.9	1.2	Positive
No. nodes (mostly secondary branches)	58.1	14.9	8.6	Positive
Aerial biomass (g)	45.5	28.3	4.6	Positive
Biomass per day of plant growth	0.69	0.50	0.12	Positive
Seed weight at 27 days after 1st flower (g)	0.85	11.3	2.69	Negative
Yield per day of plant growth (g)	0.008	0.201	0.30	Negative
Harvest index (%)	2	39	4	Negative
Seed number	3.0	33.0	7.3	Negative

allele of each gene for which the parents differed (Table 4.4). A bimodal distribution of the DTF of the F_8 progenies showed the larger range of the DTF to be inherent to the homozygous photoperiod-sensitive genotype (Figs 4.3 and 4.4). A 1:1 bimodal distribution of the progenies having narrow versus broad distributions of the DTF indicated control over response to the *P* to be by a single photoperiod gene. The dominant allele of this photoperiod gene implements the photoperiod-caused delays of the DTF, so it was assigned the gene symbol *Ppd*. The recessive and photoperiod insensitive allele was assigned the symbol *ppd* (Wallace *et al.*, 1993b).

Before the demonstration of control over the DTF by a single photoperiod gene – which required using F_8 randomly derived inbred lines – attempts to determine inheritance of the photoperiod sensitivity using F_2 segregation ratios and verifying them with F_3 progeny tests failed. They failed because quantitative variation of the apparent segregation ratio resulted from a different mean growing *T* for the growing season of the F_2 generation and the F_3 generation (Fig. 4.4). Change of the *T* rendered the F_3 progeny tests ineffective for verifying the F_2 segregation. Even more confusing, the expressed genetic dominance was reversed between the field and the controlled environments (Fig. 4.4). The late flowering and photoperiod sensitive genotype was dominant in the field whereas, on the contrary, the early and insensitive genotype was dominant in the controlled environment with long days (Fig. 4.4). However, both genotypes flowered early and simultaneously in the controlled environment with short daylength (Figs 4.3 and 3.4) and also in winter-time greenhouses. This early flowering of the photoperiod sensitive genotype under short-day verified that the early versus late DTF

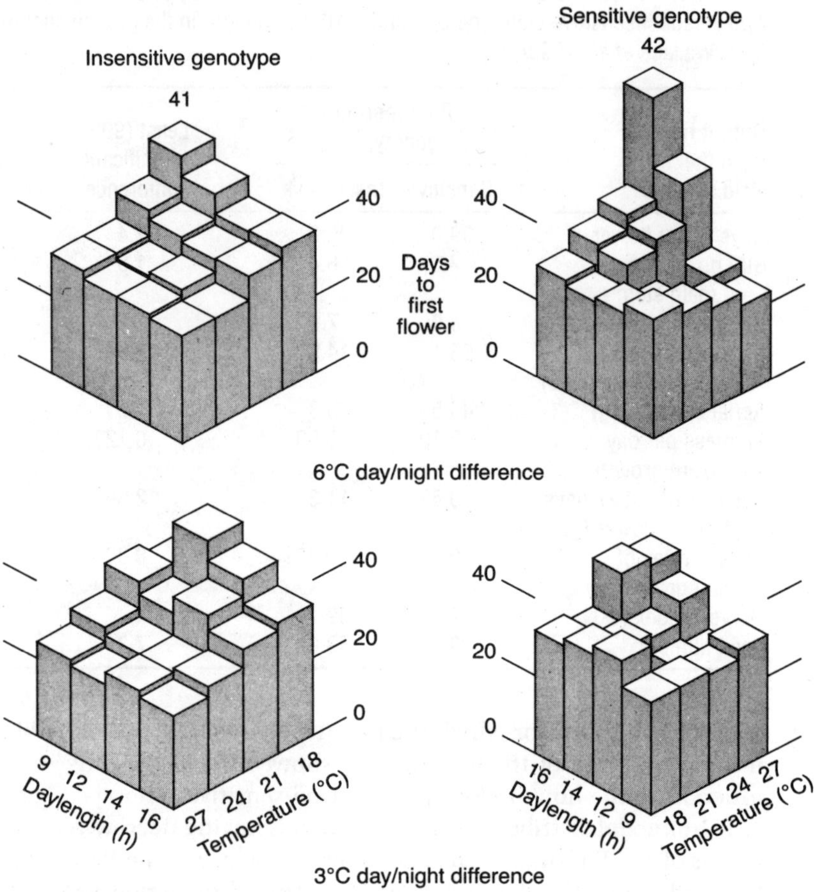

Fig. 4.1. Response surfaces for an early-maturing photoperiod-insensitive and a late-maturing photoperiod-sensitive bean genotype grown in the 32 environments of a factorial set of 4 daylengths × 4 night temperatures × 2 night/day differences in temperature. (Reprinted from Wallace and Enriquez, 1980.)

under long-daylength was controlled by the two alleles of a single photoperiod gene. The environmental difference(s) between the field and growth chamber responsible for reversal of the dominance have not been identified.

4.2.3 Photoperiod-gene activity pleiotropically controls harvest index plus the days to flowering and maturity

After demonstrating the delay of DTF by allele *Ppd* and lack of delay by allele *ppd* of a single photoperiod gene (Wallace *et al.*, 1993b), the hypothesis was tested that under any *P*, the lack of photoperiod-gene activity of photoperiod insensitive allele *ppd* allows inherent partitioning of the photosynthate predominantly toward continued growth of the earliest potential buds, flowers, pods and seeds. Hypothesized also is that delaying *P* causes photoperiod-gene activity of the sensitive allele *Ppd* that establishes quantitatively predominant partitioning of the

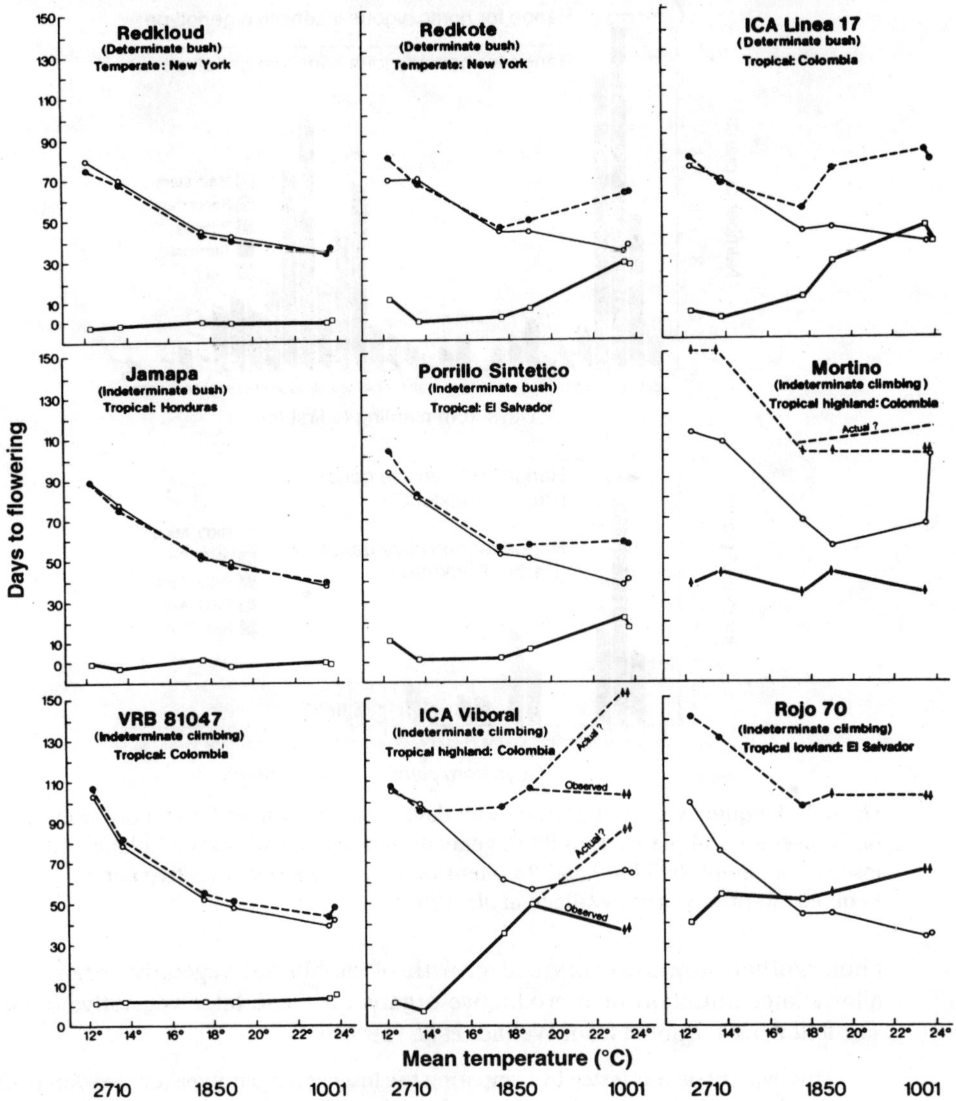

Fig. 4.2. Days to first flower and the delay of flowering caused by long daylength for nine diversely adapted bean genotypes grown under 12.3 h and 18.0 h daylength during two successive growing seasons at mean temperatures near 12°C, 18°C and 24°C. A rating for sensitivity to photoperiod is in parentheses adjacent to the cultivator name of each genotype; the sensitivity increases progressively from 1 to 8. DTF under 12.3 h natural daylength ○——○. DTF under incandescent-lamp extended 18 h daylength ●- - - -●, days delay by the long daylength □——□. (Reprinted from Wallace *et al.*, 1991.)

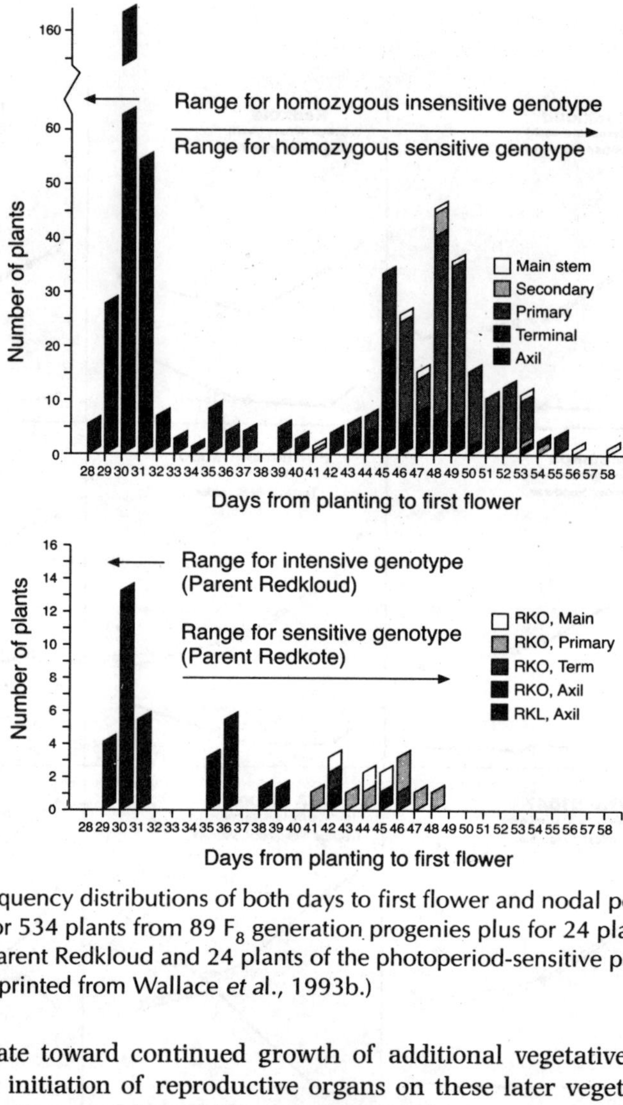

Fig. 4.3. Frequency distributions of both days to first flower and nodal position of the first flower for 534 plants from 89 F_8 generation progenies plus for 24 plants of the insensitive parent Redkloud and 24 plants of the photoperiod-sensitive parent Redkote. (Reprinted from Wallace *et al.*, 1993b.)

photosynthate toward continued growth of additional vegetative organs, but allows later initiation of reproductive organs on these later vegetative organs. The following is quoted from Wallace *et al.* (1993b).

> This hypothesis was tested by comparing the insensitive genotype (cv. Redkloud) and the photoperiod sensitive bean (*Phaseolus vulgaris* L., cv. Redkote) with their F_1, and with their F_2 segregates of undetermined genotype, plus with their randomly-derived homozygous F_8 generation segregates.... Under long daylength (LD), in addition to early vs. late flowering and maturity, the two genotypes expressed opposite levels of 25 other traits that would all be changed by competitive partitioning of the photosynthate [Table 4.3]. In contrast, under short daylength (SD), both genotypes flowered and matured early, and both expressed the levels for all 25 traits that the photoperiod insensitive genotype expressed in both SD and LD. The photoperiod gene interacted with daylength to control the levels of all three major physiological components of yield: the aerial biomass, HI, and DTHM [Table 4.4]. Other traits with levels altered by daylength-modulated photoperiod-gene activity were: the number of

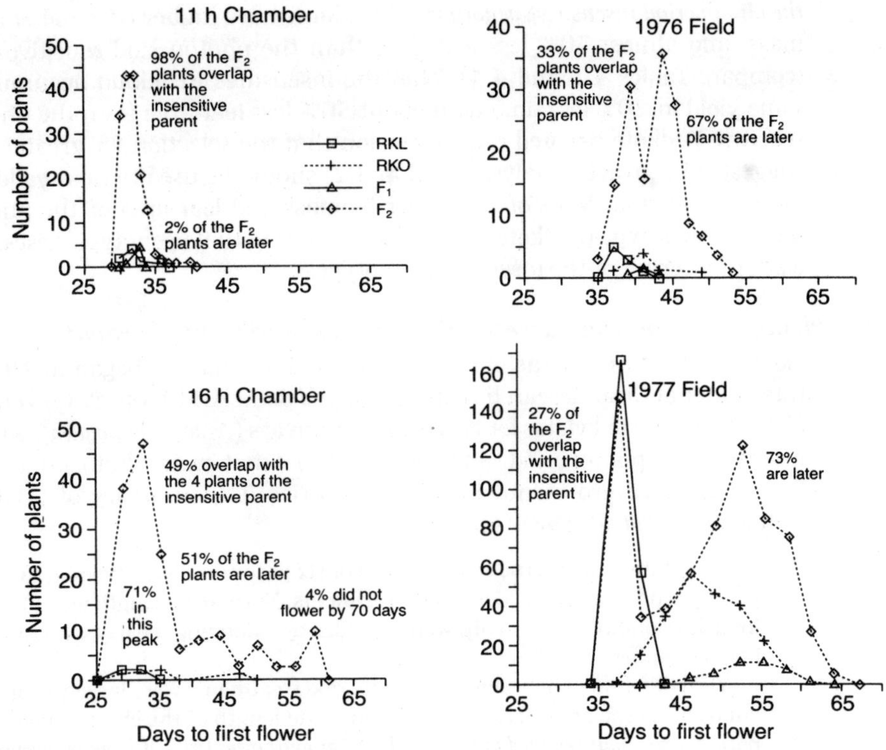

Fig. 4.4. Frequency distributions in four field environments of the days to first flower for plants of the insensitive parent Redkloud, the photoperiod sensitive parent Redkote, and their F_1 and F_2 generation progenies. (Reprinted from Wallace *et al.*, 1993b.)

branches, nodes, leaves and leaf area, the rate of yield accumulation, and the sink activity [Tables 4.3, 4.4].

The conclusion is that activity of photoperiod gene *Ppd* causes partitioning of the photosynthate toward the vegetative organs, while *ppd* causes the partitioning to be toward the reproductive organs. This agrees with earlier conclusion by Murfet (1985) that the photoperiod gene *dn* of pea controls the partitioning, and the conclusion of Beveridge *et al.* (1992) that the photoperiod response controls dry matter distribution among the plant organs of *Lathyrus odoratus*.

4.2.4 Photoperiod-gene activity also pleiotropically controls the leaf area and aerial biomass of bean

Simplified growth analysis in yield trials of Redkloud and Redkote and progenies from crosses between them, plus progenies from yearly crosses until the present time, have advanced insight about interconnections (Fig. 2.1) among the three major physiological components of accumulating crop yield. These comparisons included many early and late flowering and maturing segregates observed annually from 1969 to the present. *The following is established as a syndrome of traits for the photoperiod insensitive genotypes*. The plants have higher HI, smaller aerial biomass, and almost 50% less leaf area than the photoperiod sensitive genotype (compare Tables 4.3 and 4.4). That the insensitive Redkloud accumulated the same yield in 20% less time with about 50% less leaf area than the photoperiod sensitive Redkote negated the conclusion that the selection for greater leaf area suggested by growth analysis (section 3.2) should be used to raise yield through indirect selection. Most of the larger biomass and leaf area of the photoperiod sensitive genotype (Redkote), and of segregates with its genotype arises, after the earlier flowering of the insensitive genotype.

4.2.5 Photoperiod controls harvest index, biomass and yield of peanut

Field investigations of the effect by *P* on yield of peanuts, began in 1982 at the International Crop Research Institute for the Semi Arid Tropics. Witzenberger *et al.* (1985) reported yields for two peanut cultivars (*Arachis hypogaea*), which were insensitive to photoperiod, and four cultivars that were photoperiod sensitive. Peanut is called groundnut in many countries. The summary of a later paper (Witzenberger *et al.*, 1988) follows.

> The vegetative and reproductive growth of six groundnut genotypes in two photoperiods were studied in field experiments. Normal daylength was 11–13 h, and long day conditions (15–16h) were provided by extending the day with tungsten-filament bulbs.
>
> Yield differences between the photoperiod treatments were largely explained by changes in crop growth rate, partitioning, and length of the effective pod-filling period. *Long days resulted in increased crop growth rates [rate of biomass accumulation] but generally decreased partitioning [to the reproductive organs that become the yield], and the duration of the crop's pod filling phase.* However, it was dependent on the genotype which of the yield-determining processes had been more influenced by the day length. In some cases, [i.e. for some genotypes] partitioning contributed most to yield differences; in others, the duration of the effective pod-filling phase contributed most.

The following is from the discussion section within Witzenberger *et al.* (1988).

> Although the long day length increased the shoot growth rate by 42% and 19%, respectively, for cultivars Robust 33–1 and TMV2, the decrease in the pod fill rate for both genotypes and the moderate shortening of its duration for Robust 33–1 countered the effects on the shoot growth rate in short days, resulting in the yields due to the photoperiod treatments being not different. For those genotypes with greater final yields in the short day treatment, decrease of the shoot growth rate under short day conditions ranged from 6% to 20% below those observed in long days; however, the ratios showed that the rate of pod fill and its duration were specific for genotypes. The 37% greater partitioning in short day was largely responsible for the higher final yield of cultivar S7-2-13. In contrast, the pod fill rate of M 13 was not modified by the daylength, and the greater final pod yield could be attributed to the doubling of the duration of the seedfill in the short day treatment. In both valencia genotypes (Krapovickas St. 16 and NC Ac. 17090), higher final yields in short daylength were associated with moderate increases in both pod fill rate and its duration.

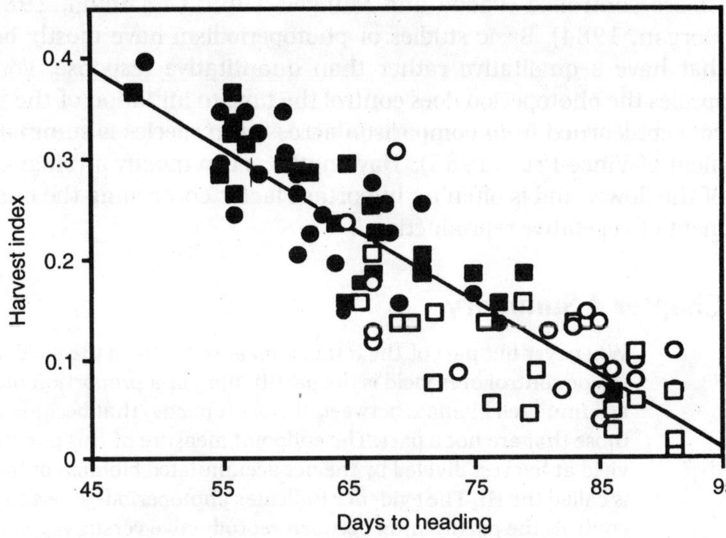

Fig. 4.5. The association between harvest index and days to heading in semi-tropical sorghum hybrids and varieties tested under two levels of drought stress at Bet Dagan. (Reprinted from Blum *et al.*, 1992.)

In addition to the above results for bean and peanut, the citation that follows from Wallace *et al.* (1993c) indicates data for 22 additional crops agree with the concept that photoperiod gene activity controls partitioning between the reproductive (yield) and vegetative organs. The original 25 citations for the 22 crops listed below can be accessed on p. 24 of Wallace *et al.* (1993c).

> Recent publications show association of early maturity with high HI, and/or show photoperiod control over partitioning toward reproductive organs of yield or storage organs of yield such as tubers, bulbs and roots. Such effects occur for: soybean, sweet pea, pea, wheat, oat, pearl millet, gladiolus, clover, radish, corn, barley, tobacco, rice, potato, onion, pigeonpea, mungbean, cowpea, bambara groundnut, sorghum, and two flower crops Chrysanthemum and Bougainvillea. For onion, the appropriate daylength for a cultivar causes partitioning of the photosynthate toward thickening of the leaves to form the economically valuable bulbs. Under an inappropriate daylength the photosynthate is partitioned toward growth of additional leaves. For potato, short daylength and/or low temperature cause partitioning toward tubers rather than toward more stems and leaves.

For sorghum, Fig. 4.5 shows the higher HI to be associated with shortest days to harvest maturity, and lower HI to attend the longer DTHM. These system-established interconnections apply across many crops.

The results in this section show that the *P* does not control the DTF of peanut, yet does control subsequent partitioning between continued growth of the reproductive (yield) organs versus continued growth of the vegetative organs. The control for bean differs: again, the *P* does not control days to the initiation of flower buds, but both the DTF and the partitioning as measured by the

HI are controlled (Padda and Munger, 1969; Greenham, 1982; Morgan and Morgan, 1984). Basic studies of photoperiodism have mostly been with plants that have a qualitative rather than quantitative response. For most of these species the photoperiod does control the time to initiation of the flower buds. The concept learned from comparison across crop species is summarized by a statement of Vince-Prue (1983): 'Daylength . . . can modify all stages of development of the flower and is often an important factor controlling the onset and development of vegetative reproduction.'.

Chapter 4 Summary

Whenever but part of the total biomass constitutes the yield, the second major control over yield is the partitioning of a proportion of the net accumulated biomass between the plant part(s) that become the yield versus those that are not a part. The endpoint measure of this partitioning is the yield at harvest divided by the net accumulated biomass at harvest. This ratio is called the HI. The evidence indicates photoperiod-gene activity strongly controls the partitioning between reproductive versus vegetative organs – as measured by the HI of a seed crop. Indicated also is that the level of this photoperiod-gene activity is directed by the allele of the photoperiod gene, and it is modulated by both photoperiod and the temperature. Depending on the plant species, the photoperiod-gene activity may begin early enough in development to: (i) control the time of initiation of flower buds followed by consequent control over the DTF and the HI; or (ii) may begin later to control only the DTF and the HI; or (iii) may begin only after flowering to control only the partitioning as measured by the HI and to cause negative correlation of the HI with the post-flowering accumulation of additional vegetative growth and biomass.

For cereal crops, the partitioning as measured by the HI is controlled additionally by the dwarf genes that shorten the plant height. The biomass not partitioned to elongation of the internodes is partitioned to the seeds to raise the HI and the yield. Photoperiod also controls partitioning to vegetative organs that become the yield, such as potato tubers, and roots and bulbs of some crops. It is established that continued selection for higher HI will raise it, while reducing both the yield and the DTHM. Similarly, continued selection for higher biomass will raise it, while lowering the yield and enlarging the DTHM. Thus, the accumulation of crop yield involves system-established interconnections among the component processes. These interconnections cause negative or positive correlations among the levels of the aerial biomass, HI and the DTHM.

All the above conclusions came with no initial intent of studying photoperiodism. Within a bean breeding programme combined with study of the physiological-genetics of yield, an early cultivar with a high HI and the smaller biomass was crossed with a late maturing cultivar. The intent was to combine the high HI with the larger biomass, the hoped for result being higher yield. The same yield as that of the late parent was produced in 23% less time by an early segregate, because it had a higher HI and higher rate of yield per DTHM. Biomass, HI and DTHM measured in many photoperiod-temperature combinations demonstrated pleiotropic control over all three plus about 20 other traits to be due to the photoperiod-gene activity and the

modulation of its level by the photoperiod and also by the temperature. The other pleiotropically controlled traits include rate of yield and biomass accumulation, number of branches and nodes, leaf area, node to first flower and flush of flowers versus only one or two flowers.

5

Days to Maturity: The Third Major Physiological-Genetic Component of Yield

This chapter describes how partitioning a larger proportion of the available photosynthate to the cells and tissues that become the reproductive organs, and/or the vegetative or storage organs, that become the yield causes them to grow faster, as explained in paradigm shift 4. Their faster growth results in a smaller days to flowering (DTF) and/or days to harvest maturity (DTHM). Based on paradigms shifts 2 to 12, it is interpreted that a strong control over the DTHM arises through differential partitioning of the net accumulated biomass (photosynthate). The partitioning between the reproductive organs and vegetative organs is quantitatively controlled by the quantitative continuum of the photoperiod–temperature combinations. The photoperiod–temperature combination modulates zero to intermediate to strong activity of any photoperiod gene(s). This photoperiod-gene activity controls the partitioning to directly controls the both harvest index (HI) and the rate or growth or the reproductive organs. This rate of growth in turn controls the DTF as well as the DTHM. The partitioning also partially controls the net accumulated biomass.

5.1 Biological Bases for Interconnectedness among the Three Major Physiological Components of Crop Yield

Chapter 4 showed the DTHM to be usually correlated negatively with the HI, but positively with the aerial biomass (Tables 4.3, 4.4). The longer duration of growth of a larger DTHM can account for its positive correlation with the aerial biomass. A physiological explanation for why the HI is negatively correlated with both the DTHM and the aerial biomass has been less obvious. The insights summarized in Figs 2.1, 5.1 and 9.1 render it intuitively obvious.

The DTHM is a time duration. Description of it as time does not indicate it to be a process as do the terms (i) biomass accumulation and (ii) partitioning of the

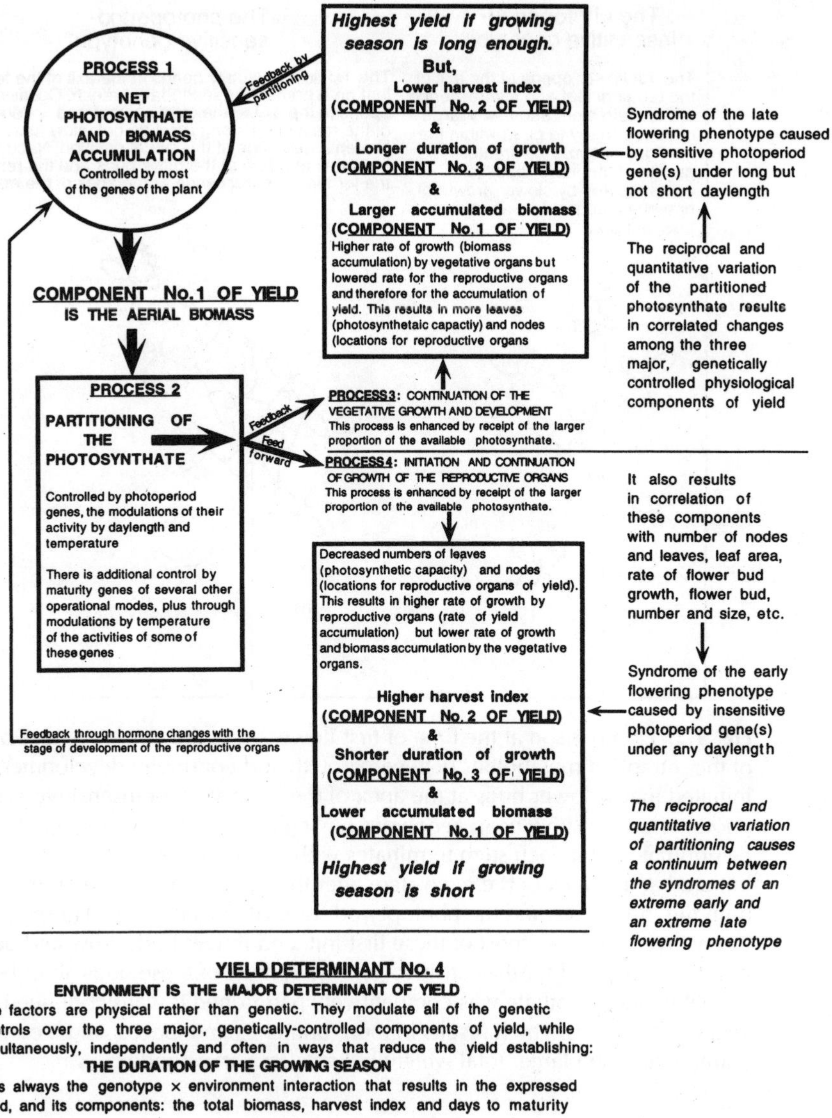

Fig. 5.1. Hypothesized role of photoperiod and other maturity genes in controlling partitioning, and thereby controlling days to flowering and maturity, cultivar adaptation, rate of yield accumulation, harvest index and the yield. (Adapted from Wallace *et al.*, 1993b.)

biomass. Without usage by any process of some span of time that process has not occurred. Thus, the DTHM is the time the genotype has used (required) to accumulate its biomass, and to partition it between yield versus the other plant organs. The genotype and environment jointly establish the time required (Figs 1.1, 2.1). From Figs 5.1, 9.1, and compare Fig. 2.1, the processes of accumulating biomass and partitioning it must share the duration of the DTHM. They must

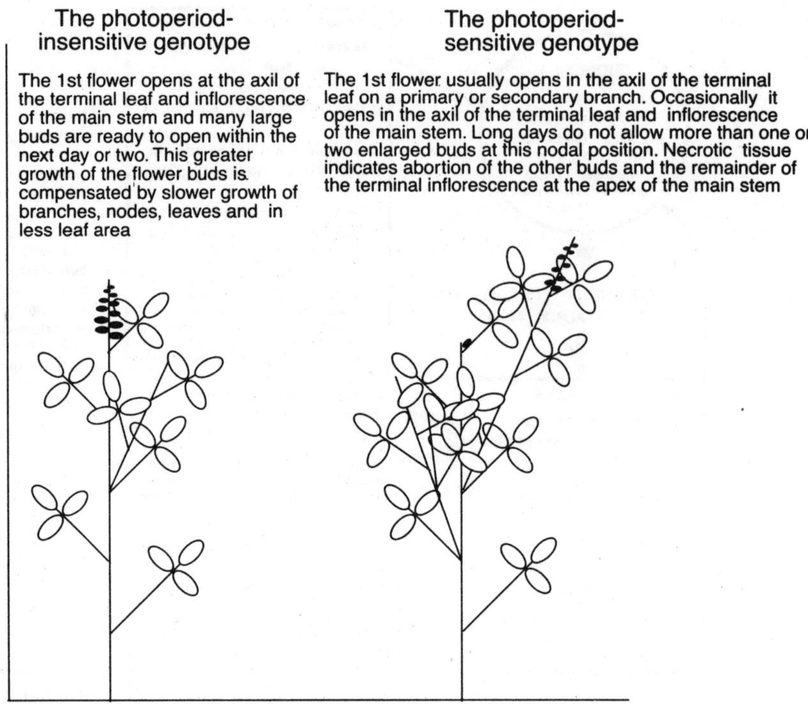

Fig. 5.2. Comparison at the time of first flowering of the consequences of partitioning of the intraplant metabolites between growth and continued development of the first-initiated many flower buds at the apex of the main stem of insensitive versus photoperiod-sensitive bean genotypes having determinate growth habit. For the determinate habit each stem terminates with an inflorescence. First flower buds are initiated at the apex of the main stem after the same number of days from planting for both the insensitive and the photoperiod-insensitive genotypes. For the photoperiod-insensitive genotype, most of these first initiated flower buds grow and develop rapidly. In contrast for the sensitive genotype, most of these buds abort because the available photosynthate was competitively partitioned toward continued growth of vegetative organs, which results in more branches and leaves, and a consequent taller plant height, and larger total weight but smaller harvest index.

also share the available photosynthate (Fig. 11.1). The proportions of both the time of the growing season used and of the photosynthate partitioned to the organs that become the yield must adequately support initiation of the yield organs followed by their continued growth. For both seed crops and crops with the yield being a vegetative organ, the HI is the endpoint result of the partitioning of the net accumulated biomass and sharing of the DTHM.

If, beginning with the cell(s) that will become a flower bud, a larger proportion of the available photosynthate is partitioned to support their growth, these cell(s) will grow faster (Figs 5.1, 5.2, 9.1, Table 4.3). Their faster growth will cause development to the bud initiation and flowering stages in fewer days. If the pods and seeds of a genotype continue to receive a larger proportion of the photosynthate than received by those of another genotype, the genotype will both flower and mature earlier than those receiving less of the photosynthate. And, this genotype will have both a higher HI in association with a higher average yield accumulation day^{-1} (compare Tables 4.4 and 9.2). This is exactly as Kooman and Rabbinage (1996) describe for potato; a more rapid growth of the tubers results in earlier maturing potato cultivars.

5.2 An Optimal Days to Maturity Will Maximize the Crop Yield of Each Genotype

If the growing season is shorter than needed by a late genotype, an early genotype is likely to give a higher yield than a late genotype that cannot grow long enough to achieve its full yield potential (Figs 5.1, 9.1). Maximization of yield of the early cultivar must arise through a higher HI, an associated higher yield per day, and in spite of a smaller total net accumulated biomass. Maximization of yield of the later cultivar will be through its longer growth, larger accumulation of biomass, in association with a smaller HI and smaller yield per day. Thus, the expressed yield of every genotype is simply the integrated result of a continuum of the compensated balances among the levels of its net accumulated biomass, HI, and DTHM, these being the three penultimate components of accumulating the crop yield. Measured ranges of the compensatorily balanced component levels and the yield, including their modulations by *P* and *T*, are described most thoroughly in section 9.3 for pigeonpea, and in section 9.2 for bean.

Growing season durations vary due to differences in moisture, and/or temperature (*T*) as altered predominantly by elevation and latitude. Stress due to lack of moisture or low or high *T* establish the beginning and end of the growing season at a site. The yield will be maximized for a genotype having the optimal duration of growth and development to harvest maturity, which will equal the duration of the growing season. A genotype that requires longer growth and development than the growing season will have much of any biomass that was partitioned to the yield wasted, because the yield organs cannot develop fully to harvest maturity.

5.3 Interdependency of Levels among the Three Major Physiological Components of Yield Dictates Requirement for an Optimal Level of Each

Partitioning a larger proportion of the photosynthate to the yield organs than is optimal will reduce the DTHM below the duration of the growing season, to thereby reduce both the yield and the aerial biomass. For an annual seed crop, an earlier completion of growth and development of the seeds will hasten plant senescence. These same results will occur for excessive partitioning toward the tubers of a potato cultivar. Partitioning a smaller than optimal proportion to the yield organs will enlarge the days needed to develop to maturity beyond the days available for the season, while simultaneously enlarging the accumulated aerial biomass. Partitioning a smaller proportion to the reproductive organs can reduce the yield through two mechanisms. The first is directly through the smaller proportion of the accumulated photosynthate that becomes yield. The second is loss of much to all of that lower proportion, because the organs that become the yield do not have sufficient time to develop to harvest maturity – before their growth and development are abruptly stopped by the moisture and/or *T* stresses that end the growing season.

These system-established interdependencies and interconnections are illustrated fully and progressively in Figs 2.1, 5.1 and 9.1. The consequence follows. Maximization of yield requires simultaneous and compensatory optimizations of each of the three major physiological components of the process of accumulating the crop yield: the net accumulated biomass, HI, and DTHM. The optimal level needed for each component is established by the environment, including the growing season duration that environment provides, and its *P* and *T*. Maximization of the yield requires genetic direction for a level of the activity of every gene that allows for the expressed environmental modulations of these gene-action levels to compensatorily optimize the levels of all three major physiological-genetic components of the process of accumulating the yield. For all three components to express their optimal levels simultaneously, the genotype-established directions must have been selected by nature or by man.

The environment of a growing season causes no less than three tandem effects on yield and its physiological-genetic components. First it establishes the beginning of the growing season. During the early to middle growing season the *P* and *T* will modulate the proportion of the net biomass that is partitioned to the yield versus is partitioned competitively to other organs. This partitioning, in turn, will control the growth duration the genotype will require to develop to harvest maturity, while also partially controlling the rate of accumulation of both the biomass and the yield, and establishing the HI. Following all these modulations of gene activities, the environment-provided moisture and/or *T* establish, somewhat randomly, independently, and totally, the actual growing season duration. Irregularity of weather always determines whether each genotype-directed and environmentally-modulated DTHM will match the actual growing-season duration. Consequently, correlation between the DTHM, HI and the accumulated aerial biomass with the yield for different growing season will vary from being large and positive to also being large and negative. The averaged correlation between the DTHM and yield across many yield trials is likely to be near zero (section 2.4).

5.4 Under Each Given Photoperiod, the Days to Flowering Responds to a Wide Range of Temperature with a U-shaped Response Curve

Temperature and photoperiod are the major environmental modulators of the DTF and DTHM (Squire, 1990; Hodges, 1991; Jordan, 1993). The genetics and physiology through which *T* and *P* jointly modulate the time a cultivar needs to complete development to its DTF and to its DTHM have not been well integrated within the published literature. Largely, the modulated controls by *T* and *P* have been independently studied.

After determination that the early DTF and DTHM of bean cultivar Redkloud are due to its photoperiod insensitivity, while the late flowering and maturing of cultivar Redkote are due to its photoperiod sensitive genotype (section 4.2), the effect of *T* on the photoperiod-gene activity was studied. Controlled-environment results showed the following: (i) Any delay of the DTF due to the photoperiod sensitive genotype of bean was exponentially enlarged by each higher *T* (Fig. 4.1). On the contrary, (ii) that same higher *T* caused a smaller and opposite change toward earlier DTF of the photoperiod insensitive genotype. (iii) The higher *T* also caused a smaller change toward earlier DTF of the photoperiod sensitive genotype – if the *P* was short enough to reduce sufficiently the photoperiod-gene activity and its delay of development that the thermal time response of earlier DTF to the higher *T* became the larger of the simultaneous but opposite controls over the DTF by the *T*. For 9 h with all higher *T* in Fig. 4.1, and for 12 h with all but the highest *T*, earlier DTF of the photoperiod sensitive genotype was caused by each higher *T*, indicating the thermal response was larger than the opposite photoperiod response.

Subsequent comparison of bean genotypes in field environments (Figs 4.2 and 5.3) also showed these opposite changes of the DTF of the photoperiod insensitive and photoperiod sensitive genotypes in response to short versus long *P* combined with high *T*. Low *T* reduced to prevented delay of DTF of the sensitive genotypes by the delaying 18 h *P*. The field studies also demonstrated a broad continuum of the genetically directed sensitivities to *P* as shown best in Fig. 4.2. In an incandescent-lamp extension of 18 h of fields at three mean *T* (elevations) in Colombia, the photoperiod sensitive genotypes expressed a U-shaped response of their DTF across the *T* range of 12–24°C (Fig. 4.2). On the contrary, under both the natural 12 h *P* of the field and the 18 h, the DTF of the insensitive genotypes decreased across all the progressively higher T. The U-shaped curve of the photoperiod sensitive genotypes occurred because, beginning from the lowest *T*, the first several progressively higher *T* decreased their DTF through the thermal-time response, while the several progressively yet higher *T* enlarged the delay of the DTF caused by the 18 h *P*. The *T* at which long *P* began to delay the DTF was generally lowest for the most highly photoperiod-sensitive genotypes, and highest for the insensitive genotypes (Figs 4.2 and 5.3). Thus, in the field, a genotype × photoperiod × temperature (GPT) interaction was controlling the DTF, as also indicated in the controlled environments (Fig. 4.1), and by Muhammad (1983).

The photoperiod insensitive genotypes also expressed a U-shaped curve of their DTF – if the *T* was higher than 24°C. This was shown in another field experiment where the DTF of ten genotypes was compared at the different *T* of multiple elevations in Guatemala (Fig. 5.3). The *P* at all elevations was near 13 h. In 1982 all the mean *T* were at elevations where beans are commonly grown. Slight delay of the DTF by the highest *T* (25.4°C) for some of the ten genotypes suggested possible beginning of the high-*T* half of a U-shaped response curve. Therefore, in 1983 the yet higher *T* of 28.3°C was compared at a 50 m elevation where beans are not normally grown. Each of the ten genotypes expressed a U-shaped curve of their DTF across the 15–29°C range (Fig. 5.3). The cultivars with independently derived scores of highest sensitivity to *P* had larger delay of their DTF by the higher *T* than the cultivars with low sensitivity. The highest *T* of 23.7°C (in Fig. 4.2) is near that giving the smallest DTF of the U-shaped response to *T* in

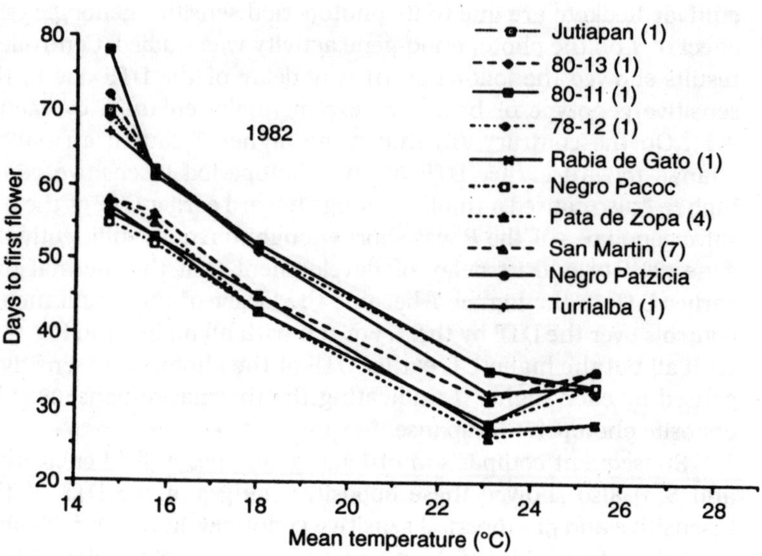

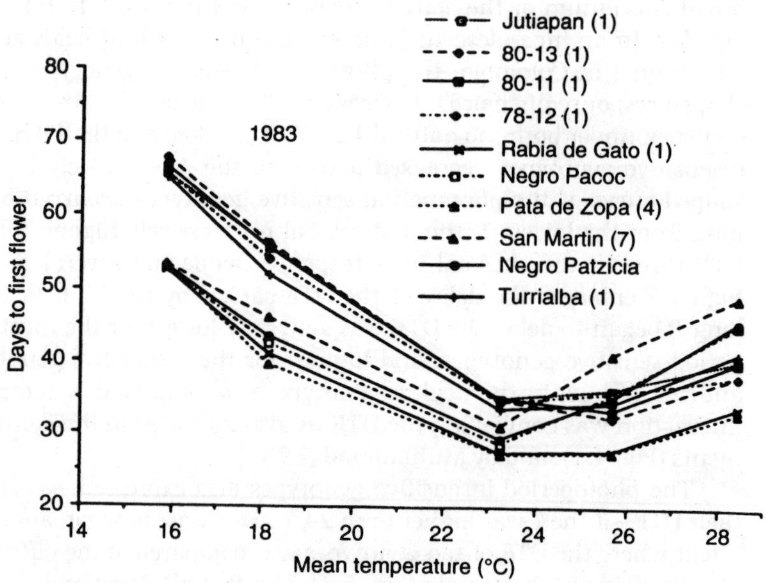

Fig. 5.3. Effect of 14.7–25.4°C in 1982 and of 15.7–28.3°C in 1983 on the days to flowering of ten indeterminate bush bean genotypes. (A rating for sensitivity to photoperiod is in parentheses adjacent to the cultivar name of each genotype; the sensitivities progressively increase from 1 to 7.) (Reprinted from Wallace *et al.*, 1991.)

Fig. 5.3. Both figures show the smallest DTF to occur at a lower *T* for the more photoperiod-sensitive than for the less photoperiod-sensitive genotypes. Both also show delay of DTF by higher *T* to be largest for the more photoperiod-sensitive genotypes.

5.5 The Physiology of Interactive Control by Photoperiod and Temperature over Time to Flowering and Maturity

Comparisons of cultivars Redkloud and Redkote and progenies from crosses between them (Fig. 4.1) led to three publications (Wallace *et al.*, 1993a, b, c). They indicate: across all *P* a genotype with insensitivity to photoperiod expresses the syndrome (section 2.4) that consists of: early DTF and maturity in association with high HI but low biomass and small leaf area (Figs 5.1, 9.1, Tables 4.3, 4.4). On the contrary, sensitivity to photoperiod under a delaying *P* results in the

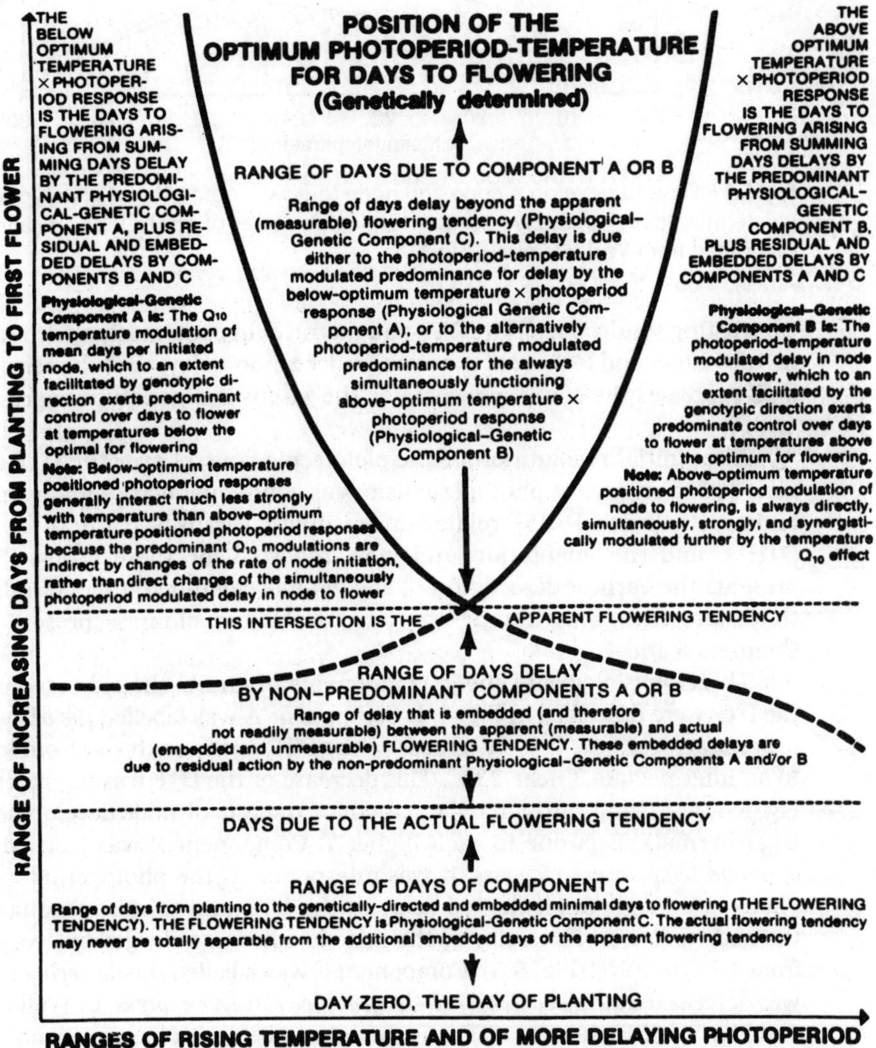

Fig. 5.4. Physiological–genetic bases for photoperiod, temperature and genotype interaction effects on days to flowering of plants. (Reprinted from Wallace, 1985.)

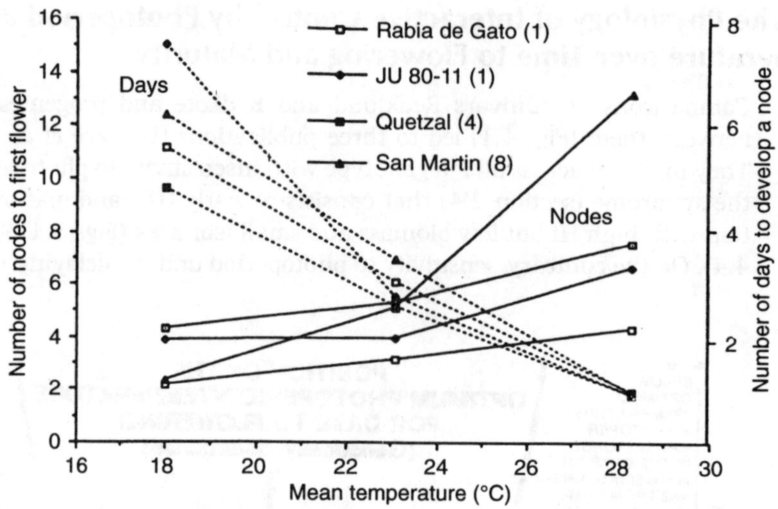

Fig. 5.5. Days to develop a node and node to flower of four bean genotypes with indeterminate bush habit grown at mean temperatures of 18.0°C, 23.1°C and 28.3°C. (Reprinted from Wallace *et al.*, 1991).

contrasting syndrome of late DTF and maturity in combination with low HI, but high biomass and leaf area. If grown under a non-delaying P, the sensitive genotype expresses the same syndrome as the insensitive genotype expresses under any *P*.

After initial recognition of the pleiotropic control over the DTF and HI by gene *Ppd*, a review of photoperiodism was conducted across many crops. That review (Wallace, 1985) related insensitivity versus sensitivity to *P* to the DTHM, and the adaptation and yield reported for these crops. Figure 5.4 presents the earliest description of the yet more detailed understanding about the interconnections among the HI, DTHM and biomass presented within Chapters 5 and 4.

Three physiological-genetic components of the U-shaped response to *T* of the DTF were described in Fig. 5.4. Component A was labelled *the below optimum temperature response*. It is a decrease of the DTF due to each rise from low (12°C) to an intermediate *T* near 23°C. This decrease of the DTF was interpreted, using Fig. 5.5, as arising through enlargement of the rate of node development by the Q_{10} (thermal) effect due to each higher *T*. Component B was labelled *the above optimum temperature response*. It was interpreted as the photoperiod-modulated, photoperiod-gene implemented delay of the DTF, and to arise through delay of the node to flower. This delay of the node to flower is enlarged by every higher *T* from 15°C to 29°C (Fig. 5.5). Component C was labelled the *flowering tendency*. It was described as the *shortest DTF the genotype can ever express*. In Wallace (1985) and Fig. 5.4 this shortest DTF was thought to be controlled by a small subset of maturity genes that do not respond to the *P*. In this book, this shortest expressible DTF is called the basic duration and symbolized as D_b. The control over the D_b is through the majority of all genes as detailed within Chapters 5, 6 and 7.

Expression of the D_b will be shown to require an optimal *T* and a simultaneously optimum *P* (Fig. 6.7). Another conclusion in the 1985 review was that the observable optimum *T* for flowering, indicated by the smallest DTF at the bottom of the U-shaped response curve, represents only an apparent optimum temperature of the DTF, as labelled in Fig. 5.4 and detailed in section 6.3. It is only an apparent optimum *T* because it is lowered by each more delaying *P*, and also by each greater sensitivity of the genotype to the *P* (Figs 4.1, 4.2).

Chapter 5 Summary

The DTHM is the time used and required by a genotype to achieve its (i) net accumulated biomass and (ii) the partitioning of this biomass between the yield and other organs of the plant. The total of the sequential and also parallel usages of days constitute component (iii) of the but three major components of the holistic process of accumulating crop yield. Usage of time is essential to accumulation of the biomass and the partitioning of part of it to the yield, as measured by the HI. The DTHM is strongly controlled by the proportion of the available photosynthate that is partitioned to the organs that become the yield. The larger this proportion the faster these organs grow and consequently the earlier the DTHM. The total accumulated biomass will be enlarged by a longer DTHM. Under a given photoperiod the DTF of each genotype responds across a range of temperatures (*T*) with a U-shaped curve of its DTF. Starting from the lowest *T*, each higher *T* causes earlier DTF, but from an intermediate *T* each higher *T* delays the DTF. The *T* that begins to delay the DTF is higher for an insensitive than a photoperiod sensitive genotype, and is made yet higher also by each decrease of the sensitivity to *P* of the genotype. This genotype × photoperiod × temperature (GPT) interaction control results in differential syndromes, which consist of alternative combinations of high versus low levels among the DTF, DTHM, HI, aerial biomass, number of branches, nodes, leaves and leaf area. The levels of as many as 26 traits all change as a consequence of the control by the photoperiod gene activity of partitioning of the available photosynthate to the reproductive organs that become the yield – versus – to continued growth of additional vegetative organs.

6

A Model of Photoperiod × Temperature Interaction Effects on Plant Development

Paradigm shifts 8, 9, 11, 12 and 13 about the photoperiod response and 14 and 16 about the vernalization response are derived in this chapter. Shift 8 is that any activity of a sensitive photoperiod gene inhibits (delays) the autonomous capability possessed by every plant to develop to flowering and all developmental stages. Shift 9 is that jointly the photoperiod and thermal responses are the largest controls over the cultivar's harvest index, adaptation and yield, in addition to their control over the days to flowering and days to harvest maturity. Paradigm shift 11 is that accurate quantification of the photoperiod response requires simultaneous quantification of (i) the Q_{10} effect on the photoperiod-gene activity, and simultaneous quantification of (ii) the Q_{10} effect known as the thermal-time response. Shift 12 is that the autonomous capability that all plants possess to develop to all their stages is implemented by the majority of all the genes of the plant, but it is delayed (inhibited) by any photoperiod and/or any vernalization gene activity. Paradigm shift 13 is that photoperiod-gene activity occurs at high but not at a sufficiently low temperature. Paradigm shift 14 is that vernalization gene activity also occurs only if the temperature is sufficiently high. Paradigm shift 16 is that any change toward earlier flowering by a low-temperature that can be replaced by a photoperiod is due to photoperiod-gene activity rather than to vernalization-gene activity. The gene actions and the modulations of them by the photoperiod and temperature are described in terms of their control over both the days to flowering (DTF) and the node to flower.

Yan and Wallace (1996) was titled: *A model of genotype × photoperiod × temperature interaction effects on plant development*. This chapter essentially duplicates two segments of that review. New insights arose while using the model in Chapter 7 to predict the DTF of 12 crops. The additional insights indicated need to change some of the original statements. The model equation remains as originally published.

6.1. Photoperiod–Temperature Modulations of Photoperiod-gene Activity

Photoperiod (*P*) and temperature (*T*) are the two major modulators of plant development (Halse and Weir, 1970; Halloran, 1975; Squire, 1990; Hodges, 1991; Jordan, 1993). The model for explaining modulations of photoperiod-gene activity by both *P* and *T* is derived from the following review of the control over the number of nodes (or leaves) to flowering.

6.1.1 Response of node to first flower to temperature and photoperiod in pea

Six pea cultivars with early to late flowering were grown by Berry and Aitken (1979) in 16 combinations of four mean *T* (6°C, 12°C, 18°C and 24°C) and four *P* (8 h, 12 h, 16 h and 24 h). Both number of nodes to first flower (NTF) and the DTF were recorded. The response surface of the NTF for each cultivar (Fig. 6.1) provided the following interpretations.

1. The NTF of the photoperiod insensitive cultivar Alaska is independent of both *T* and *P*. But, for the five photoperiod sensitive cultivars their NTF is enlarged by both each higher *T* and each shorter *P*. This suggests the photoperiod-gene(s) control responses of the NTF to both *P* and *T*.

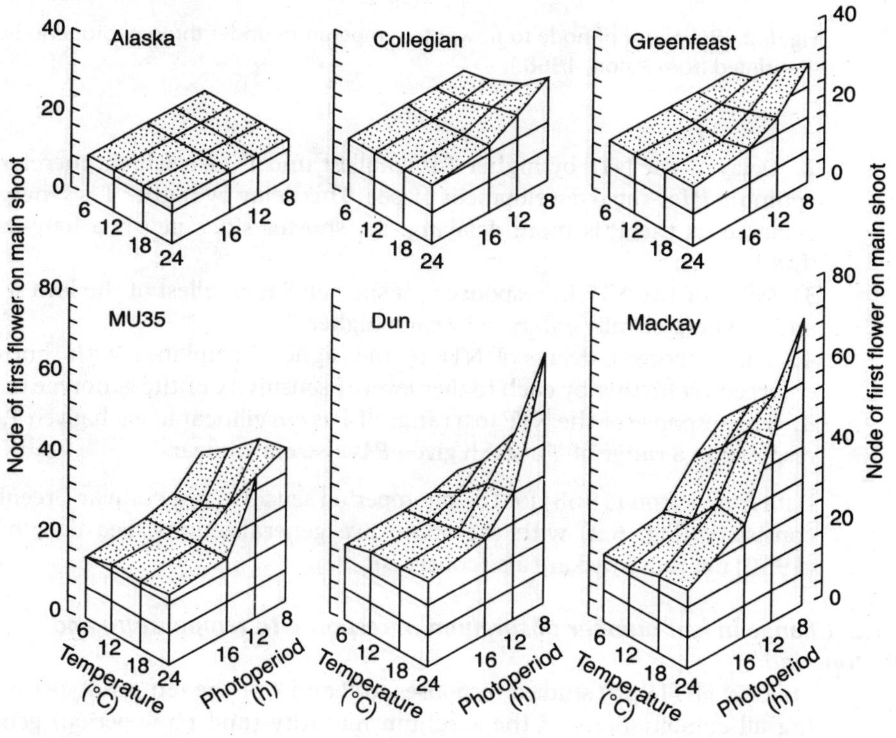

Fig. 6.1. Response surfaces showing the effect of factorial combinations of mean daily temperature (6°C, 12°C, 18°C or 24°C) and photoperiod (8 h, 12 h, 16 h or 24 h) on the node of first flower of six pea varieties of increasing maturity. (Reprinted from Berry and Aitken, 1979.)

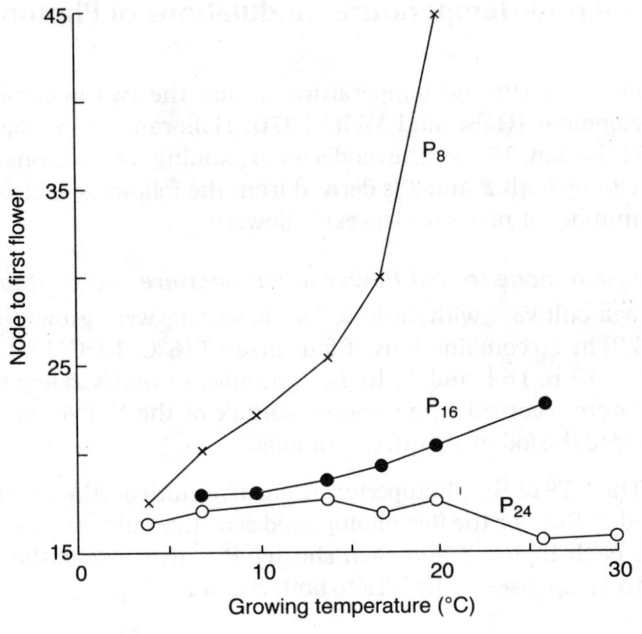

Fig. 6.2. Response of node to flower to temperature under three photoperiods. (Reprinted from Paton, 1968.)

2. Delay of the NTF by higher *T* is smallest under 24 h which, therefore, is the optimum *P* for rapid development of pea. This delay by higher *T* is synergistically enlarged as the *P* is more delaying, i.e. shorter since pea is a long-day plant (LDP).
3. Delay of the NTF in response to a shorter *P* is smallest at the lowest *T* (6°C), and is synergistically enlarged by each higher *T*.
4. The synergistic delays of NTF by the higher *T* combined with shorter *P* are enlarged yet further by each higher level of sensitivity of the genotype to the *P*.
5. The response of the NTF to a range of *P* is curvilinear at each given *T*, and the response to a range of *T* at each given *P* is also curvilinear.

Data from Paton (1968) for the photoperiod sensitive pea cultivar Greenfeast are consistent (Fig. 6.2) with the above five generalizations. Pea data of Barber (1959) not presented are also consistent.

6.1.2 Change in leaf number of sorghum in response to temperature and photoperiod

Major *et al.* (1990) studied responses to *P* and *T* of the ten genotypes representing all combinations of the sorghum maturity (and photoperiod) genes *Ma1*, *Ma2* and *Ma3*. Leaf number to tasselling (NTF) is presented for each of the ten genotypes at 20°C, 25°C and 30°C in combination with *P* of 12 h and 14 h (Fig. 6.3). The sorghum genotypes expressed synergistic increase of their NTF in response to both a higher *T* and a more delaying *P*, just as for pea. The delay for

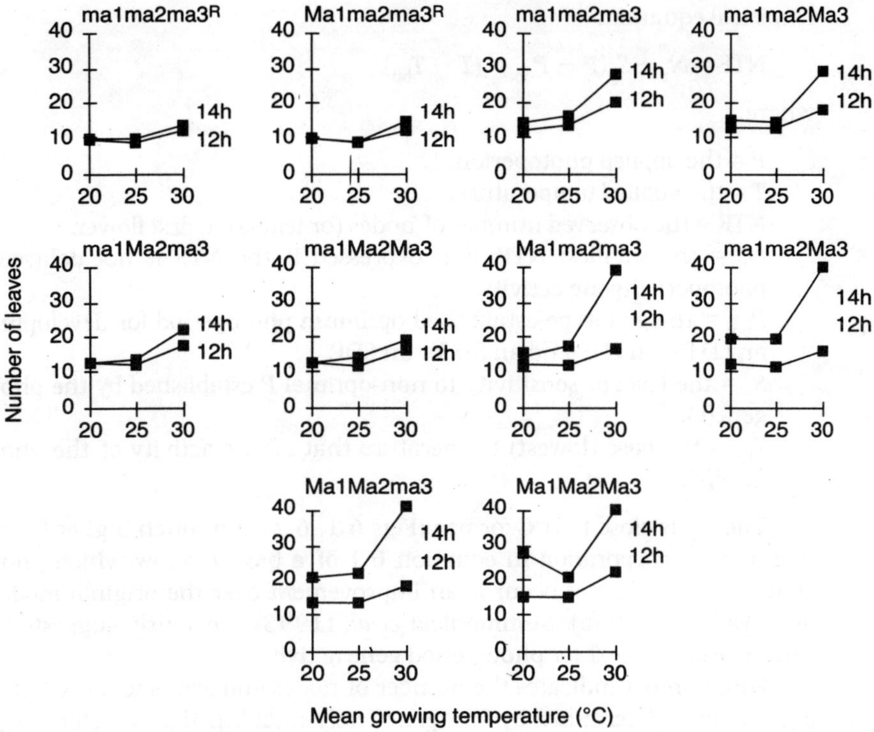

Fig. 6.3. Leaf number of ten sorghum maturity genotypes grown under mean temperatures of 20°C, 25°C and 30°C and photoperiods of 12 h and 14 h. (Data are from Major *et al.*, 1990. Reprinted from Yan and Wallace, 1996.)

sorghum resulted from longer *P* in combination with higher *T*, because sorghum is a short-day plant (SDP). Figure 6.3 provides the following interpretations.

1. For all ten genotypes the NTF tended to be larger under the more delaying 14 h than 12 h.
2. For all genotypes also, the NTF tended to be larger at the progressively higher *T*.
3. The delays described in 1 and 2 occurred only at *T* higher than a base temperature (base-*T*) of about 25°C for the two genotypes with the gene *ma3R*, 20°C for the four with *ma1*, but below 20°C for the four *Ma1* genotypes.
4. Below these base-*T*, there is no delay of the NTF by a higher *T* and/or delay by the longer 14 h. Above the base-*T*, the increase in NTF by 14 h is synergistically enlarged by each higher *T*.
5. There is no information about curvilinearity versus linearity of the response to *P*, since there are only two levels of *P*. The three *T* establish, however, that the increase of the NTF in response to higher *T* is curvilinear.

6.1.3 The model for photoperiod-gene control of the node to flower

The genotype × photoperiod × temperature (GPT) interaction effect on the NTF of pea (Figs 5.1, 5.2) and sorghum (Fig. 6.3) can be described by the following empirical equation:

$$\text{NTF} = N_b + S_p(P - P_{opt})^2 (T - T_{bp})^2 \quad [6.1]$$

where:

P = the applied photoperiod.
T = the applied temperature.
NTF = the observed number of nodes (or leaves) to first flower.
N_b = the smallest NTF; it is expressed if the NTF is not delayed by any photoperiod-gene activity.
P_{opt} = the genotype-established optimum photoperiod for development. It is an LD for an LDP and an SD for an SDP.
S_p = the level of sensitivity to non-optimal P established by the photoperiod gene(s).
T_{bp} = the base (lowest) temperature that allows activity of the photoperiod gene(s).

The T_{bp} is close to 0°C for pea (Figs 6 1, 6.2), but much higher for sorghum (Fig. 6.3). Incorporation in equation 6.1 of a base-T below which photoperiod gene activity does not occur is an improvement over the original model in Yan and Wallace (1995b). Summerfield *et al.* (1993) previously suggested possible existence of a base-T for photoperiod gene activity.

Equation 6.1 indicates the number of nodes and leaves to flowering (NTF) of a genotype is determined by four genetically-established parameters (N_b, P_{opt}, S_p and T_{bp}). These parameters interact with both the applied T and applied P to control the NTF.

6.1.4 Interpretations from the node to flower model

Equation 6.1 is a GPT interaction model of the photoperiod gene activity and it implies:

1. The NTF is controlled by the expressed level of the photoperiod gene activity. Each higher level of this activity delays the node (i.e. leaf axil) at which the first flower develops.
2. Delay of flowering by the photoperiod gene(s) occurs only when the plants are under non-optimal P (i.e. when $P > P_{opt}$ for a SDP or $P < P_{opt}$ for an LDP), and when the T is higher than the base-T (T_{bp}) that allows activity of the photoperiod gene(s).
3. The delay of NTF due to a non-optimal P is enlarged by each higher T above the T_{bp}. Alternatively stated, each higher T above the T_{bp} always modulates a larger NTF, if the accompanying P is non-optimal.
4. The responses of NTF to both T and P are curvilinear. A quadratic function is used to account for the curvilinearity.

6.1.5 Physiological explanations of photoperiodism

1. The smallest NTF results from the P_{opt}, because it prevents all photoperiod gene activity that would delay plant development.
2. A photoperiod response is inevitably a GPT interaction caused response.

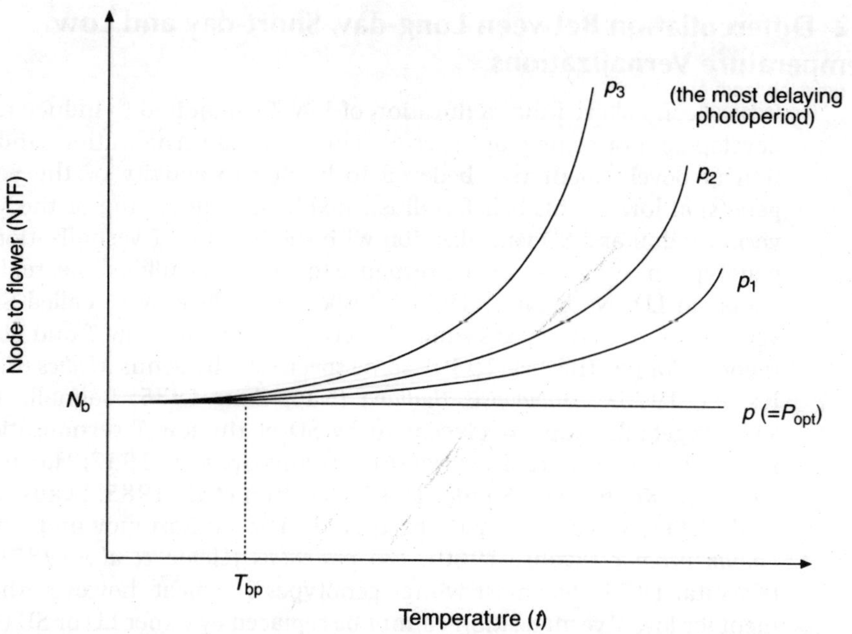

Fig. 6.4. Simulation by equation 6.1 of the photoperiod-gene directed control over node to flower (NTF) in response to temperature under four progressively more delaying photoperiods. Illustrated are: (i) The photoperiod gene controlled response is always a photoperiod × temperature interaction effect. (ii) There is no response of the NTF to temperature under the optimum photoperiod. (iii) There is no change of NTF in response to either photoperiod or temperature, if the temperature is low (when $t \leq T_{bp}$). (iv) Above the T_{bp}, the NTF becomes larger as the photoperiod is more delaying and the temperature is higher. And (v) each of the effect by a delaying photoperiod and the effect by a higher temperature synergistically enlarges the effect by the other.

3. The inevitable modulation of photoperiod gene activity by T was attributed to the Q_{10} effect (Wallace, 1985). Through its Q_{10} effect, each higher T enlarges the delay of the NTF (of development) caused by a delaying P. Reversely stated, with each reduction of the T, this Q_{10} effect depresses the photoperiod gene activity and the consequent delay of NTF by a non-optimal P. No photoperiod-gene activity occurs at any T lower than the T_{bp}.

4. A duration of low T can cause the same lack of delay of the NTF as the P_{opt}, if the T is low enough to prevent all photoperiod-gene activity and its delay of development. For equation 6.1, the low T refers to any T between biological zero and T_{bp}.

Simulations (Fig. 6.4) using equation 6.1 show lack of delay of the NTF by progressively higher T when either (i) $T < T_{bp}$; or (ii) the P is optimal, i.e. when $P = P_{opt}$. Also shown is that (iii) under each progressively more delaying P, the delay of the NTF by each higher T above the T_{bp} becomes progressively larger.

6.2 Differentiation Between Long-day, Short-day and Low Temperature Vernalizations

It has been believed that a duration of low *T* is required to induce reproductive development of some plant species. This is called vernalization, and its promotion of development was believed to be due to activity of the vernalization gene(s) at low *T*. This belief is illustrated by a confounding of the so-called LD vernalization and SD vernalization with the true low-*T* vernalization. For some genotypes of some species the response to low *T* was fulfilled (i.e. replaced) by an SD or an LD (Napp-Zinn, 1984). Respectively, these were called SD- and LD-vernalization. Genotypes seemingly vernalized by both low *T* and LD have been reported for pea (Barber, 1959), some species of the genus *Medics* (Clarkson and Russell, 1975), *Arabidopsis thaliana* (Napp-Zinn 1985; Bagnall, 1993), and some vegetable crops. Replacement by SD of the low-*T* vernalization requirement has been reported for rye (Purvis and Gregory, 1937; Hartman, 1964), wheat (McKenney and Sando, 1935; Davidson *et al.*, 1985; Evans; 1987; Miao *et al.*, 1991, 1992), barley (Roberts *et al.*, 1988a; Razumov and Limar, 1974), *Lolium perenne* (Evans, 1960), *Poa pratentsis* (Heide *et al.*, 1987) and *Pinus* (Katsuta, 1975). For most winter genotypes of wheat, however, their requirement for low-*T* vernalization cannot be replaced by either LD or SD (Davidson *et al.*, 1985; Miao *et al.*, 1991). The physiology of the replacement of the requirement for low-*T* vernalization by LD or SD has not been adequately explained. Equation 6.1 provides an explanation.

6.2.1 *Long-day vernalization*

We generated a hypothesis from the $P \times T$ modulation of the photoperiod-gene activity described by equation 6.1. The hypothesis is that the so-called LD vernalization of LDP is not a vernalization-gene-directed control over development. Rather, it is a photoperiod-gene-directed requirement for a duration of an LD. It has been called LD-vernalization, because the acceleration of development by the LD can be replaced (substituted for) by a duration of low-*T*. Any acceleration of development by low *T* was assumed to be due to the activity of vernalization genes. On the contrary, equation 6.1 indicates any photoperiod gene activity and its delay are partially depressed by each lower *T* through the range $T > T_{bp}$. All photoperiod-gene-directed delay is prevented by low *T* within the range of $0°C < T \leq T_{bp}$. Many cellular activities approach their minimum rate only at 0°C.

6.2.2 *Short-day vernalization*

Equation 6.1 provides a hypothesis also for SD vernalization. It is the same as above for LD vernalization. Read that hypothesis again, but read LD as SD. Based on these assumptions, the so-called LD- and SD-vernalization are interpretable.

This is evidenced by an example from Miao *et al.* (1992). Their experiment included two short-long-day (SLD) wheat cultivars, Xingdong 2 and Feimai. Three treatments follow.

1. A low-*T* in which the germinating seeds were kept at 0–2°C for 30 days in the dark, and then planted in a greenhouse with 24 h and about 22°C.
2. An SD treatment for which immediately after emergence the seedlings were

kept for 30 days under 6 h at 20°C, and then transferred into the same greenhouse (24 h and about 22°C).

3. An LD control for which seeds with no vernalization were planted and grown in the same greenhouse.

For cv. Xingdong 2, the leaf numbers of the respective treatments were 13, 13 and 18. For cv. Feimai, they were 16, 17, and 19. Thus, the low *T* and SD reduced the NTF comparably. Measured as the days from the beginning of the seed germination to heading, however, the SD reduced the time more than the low-*T*. See Roberts *et al.*, 1988a for similar results for barley.

6.2.3 Low temperature vernalization

We suggest that true vernalization is advance of the reproductive development by low-*T*, which cannot be substituted for nor modulated by any *P*. Contrary to past interpretation, we suggest the following. All vernalization gene activity is expressed only under *T* higher than a base-*T*, above which, the vernalization gene activity is enlarged by each higher *T*. Stated alternatively, the phenomenon called vernalization is due to depression of the vernalization gene activity by a duration of low *T*. Low *T* depress the vernalization gene activity, thereby preventing its delay of the developmental advance. On the contrary, at high *T*, high levels of vernalization gene activity delay development.

6.2.4 Confounding of vernalization versus photoperiodism

Evans (1987) studied acceleration by SD of inflorescence initiation of the winter wheat cultivar Templar. Infloresence initiation of Templar occurred earlier if its early stage of development was under an SD, but at later stages the development was made earlier by an LD. Thus, Templar is an SLD genotype. Evans pointed out that, although both short-day induction (the term he substituted for SD vernalization) and low-*T* vernalization accelerate reproductive development of this genotype, the effects by SD and low *T* differ in six respects. The most convincing is that the apical meristem of the shoot detects the low *T*, and this is also where the vernalization effect occurs, but it is the leaves that perceive the *P* even though the effect is again at the shoot apex. We suggest that the earliest development of Templar requires an SD duration rather than a low-*T* duration. According to equation 6.1, this SD requirement will be substituted for (fulfilled) by a duration of low *T*, because lower *T* than the T_{bp} of the photoperiod gene will totally prevent its activity.

For pea, late cultivars respond to both low *T* and LD, and responsiveness to both is controlled by the same complementary genes: *Sn, Dne* and *Ppd* (Barber, 1959; King and Murfet, 1985; Arumingtyas and Murfet, 1994). From this, we interpret that this gene system confers a requirement for an LD, that can be at least partially replaced by a low *T*. Murfet (1985) suggested vernalization promotes flowering by at least two mechanisms. The first arises exclusively at the shoot apex, and predisposes the plant to respond to the photoperiod, whereas the second occurs at any site at which the *Sn Dne Ppd* gene system operates. Clearly, the first mechanism refers to true vernalization due to vernalization genes, but the second refers to the so-called LD vernalization. Both LD and SD vernalization are photoperiod-gene directed responses, and the gene activity is always modulated simultaneously by both the *P* and *T*. True vernalization may not be restrained to occurring at the shoot apex. Vince-Prue (1975) stated, 'It is possible that not only apical meristems but all dividing cells, including those in leaves, are potential sites for vernalization.'

Murfet's concept of vernalization seems to be any acceleration of development by a low *T*, a commonly shared view (Vince-Prue, 1975). Both the requirement for vernalization and for SD or LD can be fulfilled by a low-*T* duration. Therefore, it is necessary to learn whether the accelerated development is controlled by the activity of photoperiod- or vernalization-gene(s). This requires determination of whether response to the low *T* can be influenced by different photoperiods.

The points requiring emphasis follow. (i) True vernalization is vernalization-gene directed with no modulation by the *P*. (ii) Any process responsive to *P* must be a photoperiod-gene controlled response. (iii) In contrast to vernalization gene activity being modulated by only the *T*, all photoperiod-gene directed response is inevitably modulated simultaneously by both *P* and *T*. All three points assume application from biochemistry, as convincingly argued by Grierson (1995), of the Q_{10} concept that the levels of all enzyme (gene-directed) activities are enlarged by a higher *T*.

Prior confusion of low-*T* vernalization with acceleration of development by LD or SD arises from the following similarities between vernalization-gene and photoperiod-gene activities. (i) Both activities delay development. (ii) Both activities occur only if the *T* is higher than a base-*T*. (iii) Both gene activities and their delay of the NTF are enlarged by the Q_{10} effect of a yet higher *T*. (iv) Both gene activities are totally prevented by a sufficient duration of a sufficiently low *T*. With T_{bp} being the base-*T* that allows any photoperiod gene activity, the range $0°C < T \leq T_{bp}$ prevents all photoperiod gene activity and thereby substitutes for the P_{opt}. Likewise, with T_{bv} being the lowest *T* that permits vernalization-gene activity, the range $0°C < T < T_{bv}$ prevents vernalization gene activity. Paradoxically, judged from the common viewpoint that vernalization gene activity promotes development and occurs at low *T*, it is prevention of the vernalization-gene activity by low *T* that accelerates the development. We hypothesize that the only major difference between delays of development due to the photoperiod- and vernalization-gene activities is that vernalization-gene activity and its delay of development are not at all modulated by the *P*.

6.3 Modulations of the Days to Flowering by Temperature, Photoperiod and Genotype

The DTF is integral to cultivar adaptation and yield. Understanding the biological bases of adaptation and yield requires understanding of the integrated control over the DTF by vernalization, photoperiodism, temperature and the genotype. This section focuses on the control through GPT interaction over the DTF of crop plants that either: (i) have no vernalization requirement, or (ii) for which their vernalization requirement has been fully completed, and only their post-vernalization development is considered.

6.3.1 The days to flowering is always modulated by temperature

If a genotype is photoperiod insensitive (Fig. 6.1), or if the applied *P* is optimal for development of the genotype, the NTF is not or is marginally modulated by the *P* (Figs 6.1, 6.2, 6.3). The DTF, on the contrary, is always modulated by the *T*, regardless of the *P* (Fig. 5.3 for bean and Fig. 6.5 for pea), and of the genotype (Fig. 4.2 for bean and Fig. 6.6 for pea).

6.3.2 The response of days to flowering to temperature follows a U-shaped curve

It was documented for bean (Figs 4.1, 4.2, 5.3) that, under a single *P*, across a sufficient range of *T*, response of the DTF to *T* follows a U-shaped curve. Other crops for which this has been shown include pea (Figs 6.5, 6.6), wheat (Slafer and Rawson, 1994b), chrysanthemum (De Jong, 1989), pigeonpea (McPherson *et al.*, 1985; Omanga *et al.*, 1995), cowpea (El-Madina and Hall, 1986), mungbean (Fernandez and Chen, 1989), dahlia (Brondum and Heins, 1993), and maize (Ellis *et al.*, 1992b).

Under a given *P* and a sufficient range of *T*, an optimal *T* is observable. It causes the smallest DTF at the bottom of the U-shaped curve (Figs 5.3, 6.5). This observed but only an apparent optimum *T* divides the U-shaped curve between two contrasting responses (Wallace, 1985; Wallace *et al.*, 1991, 1995a). The half

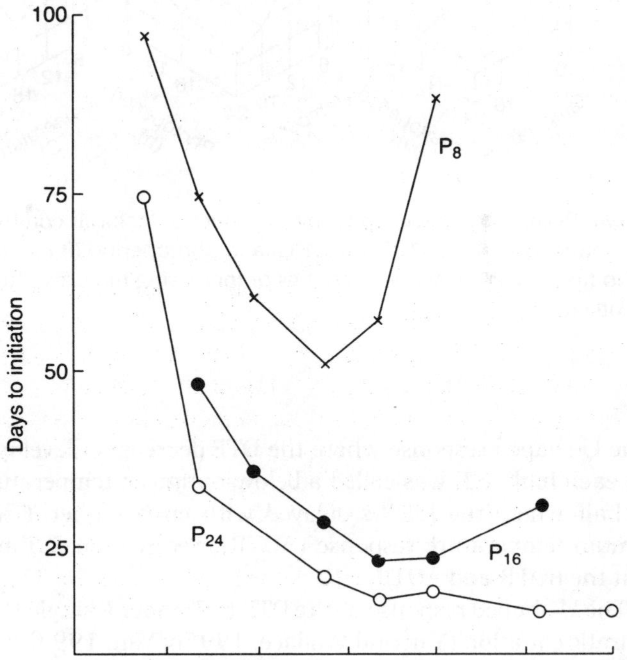

Fig. 6.5. Days to initiation of flower buds in response to temperature under three photoperiods for the photoperiod sensitive pea cultivar 'Greenfeast'. (Reprinted from Paton, 1968.)

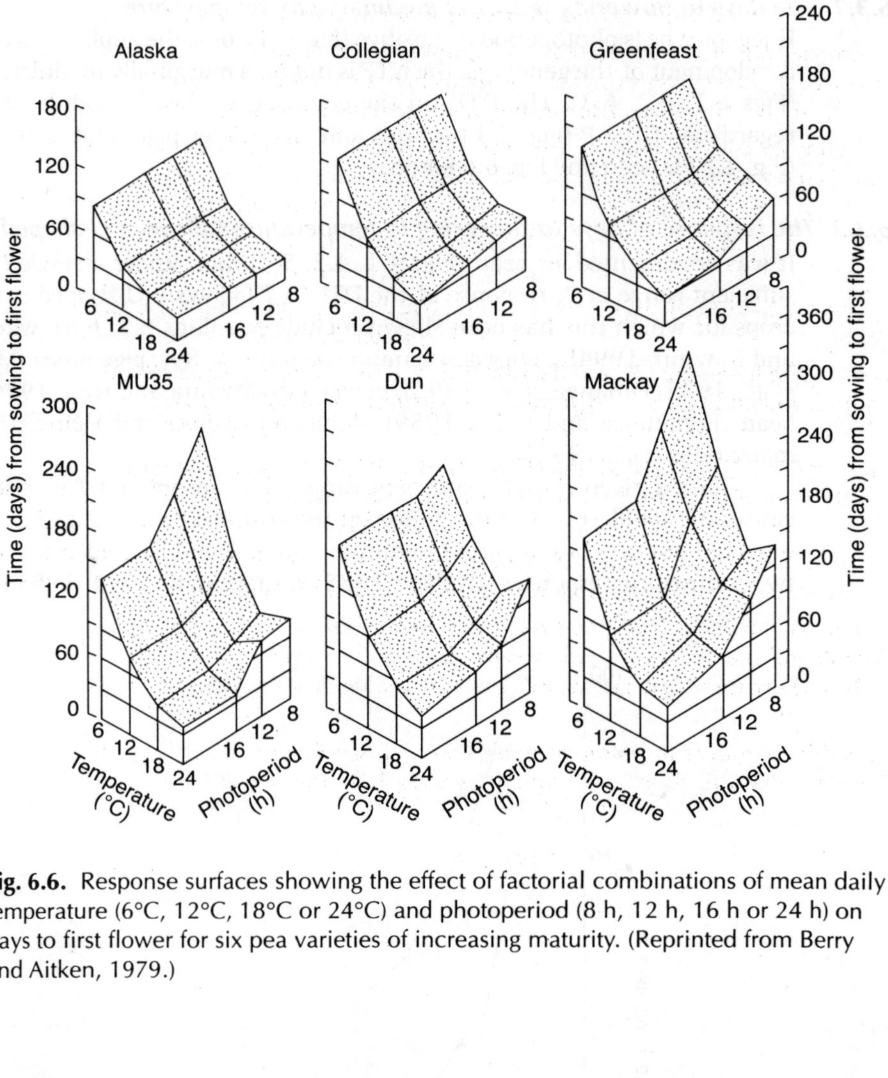

Fig. 6.6. Response surfaces showing the effect of factorial combinations of mean daily temperature (6°C, 12°C, 18°C or 24°C) and photoperiod (8 h, 12 h, 16 h or 24 h) on days to first flower for six pea varieties of increasing maturity. (Reprinted from Berry and Aitken, 1979.)

of the U-shaped response where the DTF decreases (development is accelerated) with each higher *T*, was called a below-optimum temperature response (BOTR). The half where the DTF is delayed with each higher *T* was called an above-optimum temperature response (AOTR). See section 5.5 and Fig. 5.4 for more about the BOTR and AOTR.

The U-shaped response of the DTF to *T* under a single *P* can be modelled by a quadratic equation (Yan and Wallace, 1995b; Yan, 1993b):

$$\text{DTF} = D_{ab} + S_{at}(T - T_{aopt})^2 \quad [6.2]$$

where:

DTF = the days used to develop to flowering (or other specified stages).
T = the applied temperature.

T_{aopt} = the apparent optimum temperature at which the DTF is smallest under the given P

D_{ab} = the apparent basic duration (shortest duration) of the DTF under the applied P. The D_{ab} is expressed at the bottom of the U-shaped curve where $T = T_{aopt}$.

S_{at} = the apparent sensitivity of the DTF to deviation of T from the T_{aopt}; it is the steepness of the side(s) of the U-shaped curve.

When $T < T_{aopt}$, each higher T accelerates development; this is the BOTR. When $T > T_{aopt}$, each higher T delays development; this is the AOTR. The BOTR is better known as the 'thermal time response'. The term BOTR indicates that the thermal time response is observable only if the T is lower than the T_{aopt}, even though the thermal response is always functioning. The AOTR and BOTR (i.e. the U-shaped response of the DTF to an adequate range of T) have not been adequately recognized. Summerfield *et al.* (1991) and others consider the thermal time concept to be legitimate only if the *rates* of development are *linearly* related to the T.

6.3.3 The U-shaped response of days to flowering to temperature is modified by the photoperiod

All three parameters in equation 6.2 (D_{ab}, T_{aopt} and S_{at}) are labelled as *apparent*, because each is altered by any change of the P. This is illustrated by a larger D_{ab}, a lower T_{aopt} and a higher S_{at} with each more delaying P for the pea cultivar Greenfeast (Fig. 6.5), and for three insensitive, three moderately sensitive and three highly photoperiod sensitive bean genotypes (Fig. 4.2). In Fig. 4.2, the three photoperiod insensitive genotypes of bean show only the BOTR half of the U-shaped response to T as high as 24°C. But, even insensitive bean genotypes express the complete U-shaped curve if the T range extends to as high as 28°C (Fig. 5.3).

The observed response of the DTF to T can be a typical U-shaped curve or only part of the curve, depending on: (i) the range of the tested T; (ii) the applied P; and (iii) the sensitivity to P of the genotype. If the genotype is photoperiod-insensitive (Figs 4.2, 6.6), or if the P is near-optimal (Figs 6.5, 6.6), or if the range does not extend to an adequately high-T (Fig. 6.6), the response can be only the BOTR half of the U-shaped curve. On the contrary, if the genotype is highly photoperiod sensitive and if the P is highly delaying, the temperature response can be only the AOTR half (Fig. 4.2). In conclusion, modulations (i) of the photoperiod-gene activity by both P and T and (ii) of the thermal response by only the T are jointly causal of the U-shaped response of the DTF to T.

6.3.4 The U-shaped response of days to flowering to temperature is controlled by two oppositely acting physiological components

The U-shaped response to T of the DTF was interpreted as follows, based on Fig. 5.5. The T simultaneously modulates the DTF through the (i) NTF and (ii) the days needed to develop a node. A higher T accelerates development toward flowering by reducing the time needed to develop a node (acceleration of the vegetative development), but simultaneously delays development to flowering by enlarging the NTF to delay the reproductive development. Both the delay and acceleration of DTF always occur simultaneously across the full range of tested T. But, only that

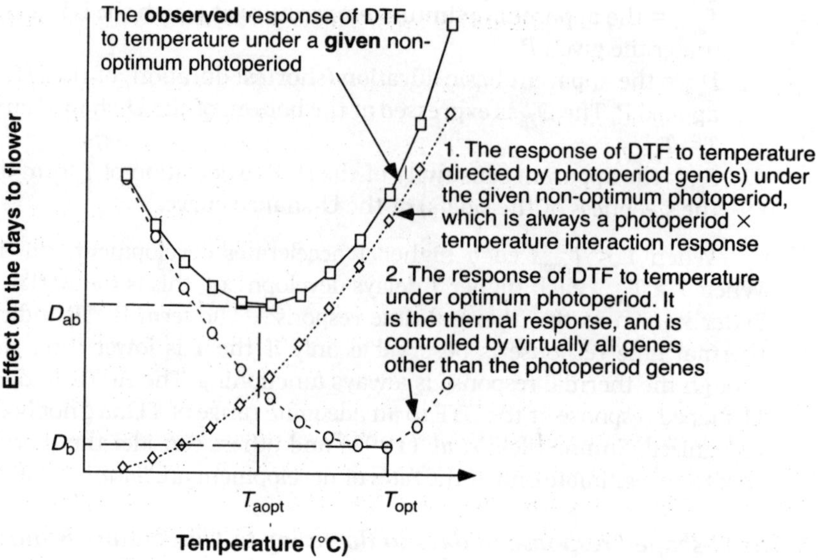

Fig. 6.7. Simulation by equation 6.3 of the U-shaped response of the days to flowering to a sufficient range of temperature under a non-optimal photoperiod. The U-shaped response curve results from two subsets of gene control that result in opposite effects on the DTF in response to changes of the temperature. The controls are implemented by: (i) the photoperiod genes, this always being a genotype × photoperiod × temperature interaction effect on the DTF for which a higher temperature enlarges the DTF, and by: (ii) the majority of all the other genes, this being the thermal response (to the temperature only) for which each higher temperature decreases the DTF until it exceeds the T_{opt}. This T_{opt} is a fairly high temperature where most species are grown. This explains why for many species development is progressively more rapid across a wide range of higher temperatures – provided the higher temperatures are not delaying the development through any vernalization-gene and/or photoperiod-gene activity. The apparent optimum temperature (T_{aopt}) is always lower than the true T_{opt}, and the apparent basic duration of the DTF (D_{ab}) is always larger than the true basic duration of the DTF (D_b). (Modified from Yan and Wallace, 1996.)

part of the larger change of the DTF caused by a change of the *T* that is not cancelled by the smaller opposite change is observable as a cause of the change of DTF. Under delaying *P*, the NTF of bean is delayed (enlarged) by each higher-*T* (Fig. 5.5) as shown for both pea in Figs 6.1, 6.2, and for sorghum in Fig. 6.3. Simultaneously, the rate of node development is accelerated across the full range of higher *T*. Interpretation follows. For the AOTR half of the U-shaped curve of the DTF (Fig. 5.4), delay of the NTF by each higher-*T* is the larger of the simultaneous delaying and accelerating controls over the DTF. On the contrary, for the BOTR half, acceleration of the rate of node development by each higher *T* is the larger effect by the *T* on the DTF (Wallace *et al.*, 1991). The observed T_{aopt} indicated by the smallest DTF along the U-shaped response occurs (Figs 6.5, 6.6, 6.7) when the two opposite effects on the DTF by the *T* are equal. Being equal but opposite, each exactly cancels the other.

Now we interpret that (i) observable delay of the NTF by a higher-T, due to its larger control of the DTF across the AOTR half of the U-shaped response curve, is implemented by the photoperiod-gene activity. (ii) The control over the rate of node development by T, which is the larger control of the DTF for the BOTR half of the U-shaped response, is due to the thermal time effect of the T (Fig. 5.5). (iii) The thermal response arises through the collective activity of virtually all the genes of the plant, excluding only (a) any photoperiod genes and (b) any vernalization genes. It is interpreted also (iv) that all control over development by T is through its Q_{10} effects, for which every gene activity is enlarged by each T higher than the base-T that allows that gene activity but remains lower than the optimum T of that gene's activity. Some enlarged gene activities accelerate development while others delay it.

6.3.5 The integrated model for the response of DTF to temperature and photoperiod

The literature review above led to the following integrated DTF model:

$$\mathrm{DTF} = D_b + S_p(P - P_{opt})^2(T - T_{bp})^2 + S_t(T - T_{opt})^2 \qquad [6.3]$$

where:

DTF = the days to flowering (or other specified stage).

D_b = the basic (shortest expressible) duration of the DTF. *It occurs only when the plant is grown under both the optimal P and optimal T established by the genotype.*

S_p, P_{opt}, and T_{bp} are as defined for equation 6.1.

T_{opt} = the optimum T of the thermal response. It is controlled collectively by the majority of all the genes of the plant – exclusive of any photoperiod-genes and any vernalization-genes.

S_t = the sensitivity of the genotype to deviation of the applied T from the T_{opt}. The S_t is also controlled collectively by the majority of all the genes of the plant, not including any photoperiod- or vernalization-genes. (See equation 7.4 in section 7.3.1.) There a definition for T_{opt} expands upon the concept that the thermal response is controlled collectively by the majority of all the genes of the plant. The most thorough discussion of control over the DTF by the majority of all the genes is in sections 7.3.2 and 7.3.3.

Equation 6.3 indicates the observable U-shaped response of DTF to T under a given photoperiod, as modelled by equation 6.2, results from the two already discussed distinct controls – the control through the photoperiod-gene directed GPT interaction $[S_p(P - P_{opt})^2(T - T_{bp})^2]$, and the control directed by the collective majority of all the other genes through the thermal-time response $[S_t(T - T_{opt})^2]$.

When the T is optimum, i.e. when $T = T_{opt}$, the DTF is exclusively controlled by the photoperiod gene activities:

$$\mathrm{DTF} = D_b + S_p(P - P_{opt})^2(T - T_{bp})^2 \qquad [6.3a]$$

Equation 6.3a is identical to 6.1, except that 6.3a focuses on the DTF while equation 6.1 focuses on the NTF. This change from NTF to DTF involves a change of unit for the S_p. But, for simplicity this same symbol is used for both equations.

When $P = P_{opt}$, the DTF is exclusively controlled by activities of the majority of all the genes, and they all respond to only the T:

$$DTF = D_b + S_t(T - T_{opt})^2 \quad [6.3b]$$

Both equation 6.3b and equation 6.2 describe a U-shaped response of the DTF to T. However, equation 6.2 describes a temperature response at any P. Therefore, all responses to T are confounded with the response to the P, so all the parameters are 'apparent' ones. In contrast, equation 6.3b describes the response to T under only the P_{opt} of the photoperiod sensitive genotypes, or under any P for truly photoperiod insensitive genotypes. Therefore, all parameters of equation 6.3b are totally determined by the genotype. Simulation of equation 6.3 is illustrated in Fig. 6.7.

Equations 6.2 and 6.3 give two relationships that are consistent with the observations. The first is:

$$S_{at} = S_t + S_p(P - P_{opt})^2$$

The apparent sensitivity of the genotype to the temperature (S_{at}) is determined by both (i), the genotype for sensitivity to non-optimal temperature (S_t) and (ii) the genotype for sensitivity to the non-optimal photoperiod (S_p). Additionally, (iii) the S_{at} is enlarged by a more delaying P. Consequently, under a delaying P, a more photoperiod sensitive genotype is seemingly more sensitive to T. The second relationship is:

$$T_{aopt} = \frac{T_{opt} S_t + T_{bp} S_p \left(P - P_{opt}\right)^2}{S_t + S_p \left(P - P_{opt}\right)^2}$$

This indicates that the apparent optimum temperature (T_{aopt}) of a genotype under a non-optimal P is always lower than the genotype's true optimum temperature (T_{opt}), with this T_{opt} being observable only when the P is also optimal. For a given genotype, the T_{aopt} becomes lower under a more delaying P. The T_{aopt} is always between the T_{opt} and the T_{bp}.

Wallace (1985) was the first to use the term apparent optimum temperature (T_{aopt}) in addition to the term optimum temperature (T_{opt}). Both were used to differentiate presence versus absence of the $P \times T$ interaction control over the DTF through any photoperiod gene activity. This interaction must be recognized for full understanding of the physiological genetic control over plant maturity. In studying this interaction, it is never sufficient to compare plants grown across a range of T in a single P, or across a range of P at a single T. Observing only a single genotype also restricts understanding of the holistic GPT interaction.

6.3.6 The predictive function of the days to flowering model

Equation 6.3 was developed from interpreting the observed GPT interaction effects on the DTF. But, it is also a predictive model. Its effectiveness for modelling DTF, based on data for 12 crops including both LDP and SDP is presented in sections 7.4 to 7.6. For all these crops, equation 6.3 gives better goodness of prediction for photoperiod sensitive cultivars than the original equation of the most commonly cited additive linear rate model developed by Summerfield and Roberts (1987) and Summerfield *et al.* (1991, 1992, 1993). For photoperiod insensitive genotypes, the two models are comparably effective. The superiority of the GPT interaction model (equation 6.3) in predicting DTF over the additive linear rate model for the more photoperiod-sensitive genotypes may result from the following three aspects.

First, the GPT interaction model is based on and predicts the DTF. In contrast, the additive linear rate model is based on and predicts 1/DTF. Summerfield *et al.* (1991) point out it is the DTF that needs to be predicted. For modelling the DTF, the additive linear rate model converts the measured DTF to 1/DTF, and the predicted 1/DTF has to be transformed back to the DTF. Therefore, high goodness of fit of 1/DTF does not necessarily mean good prediction of the DTF. The prediction is poorest when the DTF is strongly delayed by non-optimal T and/or P (Lawn *et al.*, 1995). This is amply illustrated in section 7.6.

Second, the GPT interaction model incorporates a quadratic function to account for the curvilinearity of the responses of the DTF to both T and P, while the inverse linear model accounts for this curvilinearity by using 1/DTF. Conversion to 1/DTF adequately linearizes the data across a small range of T, but does not allow modelling of the DTF when $T > T_{aopt}$. Undoubtedly, the true responses of development to T and P are not exactly 'quadratic', but assuming this has provided effective prediction across a wider range of T and P than allowed by assumptions of the original additive linear rate model.

Third, the GPT interaction model (equation 6.3) incorporates the interaction control over the DTF, whereas the inverse linear model bypasses interaction effects by modelling 1/DTF followed by use of multiple linear equations (section 7.9.2). As stated by its authors (Summerfield *et al.*, 1991; Lawn *et al.*, 1995), the inverse linear model is not valid, if the $P \times T$ interaction is large. According to equation 6.3, however, this interaction is large for all photoperiod sensitive genotypes – whenever the T is relatively high and the P is moderately to strongly delaying. Accounting for the GPT interaction is the basis for success of equation 6.3.

6.4 Integrated Control over Development by Vernalization- and Photoperiod-genes, and the Majority of all Other Genes

6.4.1 *Developmental terms not used*

In this and the later chapters, we have not used the term induction as Blondon (1972) and Heide (1984) and prior and subsequent plant scientists have, for the following reasons. Primary induction and secondary induction imply an absolute requirement for initiating or triggering development, i.e. that the development will not proceed without exposure to the specified level and duration of the low T, SD or LD. This is generally not true; the responses to different levels of P and T are quantitative rather than qualitative. Development usually proceeds but advances slowly under less favourable environments. Our second reason for abandoning the term induction is that we interpret that every plant has an autonomous capability for rapid development. Uninhibited implementation of this autonomous capability results in the shortest expressible duration (D_b) of the

DTF. But, this autonomous capability to develop rapidly is inhibited (delayed) by any vernalization-gene and/or any photoperiod-gene activity, whenever the T and/or the P are not optimal for the gene activity. The optimum T and optimum P do not 'induce' rapid development. Rather, they prevent delay of the development by activity of the photoperiod- and/or vernalization-genes, thereby allowing expression of the autonomously rapid development. Likewise, the reduced developmental rate due to low accumulation of growing degree days is a negative control over the autonomous capability of the plant to grow and develop rapidly. These suggestions of negative control over an autonomous development are backed by the following citations. Koornneef *et al.* (1991) concluded:

> *Arabidopsis* results indicate an autonomous flowering induction process, which cannot be stopped but can only be modified by promotors and inhibitors. Late flowering mutants, then, are understood to interfere with the synthesis of inhibitors, thus allowing them to act. Thus, a loss of function in the mutants explains the apparent gain of function (increased effect of vernalization).

Moreover, Koornneef and Kendrick (1994) reported:

> In all daylength sensitive species, phytochrome deficiency results in early flowering, especially in those conditions where the wild type exhibits a delay in flowering. . . . The general conclusion that can be drawn from these observations is that *Pfr* exerts an inhibitory effect on flower initiation, especially when the conditions for flowering are less than optimal.

A second commonly used term not used in this chapter is 'day neutral plants'. Others have used it in conjunction with the terms LDP and SDP species. This text always refers to plants classifiable as day neutral as being photoperiod insensitive. Our definition is that a photoperiod insensitive genotype has nil to very low response to the P. Plant species and genotypes vary quantitatively from extremely high to extremely low levels for both their optimum P (P_{opt}) and for their sensitivity to the P (S_p).

6.4.2 Phasic development

The GPT interaction model for the DTF (equation 6.3) does not account for any control through vernalization. Therefore, it cannot be used directly for the dual-photoperiodic plants (the term is from Vince-Prue, 1975), which include short-long day plants (SLDP) and long-short day plants (LSDP). Phasic development is obvious for vernalization-requiring plants, and also for the dual-photoperiodic plants of cereal and vegetable crops. For both, two developmental phases are apparent. The first phase responds to a low-T duration with an accelerated development, or responds to an SD or LD duration. The second responds only to either LD or SD. These first and second phases were called the primary induction and secondary induction phases by Blondon (1972) and Heide (1984). Both phases are essential for rapid development; they occur successively, and each complements rather than replaces the other.

The relation between the primary and secondary phases provides a key to a practical model. It has been believed that the secondary phase began immediately after, and could not begin before, the primary phase is completed. Recent evidence is that both phases have a quantitative continuum, with no unambiguous morphological indicator for completion of the primary phase due to either vernalization or the applied SD or LD. Study of wheat demonstrated the second phase (an LD response) begins before the first phase (response to vernalization or to an SD) is fully completed (Miao *et al.*, 1990, 1992). The LD response increases as the duration of vernalization is longer, and it stabilizes after a sufficient duration (unpublished data of Miao *et al.*). Vernalization makes the plants more responsive to the LD. Largely, it is the LDP species that have a vernalization requirement, and a strong stable LD response level indicates completion of the vernalization. This vernalization by photoperiod interaction in plant phasic development is detailed in Chapter 8.

Based on the previous paragraph, the original concept of phasic development can be revised as follows. Full vernalization (the primary phase) is necessary for full expression of and, therefore, for correct evaluation of the LD response (the secondary phase). Also, correct evaluation of the vernalization response (primary phase) requires that the entire post-vernalization environment be optimal for expression of the secondary (the LD) response. More detailed research is needed of the relations between the primary phase of development (a vernalization or a photoperiod response) and the secondary phase which is always a photoperiod response.

Marcellos and Single (1971) pointed out that the duration of any developmental period of wheat is a function of multiple variables including: (i) genetic factors, (ii) vernalization, temperature, and photoperiod, and (iii) miscellaneous factors like inadequate or excess nutritional and water availabilities. A complete model must account for all such factors. As an approximation, Syme (1973) suggested that the days to heading of wheat has three major components: the basic duration to heading established by the genotype; any modulation of the development by vernalization gene activity; and, any modulation (delay) by photoperiod gene activity. Incorporating also our interpretation of phasic development and of inhibition of development by any photoperiod or vernalization gene, we point out here that expressions of both the NTF and DTF result from three components: (i) the intrinsic earliness due to the genotype; (ii) any delay during the primary phase due to vernalization-gene activity – or due to the photoperiod-gene activity if the primary phase is a photoperiod response; and (iii) any delay during the secondary phase due to all the factors modelled by equation 6.3.

6.4.3 Plant classes for response to vernalization and photoperiod

The expressions of: (i) mono- versus dual-phases of development; (ii) LD versus SD responses to P; and, (iii) whether the primary phase is controlled by the vernalization-gene or the photoperiod-gene activity facilitate placing plant species into one of six classifications:

A Plants without obvious dual-phases of development
  1. Long day plants — LDP
  2. Short day plants — SDP

B Plants with obvious dual-phases of development:

B.1. The primary phase has a vernalization requirement
  3. Vernalization-long-day plants — VLDP

4. Vernalization-short-day plants (rare) VSDP

B.2. The primary phase has a short day requirement

5. Short-long day plants SLDP

B.3. The primary phase has a long day requirement

6. Long-short day plants (rare) LSDP

Of the six classes, the two with a single phase, the LDP and SDP, have long been recognized as major response classifications (Salisbury and Ross, 1969; Vince-Prue, 1975). For both, equation 6.3 models with excellent fitness (Chapter 7) responses to the *P* and *T* of both the NTF and DTF.

After fulfilment of the vernalization requirement of the VSDP class, or after a required duration of the SD of the primary phase of the SLDP, these plants respond like the LDP. Therefore, equation 6.3 also successfully models the effects of *P* and *T* on the development that occurs after completion of the primary phase.

Of the four classes with dual-phases, the VSDP and the LSDP are rarely reported; while the VLDP and SLDP are the more recognized, and strong interactions between their primary and secondary phases were reported (Heide, 1984; Davidson *et al.*, 1985; Griffiths *et al.*, 1985; Heide *et al.*, 1987; Miao *et al.*, 1991, 1992). Precisely-controlled *P* and *T* and their durations are essential to the experimental procedures needed to isolate and quantify the sequential effects of the vernalization- and photoperiod-genes. Prediction of the DTF will require models that incorporate the interaction effects arising from the gene activities of both the primary and secondary phases. Such a model requires understanding of the interactions between vernalization, photoperiodism and thermal time. Chapter 8 presents our thinking toward a model to combine all these interacting controls.

The six response classes are far fewer and more precisely defined than the 777 possible response classes suggested by Salisbury and Ross (1969) and the about 30 classes of Vince-Prue (1975). The six less complex and more specific classes are possible because of the following understandings: (i) The genetically determined sensitivity to the photoperiod among plants and genotypes both (a) varies quantitatively and (b) the observable (and apparent) sensitivity is modulated quantitatively by both the *P* and the *T*. Thereby, the *P* and *T* always interactively control any photoperiod gene activity. (ii) The genetically determined sensitivity of the vernalization genes of plants and genotypes both (a) also varies quantitatively but (b) this gene activity is modulated quantitatively by only the *T*. (iii) The holistic GPT interaction incorporates the well known Q_{10} effect taught in biochemistry. Through it, each higher *T* than the base-*T* that allows any gene activity: (a) enlarges the photoperiod-gene activity to delay development to flowering, (b) enlarges the vernalization-gene activity to delay development to flowering, and (c) enlarges all the enzyme catalysed gene activities of that majority of all the genes that implements an autonomous development of the plant to all of its stages. (iv) The quantitatively different responses of even the same allele of the photoperiod- and of the vernalization-genes occur, because the level of these gene activities can be altered by the variations in the alleles constituting the majority of all the genes which function as the background genes within which the photoperiod and the vernalization genes function.

Chapter 6 Summary

Six conclusions follow:

1. Control by photoperiod-gene activity over the DTF is inadequately quantified by comparing a range of photoperiods at just one temperature.
2. Each photoperiod gene and allele has a base (lowest) temperature below which its activity will not occur.
3. Whenever the temperature is higher than this base-*T*, the photoperiod-genes cause a curvilinear GPT interaction control over the DTF, because
4. Any delay of flowering by a non-optimal photoperiod is always enlarged by the Q_{10} effect due to each higher than the base-*T* that allows any photoperiod-gene activity.
5. The holistic control over the DTF is by the photoperiod response interacting with the simultaneous control through the thermal response.
6. A consequence of these simultaneous controls over the DTF follows. Any temperature that causes the smallest DTF under a given photoperiod other than the P_{opt} is always only an apparent optimum temperature. It will be altered by a different photoperiod.

Equation 6.3 is a physiological–genetic description of the holistic GPT interaction control over the DTF through (i) the photoperiod gene activity described by conclusions 1 to 6, plus through (ii) the well recognized and always simultaneous thermal-time control over the DTF. A seventh conclusion is: The two distinct subsets of genes that implement the photoperiod control and the thermal control over the DTF are, respectively, any photoperiod genes and the majority of all the other genes of the plant. Conclusion 8 is that the Q_{10} effects by the same temperature cause opposite changes of the DTF through the activities of the photoperiod genes and of the majority of all the genes. A higher-*T* causes earlier DTF through the thermal response and the autonomous capability for development, both of which are implemented by the majority of all the genes. But, that higher *T* simultaneously enlarges delay of the DTF through the photoperiod response. An observer can directly see and quantify the holistic control over the DTF, but cannot directly separate and quantify the opposite thermal and photoperiod response effects on the DTF. Nevertheless, the genetically established parameters of both opposite responses are quantified by applying equation 6.3 to the DTF expressed across multiple photoperiod–temperature combinations. Equation 6.3 simulates the DTF by assuming for all genotypes that an optimum photoperiod is established by any photoperiod gene(s) and an optimum temperature of the thermal response is established by the majority of all the other genes exclusive of any vernalization genes.

Conclusion 9 is that any acceleration of the DTF by a low temperature that can be replaced (substituted for) by a long or short photoperiod is due to control by photoperiod gene(s) rather than to control by vernalization gene(s). A tenth conclusion is: For some genotypes that have both a vernalization response and an LD response, the LD response arises and is intensified as the response to vernalization decreases while it is progressing toward completion of the vernalization requirement. This calls for renewal and revision of the concept of phasic development and an 11th conclusion. Expressions of: (i) mono- versus dual-phases of development, of (ii) LD versus SD responses to photoperiod, and also; (iii) of whether the primary phase is

controlled by vernalization-gene activity or photoperiod-gene activity places all plant species into one of six photoperiodism-vernalization response classifications.

7

Prediction of Phenology by the Genotype × Photoperiod × Temperature Interaction Model

This chapter applies the conclusions of Chapter 6 and equation 6.3, which are based on paradigm shifts 10 to 13, to the days to flowering (DTF) observed across multiple combinations of photoperiod and temperature. This simulates the observed DTF. High predictive fits occurred between the simulated DTF and the DTF expressed in environments not used to derive the model's parameters for the genotype. These parameters include the sensitivity to photoperiod and to temperature, a lowest temperature that allows activity of the gene, and the shortest DTF the genotype can express in any environment. The simulation results support all of paradigm shifts 2 to 17.

Plant phenology provides a major control over cultivar adaptation and yield (Wallace, 1985; Summerfield *et al.*, 1991). Understanding controls over plant phenology by the genotype (*G*), photoperiod (*P*) and temperature (*T*) is essential to understanding the biology of yield. The *T* and *P* are the predominant environmental controls, but water and mineral availabilities also influence the phenology (Marcellos and Single, 1971). Unlike water and mineral nutrient availabilities, agriculturists cannot modify the *P* and *T* on a large scale. Many models have been used to predict the DTF responses to *T* and/or *P* (Marcellos and Single, 1971; Major, 1980; Angus *et al.*, 1981; Hammer *et al.*, 1982, 1989; Hadley *et al.*, 1983, 1984; Porter *et al.*, 1987; Summerfield and Roberts, 1987; Horie and Nagagewa, 1990; Major *et al.*, 1990; Sinclair *et al.*, 1991; Grimm *et al.*, 1993).

The existing models that use *P* and/or *T* as independent variables for predicting the DTF will be classified. Two models will then be compared. The first was developed by Hadley *et al.* (1983, 1984) and Summerfield and Roberts (1987), and applied to many crops with reasonable success. It is a non-interaction, additive, linear-rate model. The second is the genotype × photoperiod × temperature (GPT) interaction model of Yan and Wallace (1996) already derived in Chapter 6. Hereafter, respectively, they are referred to as the additive linear rate model and the GPTmodel. The assumptions of the two models are compared below, followed by comparison of their fits to observed data.

7.1 Categories of Models for Predicting the Days to Flowering

Models for predicting DTF, as modulated by P and T, can be divided between statistical and biological models. Both use a mathematical equation. The following is from Summerfield *et al.* (1991):

> Models which seek to correlate dates of (or days to) flowering with various states of environmental factors such as temperature and photoperiod using formal statistical procedures seem, inevitably, destined towards increasing complexity; these models can become so bulky and unwieldy that, ultimately, they may suffer the same fate as the dinosaurs.

Focus here is on models with biological assumptions. Some are for only the effect of T on the DTF, and others for only the effect by the P. Models dealing with only the effect of T are applicable only for genotypes that are insensitive to the P, or if the P is near its optimum level for the genotype. Models that predict the DTF based on only control by the P are applicable only for DTF under an unchanging-temperature; they require assumption that there is no $P \times T$ interaction control over the DTF. Models that include effects on the DTF due to both P and T are more valid and, therefore, more practically useful. Assumption there is or is not a $P \times T$ interaction effect on the DTF is a major difference among these models. Three categories regarding this assumption follow.

The first assumes there is a $P \times T$ interaction effect on the DTF. These models multiply a function of the P by a function of the T:

$$R = F(T) \times F(P)$$

where R is the rate of development to flowering, this rate being the reciprocal of the DTF, i.e. 1/DTF. The $F(P)$ and $F(T)$ functions can be a linear, a linear-plateau, a quadratic, an exponential, or a power function (Angus *et al.*, 1981). Angus *et al.* (1981) used this approach for wheat; Major *et al.* (1975), Sinclair *et al.* (1991) and Grimm *et al.* (1993) for soybean; and, Horie and Nagagewa (1990) and Horie *et al.* (1986) for rice. These models assumed the $P \times T$ interaction totally determines the DTF. No biological explanation of the $P \times T$ interaction was assumed.

Category two is the additive, non $P \times T$ models. They are also generalized as:

$$R = F(T) + F(P)$$

These additive models assume that, across a limited range of the P and T, each independently modulates the development. To simplify the modelling procedure, the models assume there is no $P \times T$ interaction effect on the development. The best recognized is the additive linear rate model first described by Summerfield and Roberts (1987), based on research by Hadley *et al.* (1983, 1984). This model has been applied to many crops. It is by far the most published model for predicting the DTF (Summerfield and Roberts, 1993). Perry *et al.* (1987) independently described this model for wheat.

Table 7.1. A classification of models for predicting the days to flowering.

Model classification	Sources
A. For response to only temperature or only photoperiod	
1. Linear-temperature	Hadley *et al.*, 1983
2. Linear plateau-photoperiod	Major, 1980; Major *et al.*, 1990
3. Broken linear-temperature	Summerfield *et al.*, 1991; Olsen *et al.*, 1993
B. For responses to both temperature and photoperiod	
1. Multiplicative models:	$R = F(P) \times F(T)$
1.1 Linear plateau	Major *et al.*, 1975; Grimm *et al.*, 1993
1.2 Exponential	Angus *et al.*, 1981; Horie *et al.*, 1986; Sinclair *et al.*, 1991
1.3 Quadratic threshold	Angus *et al.*, 1981
2. Additive models:	$R = F(P) + F(T)$
2.1 Linear: a single plane	Hadley *et al.*, 1984; Summerfield and Roberts, 1987
2.2 Linear: three planes	Summerfield *et al.*, 1991
3. Additive plus $P \times T$ interaction	DTF (or RATE) = $F(T) + F(T \times P)$ Yan and Wallace, 1995b; Chapter 5 of this book

Abbreviations and symbols: DTF = days to flowering; RATE = 1/DTF; $F(T)$ = function of temperature; $F(P)$ = function of photoperiod; $F(T \times P)$ = function of the temperature × photoperiod interaction.

The third category incorporates: (i) a multiplicative $P \times T$ interaction control over the DTF through the photoperiod-gene activity, which is additive to (ii) the thermal effect on the DTF by the T through all the other gene activities. The latter is the commonly recognized thermal time effect, if the applied T is lower than is optimum for development. This model is illustrated as:

$$\text{DTF} = F(T) + F(P \times T)$$

The GPTmodel derived in Chapter 6 as equation 6.3 belongs to this category. Models that predict DTF, or the rate of development to flowering (1/DTF) are classified in Table 7.1.

7.2 The Best Known Non-interaction Model

Photoperiod and temperature controls over the DTF of many crops have been studied since 1971 at the Plant Environment Laboratory of Reading University, England. This research has resulted in more than 260 publications (Summerfield and Roberts, 1993). The basic equation is:

$$1/\text{DTF} = \text{a} + \text{b}T + \text{c}P \qquad [7.1]$$

where a, b, and c are fitted constants, T is the mean temperature from sowing to flowering, and P is the mean photoperiod. The assumptions follow. Across a limited range of P and T, the rate of development to flowering (1/DTF) is modulated linearly and independently by the mean T and the mean P. Transformation of the expressed DTF to 1/DTF is believed to sufficiently remove the $P \times T$ interaction effect, which the authors accept always partially controls the DTF (Summerfield *et al.*, 1991).

7.2.1 Advantages of the additive linear rate model

The advantage of the additive linear rate model is its simplicity and high fit to the DTF across a limited range of T and P. Summerfield *et al.* (1991) summarized this with the following seven statements.

1. Typically, the separate responses to both photoperiod and temperature become linear over a wide range of environments, if phenology data are transformed to rates of development toward flowering (1/DTF).
2. Whereas large, significant interactions occur between temperature and photoperiod responses if data are analysed as times to flower (DTF), these interactions often disappear if the rates are used.
3. A consequence of 1 and 2 is that simple equations without interaction terms can be developed to predict the rates of progress towards flowering and therefore, indirectly the times taken from sowing to flowering (the DTF).
4. As there are no interaction terms in such equations it is possible to identify and measure the genetic control of photoperiod-sensitivity and of temperature-sensitivity.
5. The recognition that linearized rate equations may be applied to the data permits application of the concept of thermal time (for photoperiod-insensitive genotypes) or of an analogous concept of photothermal time (for the photoperiod sensitive genotypes). The use of these concepts allows prediction of times to flowering (the DTF) in environments which are not constant, e.g. in natural environments where temperature fluctuates daily and photoperiod changes systematically. It is not always realized that these concepts are only legitimate if rates of development are linearly related to temperature and photoperiod.
6. As a consequence of 3 and 4, the genotypically controlled value of the parameters which describe sensitivity to photoperiod and temperature can, where appropriate models have been developed, be estimated from the responses of only a few, controlled and carefully-chosen, environments (in some cases no more than four environments may be essential). As a consequence of 5, it may also be possible to use fluctuating field environments, or to integrate the use of controlled and natural environments to estimate these parameters.
7. A consequence of 4 and 6 is that simple and economic techniques may prove feasible for the screening of large germplasm collections for sensitivity to photoperiod and temperature in a way that allows rational genetic analysis and predictions of time of flowering in a wide range of environments.

Lawn *et al.* (1995) summarized the key strengths of the linear rate model(s). (i) The wide applicability of the approach: large variations of the DTF were successfully explained across diverse environments for both short-day plant (SDP) and long-day plant (LDP) species. (ii) A relative lack of complexity; simplicity arises from assumption of a linear and additive rate of development and of no or limited $P \times T$ interaction effect on the rate. (iii) Relatively few environments are required to define the model's parameters. (iv) Use of the diurnal mean *P* and *T* to describe the environments, and of the DTF to calculate the mean rate of development as 1/DTF, avoids need for complex time-consuming measurements to define both the photothermal environment and the plant response.

7.2.2 *Limitations of the additive linear rate model*

Application of the additive linear rate model's equation 7.1 to many crops confirmed its authors' belief that 1/DTF enables simplified modelling of phenology. The following statement occurs in various forms in virtually every paper of Summerfield and associates. Equation 7.1 is valid only when *Tb* < *T*< *To* and *Pc* < *P* < *Pce* for an SDP, and when *Tb* < *T*< *To* and *Pc* > *P* > *Pce* for an LDP. If either *T* or *P* exceeds these limits, the model will not give accurate prediction, because the true rate will neither be linear nor without significant influence by the *P* × *T* interaction. The *Pc* for a SDP is the shortest photoperiod that causes significant delay of the DTF, and the *Pce* is the longest photoperiod that causes significant delay of the DTF as measured by 1/DTF. The *Pc* for an LDP is the longest *P* that causes any delay of the DTF, and the *Pce* is the shortest *P* that causes any further delay of the DTF as measured by 1/DTF.

Note: throughout this text, all terms used in conjunction with the linear rate model (equations 7.1, 7.2 and 7.3) include no subscripts. This is as these terms were presented by the model's authors. Within this text, therefore, lack of subscripts differentiates the terms of the linear rate model from terms of the GPTmodel described below in section 7.3.

The limitations of *Tb* < *T* < *To* and *Pc* < *P* < *Pce* (or *Pc* > *P* > *Pce*) for valid use of the additive linear rate model prevent prediction of plant phenology where crops are grown but approach being unadapted. Bean, for example, is grown at locations with *T* shown to be higher than the *To* of this linear rate model (Figs 4.2 and 5.3). Neither the *Tb* nor *To* of the additive rate model is a genetically-fixed parameter; for both, their level depends on the applied *P*. Likewise, neither *Pc* nor *Pce* is a genetically fixed parameter; both are temperature-dependent (Summerfield *et al*., 1993). These interdependencies indicate *P* × *T* interaction occurs even for DTF transformed to 1/DTF.

To enlarge the environmental range to which the additive linear rate model is applicable, Summerfield *et al*. (1991, 1993), Upadhyay *et al*. (1994) and Lawn *et al*. (1995) incorporated two equations in addition to equation 7.1. One, two or all three may be needed to account for the *P* and *T* controls over 1/DTF (section 7.7.1), depending on the photoperiod-sensitivity of the genotype and levels of the *P* and *T*. The two additional equations and their assumptions follow.

When *P* < *Pc* for an SDP, or *P* > *Pc* for an LDP, the rate of development responds to only the *T*, and the equation is:

$$1/\text{DTF} = a' + b'T \qquad [7.2]$$

where a′ and b′ are fitted constants and *T* is the mean temperature. In contrast, when *P* > *Pce* for an SDP or when *P* < *Pce* for an LDP, the rate of development is a constant (d′), which is the lowest possible rate of development of a genotype.

$$1/\text{DTF} = d' \qquad [7.3]$$

Selective use of equations 7.1, 7.2 and 7.3 extended applicability of the additive linear rate model across a wider range of *P*. But, it greatly reduced simplicity, because neither *Pc* nor *Pce* is fixed by the genotype, both being temperature-dependent. The requirement for *Tb* < *T* < *To* remains a limitation. Ellis *et al.* (1992b) resorted to use of a quadratic or cubic function to model the response of 1/DTF of several maize cultivars to *T*. Omanga *et al.* (1995) incorporated a negative term for the temperature response of 1/DTF to high-*T* of several photoperiod insensitive pigeonpea cultivars. This improved the overall fit. However, in addition to the increased complexity, it seems biologically inappropriate to assume plants respond to *T* and *P* in three or more abruptly different ways, as implied by the three linear mathematical planes of the multiple-equation procedure. This three-equation procedure will be called the three-plane linear rate model. Equation 7.1 will continue to be called the additive linear rate model.

Lawn *et al.* (1995) present the following as four limitations of the three-plane linear rate model applied to soybean, mungbean, black gram, rice, bean, cowpea, lentil and barley:

> **1.** The potential temperature responses in the models were deliberately restricted to the range where plants might reasonably be considered 'adapted', that is to the range between *Tb* and *To*. . . .
> **2.** The models are based on the rates of development, so that relatively small unaccounted-for effects can translate into large differences in the time to flowering when it is delayed. . . .
> **3.** The very simplicity of the linear relations that define the response surfaces (especially the temperature response) provide the main constraint to efficiency in representing the *E* (environment) and *G* × *E* effects in diverse environments where the photothermal thresholds were transgressed. . . .
> **4.** It is possible to identify and exclude from the analyses some situations where mean diurnal temperatures between sowing and flowering fall outside the range between *Tb* and *To*. However, that approach may be of limited effectiveness in field environments where chilling (as in temperate areas) or supra-optimal temperatures (as in the tropics) are likely to occur for only part of the diurnal cycle, and perhaps then for only some days.

Point 4 is a problem for all models that use the mean *T* and mean *P*. Our interpretation of these limitations follows. Physiologically, there is an inevitable *P* × *T* interaction control over the DTF whenever there is any photoperiod-gene activity. This *P* × *T* interaction always occurs, because the photoperiod gene activity is modulated by the delaying *P*, and always; this gene activity is further enlarged by the Q_{10} effect of each higher *T*. The additive linear rate model is relatively ineffective for DTF in environments where flowering is greatly delayed by either *P* or higher *T*, because the ever-present *P* × *T* interaction becomes very large.

In agreement with conclusion of Summerfield *et al.* (1991) that using 1/DTF sufficiently reduces the *P* × *T* interaction, but only within the ranges of *Tb* < *T* < *To* and *Pc* < *P* < *Pce* for SDP, Wallace (1985), Wallace *et al.* (1991), Mmopi *et al.* (1995), and Yan and Wallace (1995b, 1996) found the *P* × *T* interaction control over the DTF to be small to non-existent at low *T* and/or under near-optimal *P*. On the contrary, the photoperiod-gene activity was synergistically enlarged by each rise of the *T*, by each more delaying *P* (shorter for LDP but longer for SDP), and by each greater sensitivity to *P* of the genotype (Figs 4.1, 4.2, 5.3 for bean and Figs 6.1, 6.5, 6.6 for pea). The linear responses assumed for the additive linear rate model have been challenged. Shaykewich (1995) recently reviewed responses of the DTF for cereal crops, primarily, corn, wheat and barley. He concluded: 'The development rate of most species is sigmoidal rather than a linear function of temperature. Consequently, phenological models assuming a linear relationship (e.g. degree-days) are inappropriate.'

7.3 The Genotype × Photoperiod × Temperature Interaction Model

7.3.1 *Description of the GPTmodel*

This model originated from physiological-genetic interpretations of the GPT interaction control over the DTF. Wallace (1985) and Wallace *et al.* (1991) observed the *P* × *T* interaction control over the DTF of bean (Figs 4.1, 4.3, 5.2, 5.4). They interpreted it as due to the Q_{10} effect on the photoperiod gene activity and its quantitative delay of the DTF (sections 3.4.3 and 6.1.5). Incorporation of a Q_{10} modulation of the photoperiod-gene activity – and assumption of such a Q_{10} effect on all other gene activities – is one of the central-most biological-bases for the GPT interaction model (equation 7.4). The DTF of bean observed in response to multiple *P*–*T* combinations was shown via literature review (Wallace, 1985; Yan and Wallace, 1995b, 1996; Wallace *et al.*, 1995a,b) to be like responses of the DTF reported for many crops in controlled environments. The bean (Figs 4.1, 4.2, 5.2), pea (Figs 6.1, 6.2, 6.5, 6.6) and sorghum (Fig. 6.3) responses illustrate this GPT interaction control over development.

Stimulus for Wallace, a bean breeder, to study the GPT interaction was observation of a U-shaped response of the DTF to *T* in controlled *P* × *T* factorial environments. The U-shaped response curve occurred for both early (photoperiod insensitive) and late (photoperiod sensitive) bean lines. These lines had been derived (Chapters 4 and 5) from breeding for higher HI while studying the physiological genetics of yield. Controlled (Fig. 4.1) and field environments (Figs 4.2 and 5.3) showed under each given *P* that the apparent optimal *T* causal of the smallest DTF (at the bottom of the U-shaped response of the DTF to *T*) was higher for photoperiod insensitive genotypes than for the sensitive genotypes. And, it was lowered by both each more delaying *P* and each greater sensitivity to *P* of the genotype (Figs 4.1, 4.2, 5.3, 6.7).

The GPTmodel is:

$$\text{DTF} = D_b + S_t(T - T_{opt})^2 + S_p(P - P_{opt})^2(T - T_{bp})^2 \qquad [7.4]$$

This was derived as equation 6.3. It has six genetically-established and thereby fixed parameters that interact with the applied average *P* and/or the applied average *T* to control the DTF. The other terms are defined as follows:

P_{opt} = the true optimum photoperiod, under which there is no photoperiod-gene activity and therefore no delay of development by the photoperiod-genes. This is the true P_{opt} because it is established by the genetic direction of the specific photoperiod gene(s).

S_p = the photoperiod-gene established sensitivity to non-optimal photoperiods.

T_{bp} = the base (lowest) temperature that allows any photoperiod-gene activity. It is also established by the specific photoperiod gene(s). When $T < T_{bp}$, there is no photoperiod-gene activity and hence no delay of development by a non-optimal *P*.

T_{opt} = the true optimum *T* for development. The T_{opt} is directed and established, collectively, by all of the genes of the plant – exclusive of any photoperiod gene(s) and any vernalization genes. This majority of all the other genes both: (i) collectively implements growth of the plant and its autonomous capability for rapid development to flowering, which was derived in section 6.4.1; and (ii) gives expression of the thermal control over the duration and rate of development to flowering. This thermal control is the equivalent of the thermal-time response until the applied *T* becomes higher than the T_{aopt} (see the next paragraph). The true T_{opt} of the thermal response is a relatively high *T* for all plants. It is always (Fig. 6.7) a higher-*T* than the T_{aopt}, and it is always high relative to where the crop is grown.

The true T_{opt} is not the same as the optimum *T* for development symbolized as *To* by the authors of the additive linear rate model. Their *To* is the T_{aopt} described in section 6.3.2.

S_t = the sensitivity to non-optimum temperature. Like the T_{opt}, it is established also through that majority of the genes that implements the autonomous development.

D_b = the basic duration (shortest expressible DTF) of each genotype. It can be expressed only under the T_{opt} in combination with the P_{opt}, both being specifically established by each genotype.

The photoperiod genes include only that small subset of genes for which the gene-activity is modulated by the *P*. The level of this photoperiod-gene activity is always modulated also by the Q_{10} effect of the applied *T*, thereby resulting in the $P \times T$ interaction control over the DTF. The base-*T* (T_{bp}) is the lowest applied *T* for which the Q_{10} effect permits any photoperiod-gene activity to occur, and it is established at different levels by the photoperiod alleles. There is no explicit assumption of a given base-*T* for the second of the but two subsets of genes for which the GPTmodel simulates their control over development, this being that majority of all the plant's genes the collective activities of which implement both the plant growth and the autonomous capability to develop to flowering as rapidly as the *P* and *T* will allow (Fig. 6.7). The base-*T* for this majority of all the genes of the plant can be assumed to be 0°C, as often assumed for the thermal time response (e.g. Rawson, 1993; Jamieson *et al.*, 1995).

7.3.2 Three central-most biological bases for the GPTmodel

The assumption of a Q_{10} effect on every gene activity of the plant is one of the central-most biological-bases for the GPTmodel (equation 7.4), as described in the first paragraph of section 3.1 above. A second central biological-basis is that the holistic control over the DTF of all plants is by just three subsets of their genes: (i) a few if any photoperiod genes; (ii) a few if any vernalization genes, and (iii) that majority of all the other genes of the plant that implement the growth and autonomous capability to develop rapidly to flowering. Some plants possess no photoperiod and/or no vernalization genes. The third central-most biological-basis follows. Both the photoperiod genes and the vernalization genes negatively control (depress the rate of) the implementation by the majority of all the genes of the autonomous capability to develop rapidly. This overview of the three gene subsets indicates two opposite functions: (i) the majority of all the genes implement the autonomous capability for the plant to grow and development; while (ii) the photoperiod and vernalization genes – at all *T* above the base-*T* that allows their gene activity to occur – negatively control (delay) the implementation of the plant's autonomous development. The retarding of development need not attend a retarding of growth. In Chapters 4 and 5 it was determined that the photoperiod-gene control over development is a consequence of the photoperiod-gene activity controlling the proportion of the available photosynthate that is partitioned to support growth and development of the reproductive organs – versus the proportion competitively partitioned to support continued growth and development of more vegetative organs.

Equation 7.4 does not account for any effect on the DTF by the third subset: the vernalization genes. Nevertheless, equation 7.4 can always be applied to both genotypes without and those with a vernalization requirement, provided the requirement has been met in full by a prior low-*T*. Equation 7.4 models only the photothermal controls over the DTF that function during the post-vernalization growth and development. Detailed discussion about vernalization is in sections 6.1, 6.3 and 6.4 and throughout Chapter 8.

7.3.3 Concepts incorporated in the GPTmodel

7.3.3.1 The optimum temperature and photoperiod for development

1. The responses of DTF to both the *T* and *P* are curvilinear, and both a T_{opt} and a P_{opt} exist (Fig. 6.7). The quadratic functions of the GPTmodel (equation 7.4) account for these curvilinearities, and their associated T_{opt} and P_{opt}. Every genotype is assumed to have a P_{opt} and a T_{opt}. If the applied *P* and *T* are simultaneously the P_{opt} and T_{opt} of a genotype, the autonomous capability of the plant to develop rapidly to flowering is fully expressed with no inhibition or delay. Equations 7.1 and 7.2 of the additive linear rate model are also curvilinear models for the DTF. But, these equations assume that the curvilinear responses of the DTF to both *T* and *P* are hyperbolic. Thereby, they fail to account for the T_{opt} and the P_{opt}.

2. Every genotype possesses the autonomous capability to develop as rapidly as its basic duration (D_b), if both *P* and *T* are optimal. The development to flowering is not induced by a required *P* and/or required *T*. Rather, the DTF will be delayed whenever the *P* is not at the P_{opt} of the genotype and/or the *T* is either lower than the T_{opt} of the genotype as it usually is – or it is higher than the T_{opt} which occurs infrequently.

3. From concepts (1) and (2), the DTF is determined by just the three components of equation 7.4. They are: (i) the basic duration (D_b) of the genotype;

(ii) any delay of the DTF beyond this D_b by a non-optimum photoperiod $[S_p(P - P_{opt})^2(T - T_{bp})^2]$, and; (iii) any additional delay of the DTF by a non-optimum temperature $[S_t(T - T_{opt})^2]$.

4. The basic duration (D_b) of the genotype will be expressed only if the entire growth and developmental duration occur under simultaneous presence of both the P_{opt} and the T_{opt}. Under the simultaneous optima the D_b will be implemented by the majority of all the genes of the plant through the autonomous development. A different D_b for different genotypes may arise simply because this majority of all the genes consists of different genes that collectively function as background genes to the strong control by the photoperiod- and/or vernalization genes. It is also possible that one or a few of all these background genes cause a large enough effect on the D_b to become recognized as a gene or subset of genes that strongly controls the D_b.

7.3.3.2 The GPTmodel accounts for the Q_{10} effects through two subsets of gene activities

The GPTmodel differs from the additive rate model by its assumption (Fig. 6.7) of an optimum P (P_{opt}) and an optimum T (T_{opt}). It differs from all prior models for predicting the DTF, by incorporation of the concept that the well recognized Q_{10} effect by T – on all enzyme-implemented biochemical processes – is applicable to every gene-directed activity. Equation 7.4 applies this Q_{10} concept specifically to: (i) the activity of the photoperiod gene(s) (this includes every gene that has its activity modulated by the P); and (ii) the integrated activity of that subset of the majority of all other genes that implements the autonomous development.

Delay of the DTF by non-optimum P occurs for photoperiod-sensitive but not for truly insensitive genotypes. Because of the Q_{10} modulation of the photoperiod gene activity, any photoperiod-caused delay is always a GPT interaction control over the DTF. Above the base-T that allows any photoperiod gene activity (the T_{bp}), the Q_{10} effect by a higher T will enlarge the delay of the DTF caused by any departure of the P from the P_{opt}. This precludes the possibility of any control over the DTF modulated by the P being independent of simultaneous modulation by the T.

7.7.3.3 The U-shaped response of the DTF to T is due to integration of the photoperiod response and the thermal response

1. The T_{opt} for rapid development is high relative to the T for almost all environments where a species will be grown as a crop. As long as T remains below the T_{opt}, through the thermal effect, i.e. across low to moderate T, each higher T causes change toward earlier DTF. But, this change toward earlier DTF will not be observable – if the simultaneous opposite change of the DTF is larger through the photoperiod-gene activity (Fig. 6.7). When a change toward earlier DTF is observable, it has been called the thermal-time response.

2. Simultaneous with the Q_{10}-caused change toward earlier DTF through the thermal response due to a higher T – if that T is above the T_{bp} – the Q_{10} modulation of the photoperiod gene activity will enlarge any delay of the DTF caused by departure of the P from the P_{opt} of the genotype. Thus, the ever-present Q_{10} modulation of any P modulated photoperiod-gene activity renders the latter an inevitable GPT interaction control over the DTF.

3. If the change toward later DTF caused through the Q_{10} effect on the photoperiod gene activity is smaller than the change toward earlier DTF caused by the simultaneous Q_{10} effect toward earlier DTF through the thermal response – then the observable change of the DTF will be: (i) toward earlier DTF with higher *T*; and (ii) interpretable therefore as the thermal-time response (Figs 4.1, 4.2, 5.3, 6.5, 6.6, 6.7, see Chapters 4, 5 and 6 for the original derivation of these concepts).

4. If the two simultaneous – but opposite Q_{10}-modulated changes of the DTF due to the same *T* are equal – each will cancel the other. When this equality occurs, the *T* is the apparent optimum temperature (T_{aopt}) for development. It is only an apparent optimum because, for each genotype and its different photoperiod-sensitivity, it will be changed as each different *P* establishes a different level of the photoperiod gene activity to be modulated additionally by the *T* (Figs 4.1, 4.2, 6.5, 7.2).

5. All rises of the applied *T* above this observable yet (from (4)) just an apparent optimum *T* (the T_{aopt}, see Figs 6.7 and 5.4) will delay the observed DTF, because the DTF is delayed more through the Q_{10} effect on the photoperiod gene activity than it is accelerated through the Q_{10} effect on the rate of development.

6. The observable optimum *T* (T_{aopt}) under any non-optimum *P* is never the true T_{opt}, and is always lower than the true T_{opt} (Fig. 6.7). That it is altered by the *P* is indicated in many papers of Summerfield and colleagues. They always designate the T_{aopt} as *To*. In a recent paper, Omanga *et al.* (1995) also used the term apparent optimum temperature for their observed *To*. Experiments using factorial *P–T* combinations will always be required to prove the T_{opt} and P_{opt} have occurred simultaneously.

The authors of the three-plane linear-rate model recently presented two correct concepts that are contrary to conventional interpretations (Vince-Prue, 1975) about photoperiod control over the DTF. (i) Summerfield *et al.* (1991) stated: 'Genes conferring photoperiod sensitivity can cause delays in flowering but cannot promote flowering.' Based on this, a non-delaying photoperiod prevents delay by the photoperiod gene activity of the plant's autonomous capability for rapid development. (ii) Summerfield *et al.* (1993) indicated there is a 'minimum temperature', above which photoperiod genes delay flowering. This is the base temperature (T_{bp}) of the GPTmodel, below which photoperiod-gene activity does not occur, but above which it does. Both concepts are integral parts of the GPTmodel.

Another paradigm shift follows. We suggest every photoperiod response is quantitative, that the interpreted qualitative photoperiod response occurs only for the most highly photoperiod-sensitive genotypes compared at extremes of the *P–T* continuum to which these genotypes can respond. Rapid expression of yes versus no initiation of flower buds and subsequent flowering was interpreted as the qualitative response. We believe researchers accepted that an appropriate SD or LD induces flowering, because basic research has predominantly used the most photoperiod-sensitive genotypes to study the molecular and biochemical events that implement the photoperiod response (Vince-Prue, 1975). Genotypes and environments giving rapid yes versus no initiation of flower buds due to extremes of the continuum of P minimize variability of the molecular and biochemical events that implement photoperiodism, thereby maximizing progress of their reductive and basic research.

7.4 Application of the Genotype × Photoperiod × Temperature Interaction Model

The GPTmodel can fulfil two functions as an application. The first is to quantify the genotype established parameters needed to interpret the physiological-genetic processes that control differences in the DTF across multiple genotypes and/or P–T combinations. We believe all six parameters of equation 7.4 are fixed by the genotype. Their quantification will elucidate understanding of the genetic controls over the DTF, and of the modulations over this control by the P and T. For this purpose, properly-designed controlled P–T or field experiments are required; they should involve enough range of both P and T to span both the P_{opt} and the T_{opt}. If either is not spanned the genotype-determined parameter levels estimated will not be reliable. The second function is simulation and prediction of the DTF of different genotypes grown in different P–T combinations. For prediction, the requirement for a broad span of P and T is less strict, but the model's quantified parameters cannot as reliably predict the DTF in environments having a T and/or P outside the range used to estimate the parameters.

Equation 7.4 is mathematically equivalent to equation 7.4a:

$$\mathrm{DTF} = \mathrm{B0} + \mathrm{B1T} + \mathrm{B2T}^2 + \mathrm{B3}(T - T_{bp})^2(P - P_{opt})^2 \qquad [7.4a]$$

where B0, B1, B2 and B3 are fitted parameters. The parameters S_t, S_p, T_{opt}, and D_b of equation 7.4 can be obtained from B0, B1, B2 and B3 through the following relations:

$$S_t = \mathrm{B2}$$
$$S_p = \mathrm{B3}$$
$$T_{opt} = -\mathrm{B1}/(2\mathrm{B2})$$
$$D_b = \mathrm{B0} - \mathrm{B1}^2/(4\mathrm{B2})$$

Parameter B2 is S_t when it is statistically larger than zero, i.e. is positive. B2 > 0 indicates the observed DTF has spanned a T_{opt}. B2 < 0 indicates the observed DTF spanned only T lower than T_{opt}, so the estimated T_{opt} and D_b will not be realistic. To generate a meaningful model for each genotype requires at least six P–T combinations.

Below, the GPTmodel (equation 7.4) will be applied to published or available data for 12 crops. The R^2 fits of the simulated to the observed DTF will be called descriptive fits, which will be compared with descriptive fits from using the additive linear rate model equation 7.1. The GPTmodel parameters for each genotype will be tested against independent DTF data if they are available, and these simulations will be called predictive R^2 fits.

7.4.1 Modelling days to flowering of bean in controlled environments

Eight bean genotypes were grown under all 16 combinations of 9 h, 12 h, 14 h, and 16 h and night/days of 18/24°C, 21/27°C, 24/30°C, and 27/33°C by Wallace and Enriquez (1980). Four genotypes were early and photoperiod-insensitive and four late and photoperiod-sensitive. Night/day irradiances and *T* changed simultaneously with the *P*. Another planting had the same *P* and night *T*, but the day *T* were 3°C, rather than 6°C higher. All DTF for two of the eight genotypes are shown in Fig. 4.1. The 6°C day/night difference is referred to as experiment I, and the 3°C difference as experiment II.

The DTF of experiments I and II and their pooled data were modelled separately by the additive linear rate (equation 7.1) and GPTmodel (equation 7.4), using the applied *P* and weighed average *T*. The descriptive fits of the regressions were usually comparably high for the insensitive genotypes, the GPTmodel being slightly higher (Table 7.2). For the photoperiod-sensitive genotypes, the GPTmodel gave descriptive fits between 77.4% and 87.2% for the pooled data, while fits of the additive linear rate model were from 9.8% to 32%.

The parameters derived by the GPTmodel using the pooled data gave the following additional information about the eight genotypes (Table 7.3). The sensitivities to $P(S_p)$ are not significant for the four genotypes known to be photoperiod insensitive. In contrast, the S_p is highly significant for all four sensitive genotypes; these genotypes differ in their P_{opt}, base-*T* for the photoperiod gene activity (T_{bp}), and their sensitivity to photoperiod (S_p). Genotype 27 has a P_{opt} of about 11 h, and a T_{bp} of about 14°C, the latter suggesting this genotype is sensitive to *P* if the T is higher than 14°C. In contrast, genotype 37 has a P_{opt} of 11.4 h and a T_{bp} of about 21°C; thus, it is photoperiod-sensitive only if *T* is higher than 21°C. Direct comparison of S_p between genotypes makes sense only when similarity occurs for their P_{opt} and also for their T_{bp}.

The estimated T_{opt} above which each higher *T* delays the DTF, rather than accelerating it through the thermal response, is similar for all eight genotypes, ranging from 27°C to 29°C (Table 7.3). The sensitivity to non-optimum *T* (S_t), is highly significant for all eight genotypes. This indicates the range of *T* spanned the genotype-established T_{opt}. The estimated T_{opt} of 27°C to 29°C indicates each higher *T* (Fig. 6.7) caused a thermal-time response (earlier DTF) until the *T* became higher than 27–29°C, depending on the genotype. Above 27–29°C, each rise of the *T* delayed rather than accelerated the DTF. There is no consistent difference between the photoperiod insensitive versus sensitive bean genotypes in their D_b or in their S_t. The D_b ranges from 25 to 30 days. The near identical responses of genotypes 38 and 39 justified their being bulked for release as the early-maturing, photoperiod-insensitive cultivar 'Redkloud' (Wallace and Enriquez, 1980). The parameters estimated (Table 7.3) by the GPTmodel for genotype 27 were used to simulate its DTF across 10–30°C combined with *P* of 11 h, 13 h, 15 h, 17 h and 19 h. The simulations (Fig. 7.1) illustrate how the *P* and *T* jointly modulate the DTF of a photoperiod sensitive genotype so that the expressed T_{aopt} indicated by the smallest DTF of the U-shaped curve is lowered by each more delaying *P*.

A predictive model requires (i) high descriptive fits to the observed DTF and (ii) for each genotype, the parameters estimated from one data set should predict

Table 7.2. Percentage descriptive fit for four photoperiod-insensitive and four photoperiod-sensitive bean genotypes as modelled by the GPTmodel and the additive linear rate model.

Genotype	Experiment I		Experiment II		Pooled data of experiments I and II	
	Additive linear	GPTmodel	Additive linear	GPTmodel	Additive linear	GPTmodel
Photoperiod insensitive genotypes						
38	52.4	75.47	84.8	82.39	64.3	71.49
39	44.5	84.8	83.3	80.26	51.1	77.71
40	79.5	91.88	79.9	80.42	68.6	82.25
41	82.4	90.01	87.1	78.23	81.4	81.6
Photoperiod sensitive genotypes						
27	33.4	90.61	36.5	90.97	32	87.18
37	17.2	88.26	7.9	78.55	9.8	77.39
42	35.2	91.06	21	70.16	27.3	82.04
44	24.2	90.42	3.1	74.1	12.1	77.73

The modelled days to flowering are from Wallace and Enriquez, 1980. The response surfaces of an insensitive and a sensitive genotype are compared in Fig. 4.1.

Table 7.3. Descriptive fit and model parameters from modelling the days to flowering of four photoperiod insensitive and four photoperiod sensitive bean genotypes by the GPTmodel.

Genotype	Descriptive fit (%)	D_b (days)	T_{opt} (°C)	S_t	P_{opt} (h)	T_{bp}(°C)	S_p
Photoperiod insensitive genotypes							
38	71.49	27.6	28.4	0.1538	–	–	Insensitive
39	77.71	27.4	28.0	0.1806	–	–	Insensitive
40	82.25	26.4	28.8	0.1603	–	–	Insensitive
41	81.60	25.6	29.8	0.1341	–	–	Insensitive
Photoperiod sensitive genotypes							
27	87.18	24.5	29.1	0.1519	11.0	13.8	0.0058
37	77.39	30.2	27.0	0.1489	11.4	20.9	0.0140
42	82.04	30.1	28.4	0.1084	10.2	20.3	0.1095
44	77.73	28.8	28.2	0.1108	11.5	17.9	0.0092

The modelled days to flowering are the pooled data from Wallace and Enriquez, 1980, for which the response surfaces of an insensitive and a sensitive genotype are compared in Fig. 4.1.

the DTF of independently measured data sets (Summerfield *et al.*, 1993). Therefore, the model parameters (not given) from simulation of the DTF of experiment I were used to predict the DTF of experiment II. The observed DTF of experiment II of each genotype are close (Fig. 7.2) to those predicted by the six modelled parameters derived from experiment I. These parameters successfully predicted the DTF of experiment II, including the opposite responses to T of the genotypes with insensitivity versus sensitivity to P. Independent data were not available for testing the predictive fits for any of the other crops discussed below.

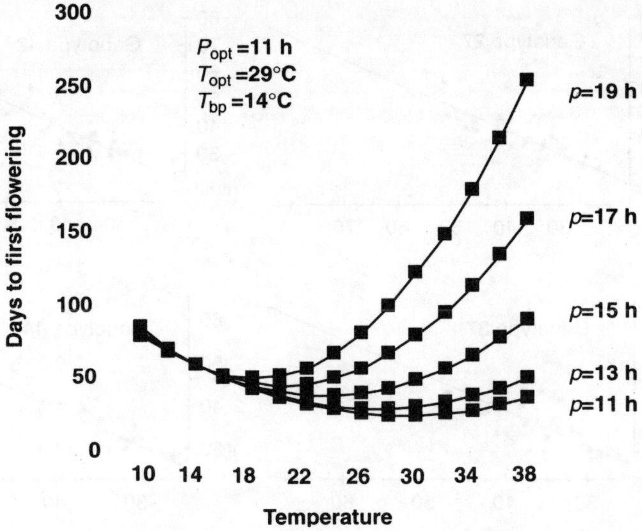

Fig. 7.1. Simulation of the days to flowering in response to temperature and photoperiod of bean genotype 27 based on the parameters (Table 7.3) derived by the DURATION version of the GPTmodel.

7.4.2 Modelling days to flowering of bean in field environments

This experiment was at three locations in Colombia (Gniffke, 1985; Wallace *et al.*, 1991). All had natural *P* near 12.3 h, but had mean *T* ranging from 23.7°C to 13.7°C due to altitudes of 1000 m, 1850 m, and 2710 m. A second *P* was by extension to 18 h using incandescent lamps. The plantings were during two successive seasons, thereby providing 12 *P–T* combinations. Sixty-eight bean genotypes of diverse origin and growth habit were grown in one or both seasons. The DTF data for nine are in Fig. 4.2. Some most photoperiod sensitive genotypes failed to flower under the 18 h, especially at the highest mean *T*.

The descriptive fits (Table 7.4) repeat results from bean grown in controlled environments (Table 7.2). (i) Both models gave high and comparable fits for the photoperiod-insensitive genotypes, being slightly higher for the GPTmodel. (ii) For photoperiod-sensitive genotypes the GPTmodel (equation 7.4) gave significantly higher descriptive fits, the average being 98.2% across all 64 genotypes. The six GPTmodel parameters for each genotype are in Table 7.4. The more sensitive genotypes tend to have a shorter P_{opt}, lower T_{bp} and a larger S_p.

7.4.3 Modelling days to flowering of cowpea

The additive linear rate model (equation 7.1) arose from studies of cowpea (Hadley *et al.*, 1983) showing 1/DTF of the photoperiod insensitive cowpea genotypes to be related linearly to the mean *T*. The 1/DTF of sensitive cowpea responded to *T* linearly also, provided the *P* was shorter than the critical photoperiod (*Pc*). But, while studying cultivar resistance to pod abortion under high-*T*, El-Madina and Hall (1986) found responses of 1/DTF to *T* of even the insensitive

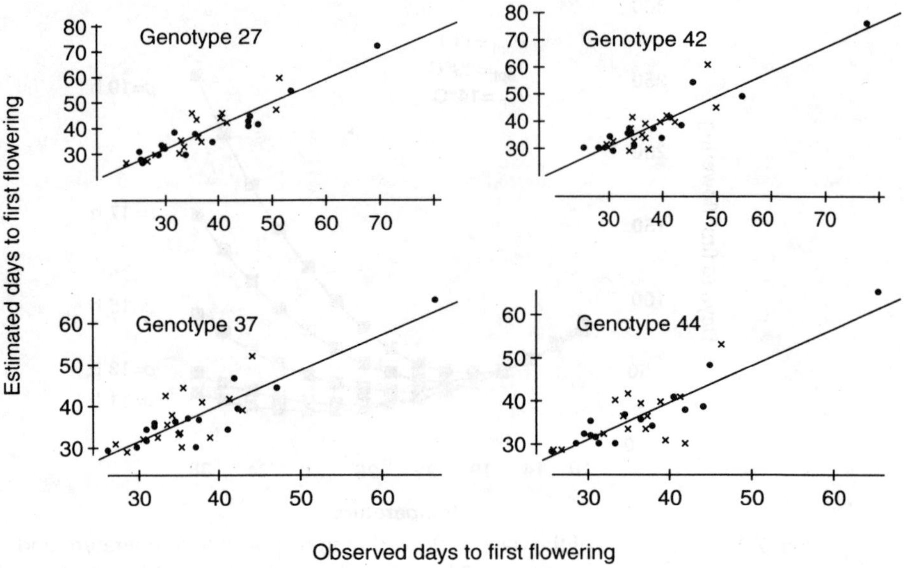

Fig. 7.2. The observed days to flowering in 16 environments plotted against the estimated days in 16 independent environments for four photoperiod-sensitive bean genotypes. The simulation was by the DURATION version of the GPTmodel (equation 7.4) using the observed DTF of experiment I (indicated by ●) plotted against the simulated DTF for experiment II (indicated by X). (The modelled data are from Wallace and Enriquez, 1980.)

genotypes to be curvilinear. They compared four photoperiod sensitive and four insensitive genotypes across seven *P–T* combinations spanning 24.8°C to 32.6°C combined with 11.5 h or 15 h. The observed DTF was used to calculate descriptive fits by the GPTmodel and the additive rate model. Fits by the GPTmodel were 89.58–99.99% (Table 7.5), in contrast to 7.1–55.9% for the additive linear rate model. Like for bean experiments I and II, the GPTmodel is highly descriptive for all cowpea genotypes. Even the photoperiod insensitive cowpea genotypes had a maximum rate of development (largest 1/DTF and smallest DTF) at an intermediate level of the 24.8–32.6°C mean *T*. This challenges validity of the linearity of the 1/DTF assumed by the additive rate model. These cowpea DTF duplicate expression of an optimal rate of development (shortest DTF) of bean at an intermediate *T* (Figs 4 1, 4.2, 5.3), and reported for pigeonpea (Omanga *et al.*, 1995).

The parameters estimated by the GPTmodel (Table 7.5) indicated the following. (i) The four photoperiod insensitive cowpea genotypes all had an insignificant response to P. All four sensitive genotypes had significant, but varying, levels of sensitivity to $P(S_p)$. (ii) The four sensitive genotypes had similar P_{opt} of about 11.3 h, and a high T_{bp} of 22–28°C. Genotype IT81D-1020 is extremely photoperiod-sensitive but has the highest T_{bp} (28.2°C), indicating it is strongly delayed by longer *P*, but only if the *T* is higher than 28°C. This seems too high for a base-*T*. But, it may be that the photoperiod sensitivity of this genotype is actually due to such a response to high-*T*. More study could clarify this. (iii) All eight

Table 7.4. Descriptive fit and the model parameter levels from modelling the days to flowering of 64 bean genotypes with the GPTmodel.

			Percent descriptive fit			Parameter levels estimated by the GPTmodel					
Genotype	Habit	Number of photoperiod-temperatures	Additive linear rate model	GPTmodel	Fit difference	D_b (days)	T_{opt} (°C)	S_t	P_{opt} (h)	T_{bp} (°C)	S_p
Puebla 172	II	12	97.8	98.6	0.8	38.7	24.2	0.3861	22.9	−10.0	0.00001 ins
Redkloud-1	IA	12	96.7	99.2	2.5	34.5	22.9	0.3900	21.0	−10.0	0.00002 ins
Rabia de Gato MG	IVA	12	96.7	99.1	2.4	38.2	23.1	0.4101	12.0	−10.0	0.00002 ins
P 326	III	12	97.6	99.1	1.5	40.2	23.1	0.4331	21.0	−10.0	0.00002 ins
ICTA Jutiapan	II	12	97.8	99.4	1.6	40.1	23.5	0.3935	16.9	−10.0	0.00004 ins
Turrialba-1	II	6	97.4	99.6	2.2	37.5	24.3	0.3882	4.8	10.9	0.00005 ins
Turrialba-2	II	12	97.4	99.4	2.0	39.5	23.1	0.4405	14.6	−10.0	0.00005 ins
PI 310668	III	12	96.2	97.9	1.7	39.6	22.9	0.4076	12.3	13.2	0.00011 ins
Ex Rico 23	II	12	97.9	98.6	0.7	34.8	23.2	0.4110	16.1	−10.0	0.00015 ins
Jamapa I810	II	12	98.9	99.7	0.8	38.7	23.7	0.3878	14.1	6.8	0.00018 ins
Puebla 152-c	III	12	94.6	97.5	2.9	42.7	23.1	0.4004	12.0	−10.0	0.00004 ins
Bat 1320	II	6	94.1	99.5	5.1	41.9	22.0	0.6013	12.2	−10.0	0.00008 ins
Jutiapa 80–11	II	12	97.2	99.4	2.1	40.0	23.1	0.4482	0.0	15.4	0.0001 ins
Negro Jamapa	II	12	97.1	99.1	2.0	38.4	23.3	0.4233	19.1	9.7	0.0001 ins
C-63	II	12	98.1	99.2	1.1	40.1	23.6	0.3931	15.7	−10.0	0.0002 ins
ICTA 78–12	II	6	99.2	99.9	0.7	38.5	23.9	0.4505	13.3	20.9	0.0073 ins
Nep-2	II	12	98.1	99.1	1.0	40.6	24.2	0.3422	0.0	23.6	0.8751 ins
ICTA Tamazulapa	II	12	95.9	99.0	3.1	38.9	22.9	0.4291	6.2	23.6	1.942 ins
E 1056	IVB	6	82.9	97.9	15.0	45.8	21.5	0.8257	10.9	22.8	10.334 ins
ICA Pijao	II	12	96.4	98.9	2.5	40.2	23.0	0.4452	0.0	15.3	0.00011
A 476	IA	5	98.6	99.7	1.1	30.3	24.2	0.4370	5.4	−10.0	0.00019
VRB 81049	IVA	6	90.9	100.0	9.1	40.5	22.1	0.6813	2.1	12.4	0.00020
Venezuela-2	II	12	81.4	98.1	16.7	34.7	22.3	0.6137	5.7	−3.2	0.00023
VRB 81047	IVA	12	93.3	98.2	4.9	39.3	22.1	0.6438	12.7	1.0	0.00033
Magdalena-3	IVA	12	88.4	96.3	7.9	45.0	22.2	0.6146	11.2	−10.0	0.00037
Sangretoro	IVB	6	83.0	99.7	16.7	37.9	21.7	0.6668	11.1	−5.2	0.00041
Sanilac	IC	12	81.6	97.8	16.2	27.1	24.5	0.3694	5.7	7.2	0.00068
#51051	II	12	76.8	95.9	19.1	37.9	21.9	0.6108	11.6	0.8	0.00102
Rojo 70	IVA	9	96.2	98.7	2.5	24.8	21.9	0.7892	9.7	−10.0	0.00121
Diacol Andino	IA	11	74.5	92.3	17.8	35.4	23.7	0.3237	9.6	4.9	0.00145
ICA L 33003 M(4)	IVB	6	74.4	96.7	22.3	16.8	22.8	0.6723	5.9	−2.9	0.00155
N 555	II	12	74.2	97.8	23.6	38.3	22.1	0.6257	10.7	4.1	0.00165

(*Continued*)

Table 7.4. *Continued*

Genotype	Habit	Number of photoperiod-temperatures	Percent descriptive fit: Additive linear rate model	Percent descriptive fit: GPTmodel	Percent descriptive fit: Fit difference	Parameter levels estimated by the GPTmodel: D_b (days)	T_{opt} (°C)	S_t	P_{opt} (h)	T_{bp} (°C)	S_p
ICA Tui	II	12	82.9	97.7	14.8	38.5	22.3	0.5694	12.5	6.0	0.00171
ICA L 33462	II	5	74.4	100.0	25.6	43.2	21.5	0.5054	10.4	3.8	0.00181
Antioquia-8	IA	11	51.5	98.7	47.2	20.6	26.5	0.2710	4.9	12.3	0.00207
ICA Tundama	II	10	92.7	98.2	5.5	40.3	22.9	0.4285	10.6	4.0	0.00214
Ecuador 299	IVA	10	94.1	93.4	-0.7	34.8	21.1	0.7355	12.0	−10.0	0.00220
Porrillo-1	II	12	80.9	98.0	17.1	38.0	22.2	0.6375	11.8	8.6	0.00220
Porrillo Sintetico	II	12	82.2	97.8	15.6	39.5	22.6	0.5299	12.7	7.0	0.00233
ICA Linea 17	IA	12	60.8	92.9	32.1	29.4	24.5	0.2946	8.6	9.0	0.00256
GN Nebraska 1–27	IC	11	80.5	96.3	15.8	33.1	23.0	0.4548	12.2	10.1	0.00298
Rojo de Seda	III	11	85.0	96.0	11.0	36.8	22.4	0.5675	12.3	12.8	0.00327
Trujillo-3	IVA	12	74.4	97.2	22.8	41.2	21.4	0.6909	12.2	9.5	0.00331
Brazil-2	IC	12	81.1	98.8	17.7	39.1	22.9	0.4177	12.3	13.7	0.00441
Limoneno	IA	12	64.9	98.8	33.9	35.2	22.5	0.3563	12.0	10.0	0.00446
Ian 5091	III	11	77.2	95.8	18.6	41.0	22.9	0.3705	13.3	4.2	0.00463
ICTA Quetzal	III	12	74.8	97.4	22.6	38.7	22.5	0.5440	12.3	13.2	0.00582
Accession 274–3	IA	12	64.7	95.2	30.5	36.1	22.4	0.4238	12.7	9.1	0.00592
Great Northern Ul1	IC	11	83.0	99.2	16.2	33.5	23.6	0.4089	12.5	12.3	0.00620
Puebla 152-n	III	12	66.9	96.7	29.8	41.5	22.2	0.5214	12.3	12.9	0.00664
Pompadour-2	IA	12	62.2	94.9	32.7	34.3	22.4	0.3984	12.3	12.8	0.00671
ICA L 33341	II	5	89.2	100.0	10.8	42.4	21.2	0.6941	12.7	7.4	0.00688
ICA L 33411	II	10	90.3	99.4	9.1	44.1	22.0	0.4863	12.8	8.9	0.00755
P523	IVA	6	65.4	100.0	34.6	36.0	25.6	0.2246	12.5	13.8	0.00999
Liborino Voluble	IVB	5	50.0	100.0	50.0	55.1	19.5	1.3157	13.1	7.8	0.01567
Jutiapa 80–13	II	12	97.5	99.7	2.2	37.8	23.9	0.4035	0.0	23.6	1.01030
ICA L 32983	IVB	7	73.2	99.0	25.8	Unrealistic	24.4	0.6173	5.9	−5.8	0.00193
Diacol Calima	IA	12	59.0	96.7	37.7	Unrealistic	Unrealistic	0.1619	0.0	12.8	0.00142
ICA Linea 24	IA	5	81.1	100.0	18.9	Unrealistic	Unrealistic	0.0976	9.0	12.9	0.01203
Redkote-3	IA	12	58.6	95.9	37.3	Unrealistic	Unrealistic	0.3622	0.0	15.5	0.00234
ICA Viboral	IV	10	65.8	98.6	32.8	Unrealistic	Unrealistic	0.1323	1.3	11.6	0.00560
Pata de Zope	IVA	6	63.8	100.0	36.2	Unrealistic	Unrealistic	0.0540	8.6	16.8	0.00834
ICA L 32980 M(8)	IVB	6	74.7	99.8	25.1	Unrealistic	Unrealistic	0.4078	14.7	−1.3	0.05320
E525	IVB	5	63.4	99.3	35.9	Unrealistic	Unrealistic	0.0270	13.2	20.8	7.94790
Mean			83.1	98.2	14.7						

... from the data of Gniffke, 1985. ins, Insensitive to photoperiod.

Table 7.5. Descriptive fit of the days to flowering for eight cowpea genotypes derived by two models and the model parameters derived by the GPTmodel.

Genotype	Descriptive fit GPTmodel model (%)	D_b (days)	T_{opt} (°C)	S_t	S_p	P_{opt} (h)	T_{bp} (°C)	Descriptive fit additive linear model (%)
Photoperiod insensitive genotypes								
Cv.7964	89.58	28.8	29.3	0.9427	ins	-	-	09.20
cv.750–1	90.85	30.8	29.8	0.6032	ins	-	-	35.50
Prima	97.38	33.7	29.9	0.7240	ins	-	-	42.30
IT82E-18	95.91	33.3	29.6	0.9445	0.0015	0.7	24.5	07.10
Photoperiod sensitive genotypes								
CB5	99.86	32.5	29.0	1.0926	0.3750	11.2	25.8	54.70
B-21	98.52	33.6	29.1	0.9517	0.4749	11.5	26.3	46.50
cv.58–57	96.06	47.9	28.9	1.8611	0.0274	11.4	22.3	55.90
IT81D-1020	99.99	32.8	31.5	0.8841	10.2494	11.2	28.2	30.90

The modelled data are from El-Madina and Hall (1986).
ins, Insensitive to photoperiod.

Table 7.6. Descriptive fits of the days to flowering derived by two models for three mungbean genotypes, and the model parameter values derived by the GPTmodel.

Cultivar	Descriptive fit GPTmodel (%)	D_b (days)	T_{opt} (°C)	S_t	P_{opt} (h)	T_{bp} (°C)	S_p	Descriptive fit additive linear model (%)
		Model parameter values						
VC.1973A	75.31	33.9	28.2	0.4012	12.3	11.0	0.0356	46.80
VC.1560D	69.48	33.6	27.8	0.5519	12.5	21.2	0.3964	32.60
V. 3029	52.87	31.9	28.2	0.2836	12.5	23.8	0.8846	29.00

The modelled days to flowering are from Fernandez and Chen, 1989.

genotypes have a T_{opt} near 29°C, but differ in their S_t. (iv) The genotypes vary in their D_b; it ranged from 29 days for insensitive genotype 7964 to 48 days for the sensitive genotype 58–57.

7.4.4 Modelling days to flowering of mungbean

Fernandez and Chen (1989) present the DTF from the 9th and 10th International Mungbean Nurseries for three cultivars with reduced sensitivity to *P* for 21 location-year combinations in nine countries. The *T* varied from 22°C to 30°C, and the *P* from 11.5 h to 13.2 h. The descriptive fits by the GPTmodel are again superior, being 75.3%, 69.5% and 52.8% for the three cultivars, in contrast to 46.8%, 32.6% and 29.0% for the additive linear rate model (Table 7.6). The three genotypes differed little in T_{opt} and P_{opt}, but did differ in T_{bp}.

7.4.5 Modelling days to tasselling of maize

Eighteen maize cultivars of different maturity, origin and adaptation were grown in up to 53 location-planting dates (Bonhomme *et al.*, 1994). The mean *T* ranged from 13.2°C to 29.4°C for the sowing-to-silking intervals of early hybrid LG-11. The *P* at tassel initiation of LG-11 ranged from 11.7 h to 19.5 h. Bonhomme kindly provided the days from sowing to silking (hereafter the DTF) at 16 to all 53 location-planting dates, the mean *T*, and the *P* at silking of cv. LG 11. The additive linear rate and the GPTmodel gave reasonably high and comparable descriptive fits for each of the 18 cultivars. Fits by the GPTmodel were slightly higher for most cultivars (Table 7.7). The GPTmodel parameters provided the following insights.

1. The cultivars all have a T_{opt} between 25.2°C and 29.4°C.
2. The P_{opt} ranged from 12.1 h to 13.7 h, except for the composite cultivar Largo Dia. The results support conclusion by Bonhomme *et al.* (1994) that quantification of the optimum *T* and sensitivity to *P* are essential to precise estimates of the thermal time to flowering and days to maturity at sites with $P > 13$ h. Since the P_{opt} for most of the genotypes is around 13 h, there should be little, if any, photoperiod response of their development when *P* is shorter than 13 h.
3. The cultivars differed greatly in their T_{bp}. The temperate cultivars have considerably higher T_{bp} than the tropical and especially the highland-tropical

Table 7.7. The descriptive fits and model parameter values estimated by both the GPTmodel and the linear rate model for 18 maize genotypes.

				The GPTmodel parameter values (equation 7.4)							The linear rate model values (equation 7.1)			
Cultivar name	Relative maturity	Cultivar type	No. of environments	Descriptive (%)	D_b (days)	T_{opt} (°C)	S_t	P_{opt} (h)	T_{bp} (°C)	S_p	Descriptive fit	a	b	c
Temperate cultivars														
F7 × F2	very early	hybrid	29	94.3	41.2	26.7	0.3407	–	–	–	89.9	−0.00719	0.00124	ins
B73 × Mo17	late	hybrid	50	93.1	51.7	27.8	0.4062	–	–	–	91.4	−0.00560	0.00094	ins
GL 11	very early	hybrid	53	92.6	39.6	27.6	0.2941	13.5	18.2	0.0951	90.1	−0.00507	0.00112	ins
F478 × W705A	early	hybrid	25	94.1	41.2	27.9	0.3072	13.4	18.3	0.1599	88.0	−0.00588	0.00111	ins
A632 × W117	medium-early	hybrid	28	91.5	44.3	27.3	0.3647	13.7	17.9	0.1306	90.7	−0.00379	0.00095	ins
W64A × F546	medium-late	hybrid	52	91.9	46.0	27.4	0.4064	13.5	18.8	0.1474	88.5	−0.00591	0.00102	ins
McC-6714	very late	hybrid	19	91.7	52.0	26.9	0.5710	13.7	21.4	1.0302	86.5	−0.00362	0.00079	ins
Lowland tropical cultivars														
IRAT 83	late	hybrid	37	95.8	53.6	27.7	0.5466	13.4	18.9	0.2334	90.3	−0.00226	0.00086	−0.00022
IRAT 81	late	hybrid	22	92.9	56.9	27.4	0.5864	13.3	18.3	0.3491	80.9	0.00064	0.00078	−0.00039
882 Z 013	very late	hybrid	17	95.9	55.2	28.7	0.5107	13.2	18.1	0.3197	86.7	0.00190	0.00072	−0.00043
IRAT 143	medium-late	hybrid	48	90.7	53.8	26.8	0.5701	13.2	13.8	0.0250	86.3	−0.00078	0.00086	−0.00029
883 Z 045	late	hybrid	19	95.3	51.1	28.6	0.4797	12.6	10.6	0.0129	91.7	−0.00052	0.00079	−0.00028
X 304 C	very late	hybrid	35	91.6	51.2	27.6	0.5651	12.1	−3.0	0.0033	86.6	0.00456	0.00090	−0.00082
Largo Dia	medium-late	composite	19	86.5	23.3	29.4	0.4419	0.0	−3.0	0.0002	88.7	0.00238	0.00077	−0.00037
Highland tropical cultivars														
Pool 1	early	composite	21	82.8	48.9	25.2	0.4383	13.7	−3.0	0.0038	80.3	0.00947	0.00066	−0.00058
Pool 5	early	composite	18	85.9	49.4	26.3	0.3625	13.3	−3.0	0.0032	86.3	0.00843	0.00071	−0.00060
Pool 12	late	composite	19	88.3	59.6	27.3	0.3598	12.4	−3.0	0.0027	89.4	0.00992	0.00054	−0.00064
Pool 2A	late	composite	16	87.4	60.1	25.2	0.3832	13.3	−3.0	0.0061	85.9	0.01578	0.00040	−0.00083
Mean				91.3							87.7			

The modelled days to flowering are from Bonhomme *et al.* (1994).
ins, Insensitive to photoperiod.

cultivars. Five temperate cultivars formerly regarded as photoperiod-insensitive were actually moderately to highly sensitive to *P*, but their T_{bp} are relatively high (about 18–21°C). In contrast, the highland tropical cultivars, and two lowland tropical cultivars, X340C and Largo Dia, are characterized by a very low T_{bp}. These results indicate seemingly more photoperiod-sensitive genotypes may be so because a lower T_{bp} allows the photoperiod gene activity (Table 7.7).

These model parameters and predictions for maize show the GPTmodel (equation 7.4) to be promising not just for prediction. The parameter levels constitute physiological-genetic explanations of how different genotypes respond to *P* and *T*.

7.4.6 Modelling days to flowering of sorghum

Twelve sorghum cultivars were grown under *P* of 17 h and 10 h combined with five *T* (32/29, 32/23, 32/17, 23/20, and 17/11°C day/nights), in greenhouses (Quinby *et al.*, 1973). The day/nights were 8/16 h. Descriptive fits by the additive linear rate model ranged from 13.9% to 85.2%, the average being 54.6%. Fits by the GPTmodel ranged from 86.9% to 99.1%, and averaged 95.2%. The GPTmodel's parameters (Table 7.8) indicate the cultivars differed little in their T_{opt}, which varied from 24.6°C to 27.3°C, and averaged 25.9°C. They did differ in S_t, P_{opt} and S_p. Quinby *et al.* indicate cv. 'Japanese Dwarf Broomcorn' was known to be relatively insensitive to both *T* and *P*, and 'Feterita FC 811' to express faster floral initiation under low *T*. Both cultivars were indicated by the GPTmodel to be photoperiod insensitive within the range of the *P*–*T* tested (Table 7.8).

7.4.7 Modelling days to flowering of pigeonpea

Thirty short-duration pigeonpea cultivars were grown in 10 *P*–*T* combinations in India (see section 9.3). The DTF, yield and aerial biomass were recorded. Ranges of the *T* and *P* were 21.7–29.6°C and 12.1–14.4 h. The mean *P* and *T* were highly correlated ($r = 0.90$).

The descriptive fits of three cultivars (11, 17 and 19, see Table 9.4 and Fig. 9.7) are lower than 43% for both models, suggesting strong control of their DTF by factors other than the *P* and *T*. The descriptive fits of the GPTmodel ranged from 49% to 92% for the other 27 genotypes, the average being 72%. Fits by the additive linear rate model ranged from 32% to 88%, and averaged 60%. For only genotype 26 was fit by the additive linear rate model higher than by the GPTmodel. Again, the GPTmodel is descriptively superior to the additive linear rate model. Strictly, the correlation of 0.90 between the mean *T* and mean *P* of this data set rendered it inappropriate for any regression analysis (modelling). This was indicated by unrealistic estimations of the T_{opt} and the P_{opt}, and consequent inappropriate estimates of the shortest expressible DTF (the D_b) of the genotypes.

7.4.8 Modelling days to booting of winter wheat cultivars

During 1982 to 1985, extensive ecological research on wheat in China included 30 wheat cultivars grown in 41 locations throughout continental China, with two to three planting dates at each year-location. In the first of a series of books

Table 7.8. Descriptive fit by two models of the days to flowering of 12 sorghum cultivars and the modelled parameters by the GPTmodel.

		Predictive fit (%)		Parameter level					
Cultivar	Genotype	By the linear rate model	By the GPTmodel	D_b (days)	T_{opt} (°C)	S_t	P_{opt} (h)	T_{bp} (°C)	S_p
Early Kalo	ma1Ma2Ma3Ma4	43.50	97.02	17.7	26.2	0.3847	7.0	−10.0	0.00007
Hegaris	Ma1Ma2Ma3ma4	13.90	98.66	14.8	24.6	0.5134	7.0	−10.0	0.00014
Kalo	ma1ma2Ma3Ma4	65.80	97.50	22.4	26.6	0.0096	11.8	−10.0	0.00016
M 937		62.40	86.90	13.5	27.3	0.2600	7.0	−10.0	0.00020
Temperate M35–1		48.90	99.04	29.7	25.8	1.0823	11.8	−10.0	0.00038
Early Hegaris	Ma1Ma2ma3ma4	19.80	98.74	15.3	24.8	0.4485	11.4	−10.0	0.00039
Combine Hegaris	Ma1Ma2ma3Ma4	51.80	98.94	13.1	24.7	0.7102	8.9	−7.0	0.00041
100 M	Ma1Ma2Ma3Ma4	67.60	96.36	5.4	25.9	0.9693	7.0	−10.0	0.00042
SM 100	ma1Ma2Ma3Ma4	78.20	94.62	7.8	26.6	0.8801	11.8	−10.0	0.00079
Feterita FC 811		59.20	89.81	15.8	25.8	0.9501	7.0	−10.0	0.00018 ins
Dwarf Broomcorn		85.20	97.17	31.6	26.6	0.3552	7.0	−10.0	0.00030 ins
Texas Bh. Kafir		58.50	87.24	29.3	25.9	0.4829	11.8	−10.0	0.00040 ins
Mean		54.57	95.17		25.9				

The modelled days to flowering are from Quinby *et al.* (1973).
ins, Insensitive to photoperiod.

Table 7.9. Descriptive fit by two models and the parameters by the GPTmodel for the days to booting of six winter wheat cultivars.

Cultivar	Descriptive fit by GPTmodel	Descriptive fit by linear rate model	Number of environments	D_b	T_{opt}	S_t	P_{opt}	T_{bp}	S_p
Xuzhou 17	97.87	92.41	19	231.6	−2.7	−1.5857	23.9	4.3	0.0222
Linfen 10	98.90	91.92	20	222.3	−3.3	−1.3136	23.9	4.4	0.0183
Jimai 7	97.86	94.33	18	214.9	−2.3	−1.4137	23.9	4.2	0.0172
Baitumai	96.69	88.87	37	222.0	−1.4	−1.8463	23.9	3.9	0.0215
Taishan 4	94.00	92.36	35	225.3	−2.9	−1.1601	23.9	4.5	0.0144
Feimai	87.37	82.70	37	236.2	−2.5	−1.1352	23.9	4.7	0.0111

The modelled days to booting are from Jing (1991).

(Jing, 1991), for six winter wheat cultivars, the days from planting to booting were grouped and presented at 5 or 10 day intervals, along with the average *T* and *P*. The GPTmodel gave descriptive fits (Table 7.9) of 87–99%. The average *T* ranged from −3°C to only 15°C, so no T_{opt} was detected for any cultivar. Each higher mean *T* caused faster development (the thermal time response), irrespective of the vernalization requirement of these genotypes. All six cultivars were responsive to *P*, with a P_{opt} near 24 h and a base-*T* for the photoperiod-gene activity near 4–5°C.

7.4.9 *Modelling days to flower bud initiation of pea*

The days to flower bud initiation (DTF) of pea cv. Greenfeast (from Paton, 1968) in response to 4–30°C combined with 8 h, 16 h and 24 h were presented in Fig. 6.5. The GPTmodel and linear-rate model, respectively, gave descriptive fits of 90.6% and 81.2%, indicating higher fit for yet another crop species by the GPTmodel.

The above nine examples from eight crops are sufficient to conclude the GPTmodel (equation 7.4) more successfully simulates descriptively the DTF than the additive linear rate model (equation 7.1). The GPTmodel's assumption of GPT interaction and incorporation of a quadratic function to describe the curvilinearities of the response to *T* and to *P* are part of the explanation. Some colleagues expressed concern a quadratic function would make the response curve too steep. The high descriptive simulations demonstrate this occurs minimally if at all. The GPTmodel's descriptive fits are most superior for the highly photoperiod sensitive genotypes, and for the extreme *P* and *T* for which excessive steepness of the DTF curve would most likely occur (Fig. 7.1). Two additional crops for which descriptive fit by the GPTmodel was superior to that of the additive linear rate model equation 7.1 but the results are not presented are lima bean (Wolokolie, 1984) and bambara groundnut (Linneman and Craufurd, 1994).

7.5 A Rate Version of the Genotype × Photoperiod × Temperature Interaction Model

Many modellers (Table 7.1), in addition to authors of the additive linear rate model, have modelled the DTF using its inverse (1/DTF). With no change of theoretical assumptions, the GPTmodel (equation 7.4) can be written in a rate form using 1/DTF (equation 7.5). A genotype has an intrinsically most-rapidly-expressible rate of development, which is symbolized as $1/D_b$. The D_b and therefore $1/D_b$ are expressed only when the applied *T* and *P* are the genotype's T_{opt} and P_{opt}. The rate of development is delayed by the *T* being either higher or lower than the genotype's T_{opt}, and also by the *P* being shorter or longer than the genotype's P_{opt}.

$$1/\text{DTF} = 1/D_b - S'_t(T - T_{opt})^2 - S'_p(P - P_{opt})^2(T - T_{bp})^2 \qquad [7.5]$$

The S'_t and S'_p have primes because their value is based on 1/DTF rather than on the DTF. As described earlier, the 'quadratic function' used in the last term $(T - T_{bp})^2$ is a mathematical technique to linearize the DTF data. Use of the 'rate', (i.e. 1/DTF) is another mathematical technique for linearizing the data. Either linearization technique may be sufficient, and both can be redundant. To avoid redundancy, the RATE version for the GPTmodel is equation 7.6:

$$1/\mathrm{DTF} = 1/D_b - S'_t(T - T_{opt})^2 - S'_p(P - P_{opt})^2(T - T_{bp}) \qquad [7.6]$$

As with equation 7.4, the requirement is $T > T_{bp}$, and if $T < T_{bp}$ then T is set to T_{bp}. Hereafter, equation 7.4 will be referred to as the DURATION version of the GPTmodel. Equation 7.6 is the GPTmodel's RATE version. Both versions predict the DTF using the GPTmodel software.

7.5.1 Modelling days to flower bud initiation of pea

Above (section 7.4.1) the DURATION version was applied to the pea data (Fig. 6.7) of Paton (1968). It gave a descriptive fit of 90.6%, while fit of the additive linear rate model was 81.2%. The RATE version gives a descriptive fit of 94.8%. The descriptive equation is:

$$1/\mathrm{DTF} = 1/16.9 - 0.0001(T - 25.1)^2 - 0.00001(P - 24)^2\,(T - 2.6)$$

The estimated D_b is 16.9 days when the T and P are simultaneously optimum (25.1°C and 24 h, respectively). The estimated T_{bp} for the photoperiod gene activity is 2.6°C. After 1/DTF is changed back to DTF, the R^2 fit becomes 95.3% rather than 94.8%. Even though the descriptive fit of 1/DTF by the additive linear rate model was 81.2%, its fit to the observed DTF was only 60.9% (Fig. 7.3). These comparisons illustrate that the R^2 fits may be weakly to strongly enlarged or else reduced when 1/DTF is transformed back to DTF. The DURATION version directly quantifies the fit to the observed DTF.

7.5.2 Modelling days to flowering of a short-long-day plant

This experiment by Heide *et al.* (1987) used *Poa pratensis*, a short-long day plant (SLDP). Its primary phase requires either an SD or a low-T, the requirement being fully met by 11 weeks under 8 h and 9°C. The secondary phase requires an LD. [Detailed discussions about SLDP plants are in sections 6.3, 8.1 and 8.4.]

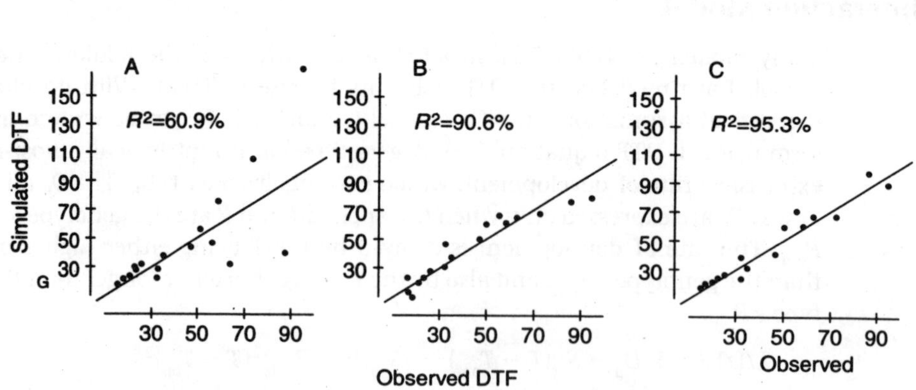

Fig. 7.3. Simulation of the days to flower initiation of the pea cultivar Greenfeast by the: (A) additive linear rate model (equation 7.1); (B) DURATION version of the GPTmodel (equation 7.4), and: (C) Rate version of the GPTmodel (equation 7.6).

After the primary phase, the plants were grown at 9°C, 15°C, and 21°C in factorial combination with 8 h, 18 h, and 24 h. The days to heading (DTF) was recorded for all nine *P–T*. The descriptive fits by equations 7.1, 7.4 and 7.6 are 94.7%, 71.4%, and 99.6%, respectively. Thus, the DURATION version of the GPTmodel gave poorer descriptive fit than both its RATE version and the additive linear rate model. This is the first data set for which the linear rate model gave higher descriptive fit than the DURATION version. The RATE version gave the highest of all fits.

The fitted model by the DURATION version (equation 7.4) is:

$$DTF = -0.68 + 0.0871(T - 30.93)^2 + 0.00014(P - 22.5)^2\,(T - (-9.95))$$

The S_t is not significant, because the 9–21°C range of the T did not span the T_{opt}. Therefore, the estimated T_{opt} and D_b are not reliable.

The fitted model by the RATE version (equation 7.6) is:

$$1/DTF = -1/271.1 + 5.907E - 5(T - (-13.57))^2 - 5.46E - 6(P - 24)^2(T - (-9.95))$$

The S_t is also not significant, so its estimated T_{opt} and D_b are also not reliable indicators of the true genotype-established values.

The descriptive fit between the observed DTF and the simulated DTF is 91.9% for the DURATION version compared with 67.3% for the additive linear rate model, and is 99.8% for the GPTmodel RATE version. These observed versus simulated DTF are compared graphically in Fig. 7.4.

7.5.3 *Modelling days to heading of oat*

Bleken and Skjelvag (1986) grew six oat cultivars at constant 9°C, 12°C, 15°C, 18°C and 21°C in factorial combination with 12 h, 16 h and 24 h. They published the days to heading (DTF). The linear rate model gave average descriptive fits of 91.3% compared with 88.3% by the DURATION version, but the RATE version gave the highest fit of 99.4%. The model parameters of the RATE version (Table 7.10) are interpreted as follows. (i) All six genotypes are highly photoperiod

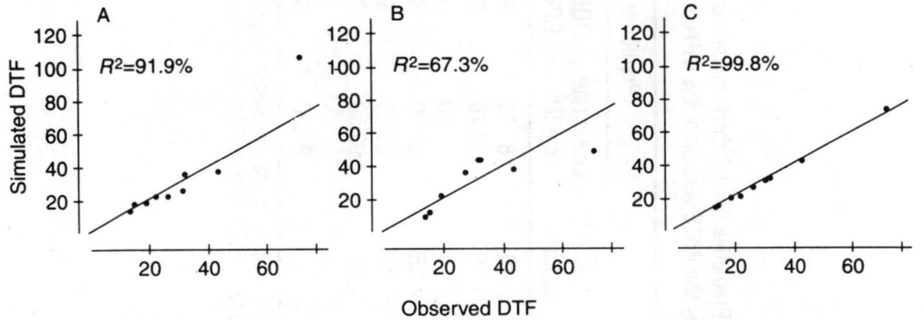

Fig. 7.4. Simulated versus observed days to flowering of *Poa pratensis* by the: (A) additive linear rate model (equation 7.1); (B) DURATION version of the GPTmodel (equation 7.4), and: (C) RATE version of the GPTmodel (equation 7.6).

Table 7.10. Predictive fits for three models of the days to heading of six oat cultivars grown under three photoperiods and five temperatures, and the model parameters by the RATE version of the GPTmodel.

		Predictive fit (%) by the			Model parameters by the RATE version (equation 7.6)					
Cultivar	No. regimes	Linear rate model	DURATION GPTmodel	RATE GPTmodel	D_b (°C)	T_{opt} (°C)	S_t	P_{opt} (h)	T_{bp} (°C)	S_p
Pol	10	87.00	91.42	99.60	*16.7*	*78.9*	−0.000010	23.3	−10.0	−0.000003
Titus	15	92.70	82.65	99.44	*14.3*	*105.8*	−0.000006	24.0	−2.9	−0.000004
Puhti	15	91.10	84.40	99.30	*13.7*	*101.5*	−0.000007	23.3	−0.4	−0.000005
Condor	15	87.90	86.94	99.55	*11.5*	*134.2*	−0.000005	23.3	1.1	−0.000005
Argentic	15	95.60	91.68	98.86	*40.0*	*33.9*	−0.000024	24.0	1.3	−0.000004
Angelic	15	93.50	92.67	99.65	*19.3*	*69.2*	−0.000011	23.3	2.7	−0.000004
Mean		91.30	88.29	99.40						

All sensitivities to temperature (S_t) are statistically insignificant. Therefore, the D_b and T_{opt} values are italicized to indicate they are invalid and most are unrealistic. However, this did not prevent the rate version of the GPTmodel from giving the highest descriptive fits of the three models. (The modelled days to heading are from Bleken and Skjelva, 1986.)

sensitive. Their P_{opt} is 23–24 h and their T_{bp} is about 0°C. Cultivar Pol has the lowest T_{bp}. Four of the six cultivars have an insignificant S_t. Obviously unrealistic T_{opt} and D_b are indicated for all six cultivars, because the T did not span the T_{opt}.

7.5.4 *Relative effectiveness of the DURATION and RATE versions of the GPTmodel*

The pea, oat and *Poa pratensis* results seem to indicate the GPTmodel RATE version simulates higher descriptive fits to the observed DTF than its DURATION version (see Figs 7.3, 7.4 and Table 7.10). Therefore, the DTF analysed for all eight crops in section 7.4 were modelled also with the RATE version. For most cultivars of all eight species, the DURATION version gave similar or better descriptive fits than the RATE version. The comparison across all crops indicated the pea, *Poa pratensis* and oat data sets, for which the RATE version gave significantly better descriptive fit than the GPTmodel's DURATION version (sections 7.5.1, 7.5.2 and 7.5.3), are all LDP. A tentative conclusion is: for LDP, the RATE version tends to give the higher fit.

Both the DURATION and RATE version gave some unrealistic parameter estimates for some data sets. This was due to poor data quality ascribable to one or more of the following problems. (i) The number of P–T combinations was too few (fewer than six, for example); (ii) the number of P and/or T was too few (as not more than two levels); (iii) the range of the T and/or P was too small to span the T_{opt} or the P_{opt}; and (iv) levels of the T and P were highly correlated, the pigeonpea data discussed in section 7.4.7 being one example. Such problems with the experimental design should be avoided in using the GPTmodel, but this may require controlled environments that introduce other problems.

Both the DURATION and RATE versions are applied by the software developed for the GPTmodel. This compares the results from both GPTmodel versions with those of the linear rate model. Comparing all three provides the most informative interpretations and conclusions.

7.6 Cross Validation of the GPTmodel

All fits above compare the simulated with the observed DTF or with the 1/DTF. These are descriptive fits; they are not predictive fits. Both the GPTmodel DURATION and RATE version usually gave description superior to the additive linear rate model. Proving the GPTmodel is most predictive requires validation using independent data. Of the modelling results above for 12 crops, only the bean DTF of Wallace and Enriquez (1980) could be validated (Fig. 7.2) using independent data.

A 'drop-one cross validation' procedure can be used to validate the results (Hamilton, 1991). Let N be the number of P–T combinations the DTF has been measured in. Use $N-1$ of them to generate the model parameters. Then, use the parameters to predict the DTF of the genotype in the P–T not used to derive the parameters. This will be repeated N times, each predicting the DTF of the genotype for one of all the environments. Two statistical parameters will be used to judge validity of the model's simulation. One is the Root Mean Square Prediction Deviation (RMSPD) calculated as follows:

$$\text{RMSPD} = \sqrt{\sum (Y' - Y)^2 / N}$$

where Y' is the observed DTF and Y is the predicted DTF. The smaller the RMSPD, the better the prediction. The second measure is the R^2 fit between the independently predicted versus observed DTF. The RMSPD is an absolute measure of the prediction, having the same unit as the observed values. The R^2 fit is a relative measure without unit. Usually, but not always, the two measures provide consistent conclusions.

Six of the 12 data sets discussed above in sections 7.4 and 7.5 were subjected to the cross validation test. The others had too few environments for cross validation.

7.6.1 Cross validation of days to flower bud initiation of pea

There are 19 P–T combinations of the data set (Fig. 6.5) of Paton (1968) for the photoperiod sensitive pea Greenfeast. The RMSPD for the GPTmodel's DURATION and RATE version and the linear rate model are 11.5, 8.9 and 32.4 days, respectively. The threefold larger RMSPD for the linear rate model indicates the predictive fits of the days to flower bud initiation are superior for both versions of the GPTmodel. Respective predictive fits of 78.3%, 89.6% and 53.9% reinforce this conclusion from the descriptive fits in section 7.5.1.

7.6.2 Cross validation of days to flowering of bean

The RMSPD and predictive fits for the eight bean genotypes grown in all 32 P–T combinations from two experiments (sections 7.4.1 and 7.4.2) are in Table 7.11 (compare Tables 7.2 and 7.3). All the results for the photoperiod sensitive genotypes 27, 37, 42 and 44 indicate both GPTmodel versions gave significantly better prediction of the DTF than the linear rate model. In contrast, for the insensitive genotypes 38, 39, 40 and 41, the cross validation indicates all three models are equally predictive, just as indicated by the descriptive fits (section 7.4.1).

Table 7.11. Root mean square prediction deviations (RMSPD) and prediction fits of the days to flowering of bean by three models.

	RMSPD (days)			Prediction fits (%)		
Genotype	DURATION GPTmodel	RATE GPTmodel	Linear rate model	DURATION GPTmodel	RATE GPTmodel	Linear rate model
27	4.4	7.0	9.5	81.6	74.2	16.3
37	4.6	4.3	7.6	61.4	68.0	0.4
42	5.6	6.3	9.1	66.7	69.9	15.3
44	4.5	4.6	7.7	64.3	69.0	2.0
38	2.9	3.1	3.2	64.7	61.2	58.5
39	2.8	3.1	3.2	69.7	63.8	59.5
40	2.5	2.7	2.8	76.9	74.0	71.3
41	2.7	2.9	2.8	74.4	71.1	72.8

The modelled days to flowering are from Wallace and Enriquez (1980), compare Fig. 4.1 and Tables 7.2 and 7.3.

Table 7.12. Root mean square deviations (RMSPD) and prediction fits of the days to flowering of 47 bean genotypes by three models.

		RMSPD (days)			Prediction fit (%)		
Cultivar	No. regimes	DURATION GPTmodel	RATE GPTmodel	Linear rate model	DURATION GPTmodel	RATE GPTmodel	Linear rate model
ICA Linea 17	12	7.2	7.4	17.7	79.93	81.21	19.15
Diacol Calima	12	5.5	4.2	14.1	85.98	91.65	24.42
Antioquia-8	11	2.9	6.1	12.8	95.47	86.41	26.74
Limoneno3	12	2.5	3.2	13.7	96.68	94.99	28.51
Accession 274-3	12	6.6	6.3	15.8	83.35	85.49	30.91
Pompadour-2	12	6.5	6.7	13.0	80.21	79.87	31.36
Redkote-3	12	4.7	3.8	12.8	89.37	93.12	31.44
Diacol Andino	11	12.4	9.6	15.4	59.88	74.31	38.06
ICA Viboral	10	8.4	13.3	17.8	86.61	68.44	38.11
Ian 5091	11	9.1	6.6	15.8	77.38	87.73	41.16
Puebla 152-n	12	6.2	5.0	14.1	88.31	92.39	42.51
N 555	12	5.3	6.1	17.4	93.64	92.21	51.90
ICTA Quetzal	12	4.9	3.1	13.8	93.83	97.45	53.98
Trujillo-3	12	6.1	4.9	15.1	92.28	94.83	54.59
Sanilac	12	4.0	2.6	12.5	94.16	97.63	56.61
51051	12	7.5	6.8	13.3	86.58	88.88	61.02
Great Northern UI1	11	6.7	3.7	12.6	88.62	96.09	62.58
Porrillo-1	12	5.5	2.9	13.4	93.82	98.18	64.41
GN Nebraska 1–27	11	6.2	5.2	10.0	85.93	90.15	67.57
Brazil-2	12	3.5	3.2	9.4	95.45	96.31	68.12
Venezuela-2	12	4.9	6.8	12.0	94.51	90.02	68.85
Porrillo Sintetico	12	5.5	3.9	11.5	92.68	96.27	69.27
ICA Tui	12	5.8	4.4	11.0	92.13	95.44	71.83
Rojo de Seda	11	7.3	4.6	9.6	84.56	94.46	74.06
Rojo 70	9	8.8	15.3	37.8	95.03	90.69	74.26
Ecuador 299	10	15.9	14.9	28.7	79.47	81.98	75.62
ICA L 33411	10	2.6	3.6	8.8	97.71	95.68	76.31
ICA Tundama	10	5.1	6.5	10.8	93.20	89.48	77.50
Magdalena-3	12	6.5	4.6	10.7	91.52	95.92	77.66
VRB 81047	12	4.8	2.6	8.8	95.70	98.73	88.34
Puebla 152-c	12	4.3	4.6	5.0	94.28	93.80	93.79
PI 310668	12	3.7	2.8	4.9	95.60	97.45	93.92
ICA Pijao	12	2.9	3.6	4.7	97.83	96.84	95.76
ICTA Tamazulapa	12	2.8	3.2	4.4	97.78	97.13	96.02
Redkloud-1	12	2.1	3.5	4.0	98.42	96.19	96.21
Rabia de Gato MG	12	2.8	2.5	3.9	97.66	98.17	96.81
Negro Jamapa	12	2.6	4.0	4.0	98.22	96.33	96.88
Jutiapa 80–11	12	2.5	3.5	4.1	98.50	97.12	97.09
P 326	12	2.7	2.6	4.0	98.04	98.27	97.10
Nep-2	12	2.5	3.3	3.2	98.16	97.11	97.33
Turrialba-2	12	2.4	2.6	3.9	98.54	98.28	97.34
Ex Rico 23	12	3.2	2.8	3.5	97.23	97.79	97.43
Puebla 172	12	3.7	3.4	3.3	96.86	97.41	97.68
C-63	12	2.4	3.0	3.2	98.51	97.76	97.82
ICTA Jutiapan	12	2.2	2.8	3.2	98.71	97.98	97.92
Jutiapa 80–13	12	2.3	2.7	3.4	98.76	98.45	97.99
Jamapa I810	12	1.5	2.2	2.5	99.40	98.82	98.78
Mean		5.0	4.9	10.5	92.05	93.00	70.06

The modelled DTF are from Gniffke (1985). Compare Table 7.4. The number of genotypes is smaller than in Table 7.4, because a larger number of regimes is required to determine the predictive fit by cross validation than is required to obtain the descriptive fit.

Table 7.13. Root mean square prediction deviation (RMSPD) and prediction fits of the days to heading of oat by three models

	RMSPD (days)			Prediction fits (%)		
Genotype	DURATION GPTmodel	RATE GPTmodel	Linear rate model	DURATION GPTmodel	RATE GPTmodel	Linear rate model
Pol	11.7	2.3	3.1	71.3	98.8	98.3
Titus	24.3	8.3	17.2	59.3	97.3	89.4
Puhti	24.7	11.9	25.8	62.5	95.8	84.7
Condor	21.6	5.3	23.1	67.5	98.6	82.8
Argent	15.3	11.5	18.5	79.4	93.4	84.2
Angeli	10.9	2.5	9.6	82.2	99.1	91.4
Mean	18.1	7.0	16.2	70.3	97.1	88.5

The modelled days to heading are from Bleken and Skjelvag (1986) (compare Table 6.10).

Forty-seven genotypes flowered in more than ten of the 12 *P–T* combinations of Gniffke's (1985) data. The DURATION and RATE versions gave significantly smaller average RMSPD (5.0 and 4.9 versus 10.5 days) and larger predictive fit (92% and 93% versus 70%) than the linear rate model (Table 7.12). Both GPTmodel versions are similarly predictive compared with the linear rate model for the photoperiod insensitive genotypes, but more predictive than it (RMSPD of 10.5 and fit of 70.06) for the sensitive genotypes.

7.6.3 Cross validation of days to heading of oat

The results (Table 7.13) indicate the GPTmodel RATE version to be far more predictive than the DURATION version and prediction by the linear rate model to be intermediate. This repeated the conclusion in section 7.5.3 (Table 7.10).

7.6.4 Cross validation of days to tasselling of maize

The RMSPD and prediction fits (Table 7.14) of the three models similarly predict the DTF, the linear rate model's predictions being slightly higher for some cultivars. This differs from the descriptive fits of the GPTmodel being slightly higher (section 7.4.5 and Table 7.7). A possible basis for this reversal follows. The mean *T* was 13.7–29.4°C. Except for one of the 16 to 53 *P–T*, the applied *T* was lower than 26°C; this exception had 29.4°C in combination with a near optimum *P* of 13.6 h. Therefore, in most environments the plants grew under a below-optimum *T*. The optimum *T* must be spanned for the GPTmodel to simulate the curvilinearity of the response to *T* and derive valid parameters. That the DTF observed at the extreme *T* can be used to predict the DTF under intermediate *T*, but not vice versa, explains why the linear rate model's equation 7.1 had slightly higher prediction fits as indicated by the cross validations. Poor prediction of the DTF for some of the maize genotypes was primarily due to deviation of the predicted DTF for only the *P–T* involving the 29.4°C.

7.6.5 Cross validation of days to flowering of mungbean

The RMSPD and predictive fits for three mungbean genotypes (Table 7.15) provide conclusions similar to that for maize. The ranges of mean T and P again provide understanding. The T range of 22–30°C just spanned the T_{opt} of near 28°C, and the P of only 11.5–13.2 h were all close to the P_{opt} of about 12.4 h (Table 7.6). This would render GPT interaction an unimportant source of the variation in the observed DTF.

The cross validation of six data sets for five crops reinforces the three conclusions from descriptive fits in sections 7.4 and 7.5. (i) When photoperiod sensitive genotypes are grown across wide ranges of both P and T, both GPTmodel versions (equations 7.4 and 7.6) provide descriptive and predictive fits superior to the linear rate model (equation 7.1). (ii) For insensitive genotypes, especially when grown across small ranges of T and/or P, the GPTmodel(s) and the linear rate model are comparably predictive. (iii) The GPTmodel versions give superior physiological genetic interpretation of the GPT interaction controls over the DTF, the RATE version being relatively better.

Table 7.14. Root means square prediction deviations (RMSPD) and prediction fits of the days to tasselling of maize by three models. (The original data are from Bonhomme *et al.*, 1994; compare Table 7.7.)

	RMSPD (days)			Prediction fits (%)		
Genotype	DURATION GPTmodel	RATE GPTmodel	Linear rate model	DURATION GPTmodel	RATE GPTmodel	Linear rate model
Temperate cultivars						
F7 × F2	5.1	5.7	5.1	92.0	90.7	92.1
B73 × Mo17	6.3	6.3	5.9	91.3	91.8	92.4
GL 11	5.3	5.1	5.0	89.9	90.6	91.1
F478 × W705A	5.8	5.8	5.9	89.2	89.3	89.6
A632 × W117	8.2	6.7	6.6	79.8	86.6	87.3
W64A × F546	6.7	7.0	6.1	88.4	87.7	90.4
McC-6714	20.4	11.5	11.6	52.7	83.9	86.6
Lowland tropical cultivars						
IRAT 83	7.0	9.6	6.2	93.1	90.4	94.8
IRAT 81	12.9	7.6	9.6	74.8	91.5	86.1
882 Z 013	13.1	12.0	14.8	83.4	88.5	85.9
IRAT 143	9.3	9.3	10.6	87.8	88.0	85.3
883 Z 045	14.9	14.0	45.3	84.7	87.6	57.6
X 304 C	11.9	11.2	10.7	85.4	87.3	89.2
Largo Dia	17.7	15.7	11	57.2	63.8	81.8
Highland tropical cultivars						
Pool 1	15.3	10.7	11.0	58.0	78.2	79.5
Pool 5	13.1	7.5	8.2	66.6	89.4	87.4
Pool 12	15.7	11.8	8.2	57.7	76.9	88.8
Pool 2A	15.6	10.7	10	63.7	81.9	84.9
Mean	11.4	9.3	10.7	77.5	85.8	86.2

Table 7.15. Root means square prediction deviations (RMSPD) and prediction fits of the days to flowering of mungbean by three models.

	RMSPD (days)			Prediction fit (%)		
Genotype	DURATION GPTmodel	RATE GPTmodel	Linear rate model	DURATION GPTmodel	RATE GPTmodel	Linear rate model
V 3029	6.2	6.1	5.2	15.4	16.1	25.2
V 1560D	5.2	5.2	5.6	42.6	42.9	30.8
V 1973A	4.1	4.7	4.4	53.3	41.5	45.8
Mean	5.2	5.3	5.1	37.1	33.5	33.9

The modelled days to flowering are from Fernandez and Chen (1989).

Model parameters derived across a range of *P* and *T* can be used to predict DTF at *P–T* combinations within this range, but will not be highly predictive for environments outside of this range. This has been ignored by the cross validations. Poor validation because of poor prediction of the DTF of only a few extreme environments should not, therefore, be considered as decisive evidence for rejecting a model.

7.7 Comparison of the GPTmodel with the Three-plane Linear Rate Model

7.7.1 Modelling procedure of the three-plane linear rate model

Sections 7.4, 7.5 and 7.6 indicate both the GPTmodel's DURATION and RATE versions to usually give higher predictive fits for photoperiod sensitive genotypes than the additive linear rate model, the latter being the most documented predictive model. It is highly recognized for simplicity due to its avoidance of the $P \times T$ interaction terms, and reasonably predictive fit to the observed DTF across its narrow valid ranges of *P* and *T*. In section 7.1 it was stated that Summerfield *et al.* (1991) extended the original additive linear rate model (equation 7.1) to improve its predictive fits using selective additional applications of equations 7.2 and 7.3. This is called the three-plane additive linear rate model. The authors (Watkinson *et al.*, 1994) released a computer software to apply the three-plane model. It applies equations 7.1, 7.2 and 7.3 selectively to the several *P–T* combinations in which a genotype has been grown. The procedure places the DTF data of each genotype into a first, second or third group based on the sensitivity to the *P–T* combinations. The groups are differentiated by the relation among the applied *P*, critical *P* (*Pc*) and the ceiling *P* (*Pce*) which the model derives for each different applied *T*. The *Pc* and *Pce* are described in section 7.2.2.

For an SDP the classifications for the three groups are:

Environment group 1: $Pc < P < Pce$
Environment group 2: $P < Pc$
Environment group 3: $P > Pce$

For an LDP, the opposite relationships are:

Environment group 1 : $Pc > P > Pce$
Environment group 2 : $P > Pc$
Environment group 3 : $P < Pce$

Re-emphasized is that for each genotype, neither its *Pc* nor *Pce* is a genetically fixed value; both vary with the applied *T*. For each genotype, a specific *Pc* and specific *Pce* is derived, but both differ with the *T*. The iterative software (Watkinson *et al.*, 1994) of the three-plane model includes the following steps:

Step 1: Apply equations 7.1 and 7.2 separately to the 1/DTF of all *P–T*. If equation 7.1 gives significantly better fit than equation 7.2, go to step 2. Otherwise, accept equation 7.2 as the superior model.

Step 2: The observed 1/DTF are iteratively placed into response group 1 or 2 and, respectively, equations 7.1 and 7.2 are re-applied. For each iteration, the simulated 1/DTF is calculated for each *P–T* combination. If the combined use of equations 7.1 and 7.2 gives higher fit across all environments than 7.1 alone, go to step 3. If it does not, accept 7.1 as the superior model.

Step 3: The observed 1/DTF are iteratively placed into response group 1, 2 or 3 and, respectively, equations 7.1, 7.2 and 7.3 are re-applied. For each iteration the simulated 1/DTF is calculated for each *P–T* combination. The joint use of all three equations is accepted if it gives significantly higher fit than use of just equations 7.1 and 7.2. If it does not, accept the joint use of equations 7.1 and 7.2 of step 2 as the solution.

7.7.2 Descriptive and predictive fits

All modelling with the additive linear rate model in sections 7.4 to 7.5 above used just equation 7.1. Below, the descriptive fit by the three-plane software application to the observed DTF will be compared with the descriptive fit by the DURATION version of the GPTmodel (equation 7.4). Such comparison follows for the DTF of the photoperiod-sensitive bean genotype 27. It was modelled in section 7.4.2 and cross validated in section 7.6.2.

Step 1: Application of just equation 7.1 across all the *P–T* gave descriptive fit of 32.08%, and applying just equation 7.2 gave fit of 1.08%.

Step 2: The observed 1/DTF were iteratively placed into response group 1 and 2. Then, respectively, equations 7.1 and 7.2 were re-applied. The joint use of equations 7.1 and 7.2 gave a fit of 84.96%.

Step 3: The observations of the 1/DTF were iteratively placed into response group 1, 2 and 3 and, respectively, equations 7.1, 7.2 and 7.3 were re-applied. This joint use of all three equations gave no higher fit than step 2. Therefore, the R^2 fit of 84.96% for step 2 was accepted.

The descriptive fit for genotype 27 of 84.96% using the combined equations 7.1 and 7.2 is lower than the descriptive fit of 87.18% by the DURATION version of the GPTmodel (Table 7.2 and section 7.4.1). The triple-equation procedure did advance insight about the photothermal response of this genotype; when $P < Pc$, the development was modulated by only the *T*. And, every higher *T* from 19°C to 31°C caused shorter DTF. When $P > Pc$, the development responded to both *T* and *P*, and each higher *T* (than 23°C) and/or longer *P* delayed DTF. This evidence that under delaying P each higher T causes later DTF of a photoperiod sensitive genotype conflicts with assumption for the three-plane linear procedure that it is modelling a linear thermal-time response across the entire but limited range of $Pc < P < Pce$ for an SDP (Watkinson *et al.*, 1994).

Similar results follow for the also photoperiod sensitive bean genotypes 37, 42 and 44. Using the same three-plane linear rate model software of Watkinson *et al.* (1994), the respective best descriptive fits were 60.8%, 71.4% and 58.0%. All are lower than corresponding fits by the DURATION version of the GPTmodel of 77.4%, 82.0% and 77.7% (Table 7.2). For all four photoperiod-sensitive bean genotypes, when $P > Pc$, the three-plane model indicated their rate of development was lowered (their DTF was enlarged), rather than made faster (their DTF reduced) in response higher T as expected for the thermal time response.

Application of the three-plane linear rate model software to the DTF of pea cv. Greenfeast (Fig. 6.7) gave a descriptive fit of 94.8% for 1/DTF. However, after transfer back to DTF, the fit was only 84.1%. This was significantly lower than the predictive fits given by both the GPTmodel's RATE (95.3%) and DURATION versions (90.6%) presented in section 7.5.1.

The cross validation procedure described in section 7.6 was used to test predictiveness for the bean and pea data of joint use of equations 7.1 and 7.2. Respectively, for photoperiod-sensitive bean genotypes 27, 37, 42 and 44, their predictive fits and RMSPD were: 81.3% and 4.67 days, 61.1% and 4.63 days, 64.4% and 5.96 days, and 37.7% and 6.06 days. Those for pea cv. Greenfeast were 66.5% and 14.52 days. These RMSPD are usually lower or at least not larger and therefore are more predictive than those by the GPTmodel.

7.7.3 The three-plane linear rate model is a mathematical way to account for the $P \times T$ interaction

Joint use of equations 7.1 and 7.2 gave much higher descriptive and predictive fits for these bean and pea cultivars than separate applications of equations 7.1 or 7.2. This indicates predictions of the DTF for bean and pea using the three-plane model can be as high as those by the GPTmodel versions (section 7.6.2). That they are as high indicates joint use of equations 7.1 and 7.2 (and occasionally 7.3) accounts for the $P \times T$ interaction. The authors of the three-plane linear rate model acknowledge existence of a large $P \times T$ interaction effect on the expressed DTF of the genotype. They believe that transformation to 1/DTF (Summerfield *et al.*, 1991; Watkinson *et al.*, 1994) sufficiently linearizes the response for effective prediction, provided the observed DTF has not spanned a T lower than the base-T or above the optimum T for development, and/or has not spanned a P that exceeds either the critical (Pc) or the ceiling P (Pce). Exceeding these limits invalidates application of the model as described in section 7.2.2.

Accounting for the $P \times T$ interaction by the three-plane linear rate model will be interpreted, by examining its Pc. The Pc is the P at which the rate of development is the same whether defined by equations 7.1 or 7.2. Therefore:

$$Pc = (a' - a) + (b' - b)T)/c$$

or

$$Pc = (a' - a)/c + \{(b' - b)/c\}T \quad [7.7]$$

All symbols in equation 7.7 are as defined for equations 7.1 and 7.2 in section 7.2. Equation 7.7 implies that Pc is a linear function of the T. Actually, the $\{(b' - b)/c\}$ measures the $P \times T$ interaction. The $P \times T$ increases with larger difference between the sensitivity to T when there is no photoperiod response (b′) versus when there is a photoperiod response (b). It increases also as the sensitivity to P(c) is increased. For the four SD bean genotypes compared above with one LD pea genotype, or any SD compared with an LD species, the b′ and b have opposite signs. For environments giving no response to P, higher T accelerates development. But, for environments with response to P, higher T delays development. Successful prediction by the three-plane model has not been described in these or other biological terms by the three-plane model's authors. They have emphasized the improved predictiveness from using all equations of the three-plane model. Recognition of this biology for the $P \times T$ interaction control over the DTF of photoperiod sensitive genotypes is described (section 7.3) as the biological basis of the GPTmodel. Magnitude and intensity of the $P \times T$ interaction and sensitivity to P are the same physiological phenomenon. This biological understanding of the $P \times T$ interaction is the underlying difference between the GPTmodel versions (equations 7.4 and 7.6) and the three-plane linear rate model's selective use of equations 7.1, 7.2 and 7.3.

The three equations of the three-plane linear rate model, when separately used, are less biologically descriptive and predictive than the GPTmodel, especially for photoperiod sensitive genotypes grown across wide ranges of T and P (sections 7.6.1 to 7.6.5). The three equations enhance the predictiveness, because their joint use at least partially accounts for the $P \times T$ interaction effect on the plant development. The GPTmodel versions incorporate the biology of the $P \times T$ interaction, so they are highly descriptive and predictive for all genotypes of any sensitivity to P, provided the model parameters are estimated from data sets of high quality as described in section 7.5.4. Among the factors of high quality, the GPTmodel requires DTF data from a broad range of P–T combinations. In contrast, the three-plane additive rate model requires the data to come from P–T combinations that do not exceed the narrow range for which use of 1/DTF removes most of the curvilinearity of the DTF across the compared P–T.

7.8 Simulating Yield and its Components Using the GPTmodel

Chapters 4, 5, 6, 7 and 9 show that photoperiod-gene activity strongly controls partitioning of the available photosynthate between continued growth of additional vegetative organs – versus – competitive growth of the reproductive (or other) organs that become the yield. Shown also is that inhibition of partitioning to the organs that become the yield enlarges the growth duration and thereby the net accumulated biomass. Control of partitioning by P–T combinations occurs also for other than sexual organs that become yield. Potato is an LDP with respect to flowering but an SDP with respect to tuberization and yield (Purohit, 1970; Owen *et al.*, 1988; Kooman and Rabbinage, 1996). For potato an LD and/or high T causes predominant partitioning to above-ground growth and development, including the flowering. On the contrary, SD causes predominant partitioning toward growth of the tubers.

In the rest of this section, the GPTmodel will be used to simulate the yield, biomass, harvest index (HI) and DTF of pigeonpea, and of some morphological components of the yield of peanut. Figure 2.1 shows both the plant growth and development to be controlled by the same two subsets of gene activity. One is the majority of all genes; they implement growth attended by the autonomous capability to develop. The effect on the growth and development by T is the thermal response, for which each higher T accelerates the autonomous development. The second subset is any photoperiod genes, the activity of which at usual growing T will delay the autonomous development, with this delay being enlarged by each higher T. The simultaneous controls by these two subsets of gene activity over both growth and development (Fig. 2.1) suggest a potential to use the GPTmodel to simulate the biomass, HI and yield, in addition to the DTF.

Table 7.16. Descriptive fits of flowering and yield traits of peanut by the DURATION version of the GPTmodel.

Trait	Cultivar	Descriptive fit (%)	Photoperiod sensitivity (sp)
Days to flower	A2 (Valencia)	96.49	Insensitive
	B2 (Virginia)	93.58	Insensitive
	C2 (Spanish)	97.13	Insensitive
Flowers plant^{-1}	A2 (Valencia)	90.71	Insensitive
	B2 (Virginia)	93.43	Insensitive
	C2 (Spanish)	97.72	Insensitive
Pegs plant^{-1}	A2 (Valencia)	97.31	Sensitive
	B2 (Virginia)	93.82	Sensitive
	C2 (Spanish)	98.25	Sensitive
Fruits plant^{-1}	A2 (Valencia)	98.35	Sensitive
	B2 (Virginia)	100.00	Sensitive
	C2 (Spanish)	99.61	Sensitive
Yield plant^{-1}	A2 (Valencia)	99.68	Sensitive
	B2 (Virginia)	100.00	Sensitive
	C2 (Spanish)	99.80	Sensitive
Biomass plant^{-1}	A2 (Valencia)	99.45	Sensitive
	B2 (Virginia)	98.65	Sensitive
	C2 (Spanish)	99.91	Sensitive
Harvest index	A2 (Valencia)	99.26	Sensitive
	B2 (Virginia)	100.00	Sensitive
	C2 (Spanish)	98.93	Sensitive
Mean		97.72	

The modelled data are from Wynne *et al.* (1973).

7.8.1 *Simulation of phenology, yield and biomass of peanut*

Wynne *et al.* (1973) grew three peanut genotypes under day/nights of 30/26°C, 26/22°C and 22/18°C combined with both a 9 h SD, and an 'LD' established by interruption of the 15 h nights with 3 h of illumination. Data were presented for DTF, flowers per plant, number of pegs, number of fruits, yield, HI and total biomass per plant. Physiologically, the '9 + 3 h LD' is not equal to a 12 h day. There were only two *P*, so we arbitrarily assume the LD to be 18 h, because the night interruption resulted in 6 h of continuous dark. This assumption is supported by an experiment on oats by Sampson and Burrows (1972). The descriptive simulations (Table 7.16) by the GPTmodel DURATION version (equation 7.4) indicated: (i) All seven traits were simulated with descriptive fits between 93% and 100%. (ii) The DTF and flowers per plant were not responsive to *P*. But, the pegs plant^{-1}, fruits plant^{-1}, and yield plant^{-1} derived from the flowers were all reduced by the LD combined with the higher *T*. The GPTmodel's simulations support conclusions of Wynne *et al.* (1973) that the photoperiod-gene activity inhibits growth of the yield organs. Such inhibition is described in section 4.2.5 for field grown trials of peanut. The biomass plant^{-1} was highly responsive to the *P*, its changes being opposite the changes for yield. Biomass plant^{-1} was increased by the LD when combined with the higher *T*, while the yield and consequently the HI were reduced.

7.8.2 *Simulation of biomass of pigeonpea*

Simulation of the control by *P*–*T* combinations over the net accumulated biomass was demonstrated for the 30 pigeonpea genotypes for which the DTF was modelled in section 7.4.7. The biomass data (Table 9.4) are from ten location-soil planting-dates. Each soil/date included irrigation versus non-irrigation, thereby giving 20 environments. The biomass (t ha^{-1}) of each genotype was simulated using the DURATION version (equation 7.4) for the irrigated, the non-irrigated environments of ten *P*–*T* combinations. Respectively, for the irrigated, non-irrigated and combined 20 environments, the descriptive fits averaged 94.8%, 94.9% and 92.8% (Table 7.17). These high fits reinforce conclusion that *P* and *T* are major environmental determinants of both plant growth and the development – even when water availability differs greatly. The model parameters are in Table 7.18.

Application of the drop-one cross-validation procedure (section 7.6) to the pigeonpea biomass showed predictive fits for the genotypes to range from 52 to 95%, the average being 82.7% (Table 7.19). The calculated RMSPD ranged from 0.9 to 5.7 t ha^{-1}, with an average of 2.0. Mathematically, RMSPD is larger than the true prediction deviation (PD). The true PD was not discussed for any DTF data above, because replicated data required to calculate it were not available. The PD is calculated as follows:

$$\mathrm{PD} = \sqrt{\mathrm{RMSPD}^2 - \mathrm{Error}^2}$$

where Error is the statistically quantified experimental error. Ten genotypes have near zero prediction deviations while the following five have PD larger than 2 t ha^{-1} deviations: genotypes 11, 17, 19, 23 and 30 (Table 7.19). They are high

Table 7.17. Descriptive fit by the DURATION version of the GPTmodel of the net accumulated biomass at harvest maturity of 30 pigeonpea cultivars grown in 20 different environments.

	Ten irrigated environments		Ten non-irrigated environments		All 20 pooled environments	
Genotype*	Descriptive fit (%)	S_p	Descriptive fit (%)	S_p	Descriptive fit (%)	S_p
ICPL4	94.38	0.7877	93.16	0.0137	91.60	5.2346
ICPL87	93.29	0.0200	98.00	22.6924	85.45	0.0193
ICPL151	88.14	0.3802	96.12	0.0213	91.62	0.0180
83006	99.01	0.2198	96.54	0.0296	96.84	0.0263
83015	92.32	0.9459	76.98	0.0002	88.66	0.9323
83019	95.54	0.1222	96.46	0.1073	95.63	0.1099
84023	92.32	0.0007	90.97	5.5717	80.44	0.0003
85010	95.98	0.0167	95.45	0.0242	91.66	0.0169
85014	93.47	0.3331	94.17	0.0236	91.94	0.0923
85024	98.07	1.6810	97.37	0.2358	95.83	0.2971
86005	91.98	96.7000	98.13	20.5713	93.90	12.4464
86009	98.22	0.1773	96.42	0.2524	96.67	0.0847
86010	96.34	0.0173	98.19	0.0234	96.61	0.0204
87095	97.57	0.6423	96.84	0.0293	95.72	0.0289
87097	95.75	0.3547	98.50	0.0326	96.20	0.2020
87098	95.76	0.0136	97.13	0.0188	94.42	0.0158
87101	93.20	0.0153	98.30	57.0041	88.88	70.4672
87104	97.70	103.1611	98.09	47.7201	96.11	58.3320
87105	95.01	165.9513	98.16	16.3538	93.85	123.5387
ICPH9	96.18	0.0500	94.55	0.0309	94.11	0.0323
UPAS120	98.18	0.0251	93.36	0.0233	94.85	0.0243
MANAK	97.31	0.0231	87.57	0.2900	92.90	0.0207
86019	98.70	0.3317	89.62	0.3881	90.43	0.0305
86020	99.52	96.2225	98.65	79.9632	97.81	75.6847
87110	93.65	0.0137	97.44	45.6259	87.95	0.0126
87111	93.26	0.0315	90.91	0.0303	91.02	0.0309
87113	88.22	0.0146	96.36	13.8121	96.16	8.9991
87117	95.03	0.0727	92.43	0.0190	89.74	0.0178
ICPH	96.22	16.9048	97.63	21.0565	94.96	19.1997
ICPH149	83.33	0.0274	94.73	17.7580	90.79	14.2129
Mean	94.79		94.94		92.76	

*The names of the 30 cultivars are given in Table 9.4. These numbers were used in order to plot the 30 genotypes on the graphs of Fig. 9.7 and label them as needed without excessive crowding.
The modelled biomass data are from unpublished data of Y.S. Chauhan, S. Johansen and L. Singh at the International Crops Research Institute for the Semi Arid Tropics, Patancheru, Andra Pradesh, India.

yielding cultivars shown to be photoperiod sensitive by responses of their DTF (Fig. 9.7A) and responsive to factors in addition to P and T such as the soil type and irrigation versus non-irrigation (section 9.3.4.1).

The model parameters (Table 7.18) provide insight about response of biomass to P and T. (i) Every cultivar had a T_{opt} for the accumulation of biomass.

Table 7.18. The model parameters by the DURATION version of the GPTmodel for the biomass at harvest maturity of 30 pigeonpea genotypes grown in 20 environments.

	Model parameters					
Cultivar (Genotype)*	D_b (t ha^{-1})	T_{opt} (°C)	S_t	P_{opt}, i.e. the least optimum h	T_{bp} (°C)	S_p
1	7.2	31.7	−0.0577	13.8	28	5.235
2	9.1	31.9	−0.714	0	27.9	0.0193
3	6.5	27.5	−0.1422	0	27.9	0.018
4	6.7	28.6	−0.1022	1.3	28.1	0.0263
5	6.7	29.2	−0.0952	12.3	28.4	0.9323
6	5.7	29.7	−0.061	8.8	28.1	0.1099
7	0.9	24.6	0.1109	10.3	−10	0.0003
8	5.2	27.9	−0.0972	0	28	0.0169
9	6.8	28.8	−0.1018	8.2	27.9	0.0923
10	3.7	26.5	−0.1149	13.7	22.9	0.2971
11	8.8	26.6	−0.3015	15.2	28.3	12.4464
12	6.5	28.9	−0.09	7.8	28.1	0.0847
13	7.5	31.2	−0.0752	0	28.1	0.0204
14	4.6	26.7	6.4624	0	28.2	0.0289
15	5.4	27.5	−0.1191	9.7	27.9	0.202
16	7.4	32.7	−0.0435	0	27.9	0.0158
17	7.6	26.8	−0.2223	14.1	28.4	70.4672
18	8.1	29.7	−0.0989	14.1	28.4	58.332
19	8.6	26.1	−0.3742	14.1	28.4	123.5387
20	9.2	32.1	−0.0636	2.4	28.2	0.0323
21	6.7	28	−0.1122	0	27.9	0.0243
22	7.8	27.9	−0.15	0	27.9	0.0207
23	6.3	26.7	−0.1739	1.2	28	0.0305
24	6.2	30.4	−0.0473	14.1	28.4	75.6847
25	6.6	30.3	−0.0666	0	27.9	0.0126
26	5.7	25.9	−0.2398	0	28.2	0.0309
27	7.6	27.3	−0.1795	15.3	28.4	8.9991
28	6.7	28.8	−0.0838	0	27.8	0.0178
29	9.4	28.9	−0.1268	15	28.4	19.1998
30	9.8	26.6	−0.2783	15.3	28.3	14.2129
Mean		28.3			26.6	

*The names of the 30 cultivars are given in Table 9.4. These numbers were used in order to plot the 30 genotypes on the graphs of Fig. 9.7 and label them as needed without excessive crowding.
The modelled biomass data are from unpublished data of Y.S. Chauhan, S. Johansen and L. Singh at the International Crops Research Institute for the Semi Arid Tropics, Patancheru, Andra Pradesh, India. Compare Tables 7.17, 7.19.

The average was about 28.5°C. This T_{opt} indicates a T above 28.5°C would be high enough to decrease the rate of biomass accumulation, because it is higher than is optimal for growth. (ii) The biomass of all genotypes was sensitive to the P, all S_p being positive. For the biomass the P_{opt} indicates a least-optimal rather than the optimum P. It ranged from 0 h for the less sensitive to 15 h for the more

Table 7.19. Prediction fit by the DURATION version of the GPTmodel, the root mean square prediction deviation (RMSPD) and the true prediction deviation (PD) for the pigeonpea biomass of 30 genotypes grown in 20 environments.

Genotype*	Prediction fit (%)	RMSPD ($t\ ha^{-1}$)	Experimental Error ($t\ ha^{-1}$)	PD ($t\ ha^{-1}$)
1	83.50	1.3	1.2	0.6
2	79.93	2.6	2.2	1.4
3	84.44	1.7	1.8	<0.0
4	94.57	1.0	1.6	<0.0
5	68.26	2.0	1.2	1.6
6	92.66	0.9	0.7	0.5
7	66.34	1.8	1.0	1.4
8	82.50	1.6	1.9	<0.0
9	86.23	1.7	2.0	<0.0
10	89.71	0.8	0.9	<0.0
11	85.28	2.5	1.5	2.0
12	94.63	0.9	1.4	<0.0
13	94.55	1.1	1.7	<0.0
14	91.22	1.3	1.2	0.6
15	93.35	1.1	1.0	0.4
16	89.38	1.4	1.4	<0.0
17	56.79	3.9	1.9	3.4
18	93.51	1.4	1.4	<0.0
19	63.89	4.3	2.1	3.8
20	90.54	1.5	1.6	<0.0
21	92.21	1.6	1.5	0.7
22	84.66	2.2	1.7	1.3
23	74.72	2.4	1.3	2.0
24	78.59	2.7	1.9	1.9
25	77.55	1.9	1.5	1.2
26	84.93	1.5	0.9	1.2
27	80.04	2.7	2.0	1.8
28	83.62	1.9	1.6	1.0
29	90.75	2.2	3.4	<0.0
30	52.00	5.7	4.0	4.1
Mean	82.7	2.0	1.6	1.1

*The names of the 30 cultivars are given in Table 9.4. These numbers were used in order to plot the 30 genotypes on the graphs of Fig. 9.7 and label them as needed without excessive crowding.
The modelled biomass data are from unpublished data of Y.S. Chauhan, S. Johansen and L. Singh at the International Crops Research Institute for the Semi Arid Tropics, Patancheru, Andra Pradesh, India. Compare Tables 7.17 and 7.18.

photoperiod sensitive genotypes, the 15 h being longer than the longest 14.4 h. (iii) The base-T (T_{bp}) for this 'photoperiod-gene activity' below which there would be no response of the biomass to the P averaged about 28°C. (iv) Above this, the response of biomass to longer P differed greatly among the cultivars (Table 7.18) The cultivars 11, 17, 18, 19, 24, 29 and 30 are the highly photoperiod sensitive genotypes for biomass, as indicated also for the DTF (compare Fig. 9.7A, and

Table 9.4). They have the highest average biomass and yields (compare Table 9.4 and Fig. 9.7D and E). Their biomass will be reduced dramatically by longer P – but, based on the T_{bp} – only if T is higher than the 28°C estimated as the T_{bp}. This is about equal to the 28.5°C at which the T would begin to reduce biomass accumulation by being higher than the T_{opt} of the thermal response. Biomass of the less photoperiod sensitive genotypes was indicated to increase slightly as the P was longer and as the T was higher than 28°C. This suggests the biomass is determined primarily by the T if it is not higher than 28°C. The estimated T_{opt} of 28.5°C and T_{bp} of 28°C are very close. *This may indicate that for this pigeonpea data set the interpreted photoperiod-caused reduction of the biomass is actually a response to* T *higher than the* T_{opt}. The correlation of 0.90 between the P and T make it difficult, as detailed in section 9.3.3.3, to separate the $P \times T$ interaction effects on the measured traits from the thermal effects on them.

7.8.3 The biological basis for successful prediction of the biomass

Successful simulation of phenology by the GPTmodel is an unexpected benefit from the long term research Wallace and colleagues have directed toward understanding the photoperiod-gene control over the HI, yield and the DTF and DTHM (Chapters 4, 5 and 9). Simulation of plant biomass by the GPTmodel is a far-more unexpected potential. Yan developed equation 7.4 (the GPTmodel) from Wallace's (1985) verbal physiological-genetic description of the GPT interaction controls over the DTF, including the simultaneous opposing control over the DTF through the thermal response now considered to be implemented by the majority of all the genes of the plant. Simulation by the GPTmodel of just these two subsets of gene activity fulfils the requirement for basic whole-system research, because the interpretations are based on the smallest number of but two near-fully integrated subsets of gene activity that collectively account in full for the expressed DTF. These two gene subsets simultaneously account for the net accumulated biomass and its partitioning (Fig. 2.1). This explains the good predictions of the biomass, HI and yield, in addition to the DTF. The GPTmodel predicts the biomass and yield using far fewer daily measures of change in level of multiple plant traits than commonly used (Boote *et al.*, 1994) by yield prediction models.

7.9 A GPTmodel Based on the Critical Photoperiod

This section was written months after comparisons above of the simulated fits of the DURATION and RATE versions of the GPTmodel with the linear rate model's equation 7.1 and also with its three-equation, three-plane version. Below, we derive a GPTmodel version based on the critical photoperiod (P_c) rather than the P_{opt}.

7.9.1 Derivation of the concept of an optimal photoperiod

The DURATION and RATE versions of the GPTmodel described above assumed a P_{opt} for the development of every genotype. Both versions assumed the difference in P_{opt} between different SDP and different LDP is quantitative. The assumption of a P_{opt} facilitated simulation of the DTF of any genotype of any plant species using an identical iterative procedure for the GPTmodel, and with or without prior knowledge about whether the genotype is an SDP or an LDP. Assumption of a P_{opt} arose through reasoning that an SDP should have a P_{opt} that is intermediate between complete dark (0 h) and continuous light (24 h). The P_{opt} would represent a compromise between the controls over development through both photoperiodism and photosynthesis. A longer or shorter P than the P_{opt} will enlarge the photoperiod gene activity of an SDP and delay its development. But, the larger incident energy of this same longer P will enlarge the accumulated biomass and may accelerate growth and development. Above, the assumed compromised P_{opt} has been shown to model the phenology effectively. But, a minimum assimilate availability is essential for any growth and development, as detailed in Chapter 11. In some plants this assimilate can come from storage organs, rather than coming directly from photosynthesis.

A compromised P_{opt} has definite conceptual problems. If the estimated P_{opt} for an LDP is shorter than 24 h, it suggests a 24 h P should delay development. Similarly, for an SDP, a P shorter than the estimated P_{opt} would delay development. Both deductions conflict with facts. A wide range of both (i) P longer than the P_c of an LDP and of (ii) P shorter than the P_c of an SDP facilitates rapid development (Major, 1980; Major *et al.*, 1990). For an SDP the P_c is the longest of all P that allows near-equally rapid development. For an LDP, the P_c is the shortest of all P that allows near-equally rapid development. These are the definitions also of Summerfield *et al.* (1991) for the additive linear rate model. *However, their 'critical photoperiod' (Pc) varies with the* T *and, therefore, is not a genotype-fixed parameter*. In recognition of these extended ranges of near equal rapid development, Major *et al.* (1990) called the P_c for SDP the 'maximum optimum photoperiod', and the P_c for an LDP the 'minimum optimum photoperiod'.

7.9.2 Quantification and comparison of optimal and critical photoperiods

The DTF of the insensitive SDP bean line 39 and photoperiod sensitive bean line 27 were earlier simulated using the GPTmodel compared with the linear rate model (Tables 7.2, 7.3 and 7.11). The DTF for the LDP Greenfeast (Fig. 6.5) was simulated and compared in sections 7.4.1 and 7.5.1. These three genotypes and assumptions of different P_{opt} using the GPTmodels will show existence of the P_c in addition to a P_{opt}. The GPTmodel was applied 49 times to the DTF data of each of the genotypes, once after designating each half hour interval of 0.0, 0.5, 1.0 and 1.5 h to 24 h as the T_{opt}. For each genotype, the R^2 fit between the simulated DTF and the observed DTF was plotted against the 49 designated P_{opt} (Fig. 7.5). The DURATION and RATE versions gave a similar result, so only that of the DURATION version is presented.

As the designated P_{opt} was progressively enlarged from 0.0 h, the R^2 fit was progressively enlarged from 74% to 88% at 10.5 h for the photoperiod sensitive SDP (Fig. 7.5). Designation of the P_{opt} as 11.0 h caused reduction of the fit to 84%, thereby identifying the Pc of the photoperiod sensitive SDP genotype. The several next longer half hours reduced the fit precipitously. At 14.5 h, sign of the sensitivity to photoperiod (S_p) reversed from being positive as required for any fit of the model to being negative. The negative S_p indicates misfit rather than fit. Each designated P_{opt} longer than 14.5 h progressively enlarged this misfit of the photoperiod sensitive SDP.

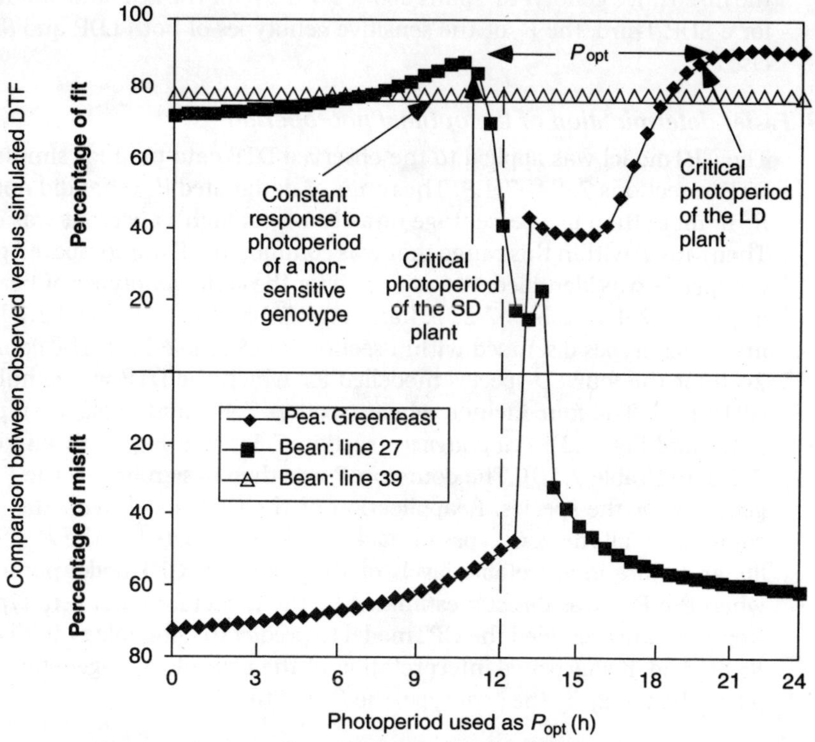

Fig. 7.5. The percentage fit or misfit between the observed days to flowering and the days simulated by the GPTmodel's DURATION version for 49 photoperiods designated one at a time as the 'optimum photoperiod'. For the photoperiod-sensitive short-day bean line 27, a 9.5–11.5 h range of designated P_{opt} all gave a near-optimum fit. All 49 designated P_{opt} gave fits near 79% for the photoperiod-insensitive bean line 39. For the photoperiod-sensitive pea cultivar Greenfeast, an LDP, the P_{opt} spans from about 20–24 h. The misfits indicate the sign for the sensitivity to photoperiod (S_p) has changed to being negative rather than positive as required by the GPTmodel. The critical photoperiod (P_c) of the short-day plant is about 11 h while that of the long-day plant is about 20 h.

For the photoperiod sensitive LDP, designation of 0.0 h as the P_{opt} gave the largest of all misfits of 74% (Fig. 7.5). Enlarging the designated P_{opt} from 0.0 h to 12.5 h, progressively decreased the misfit to near 50%. Designation as 13.0 h abruptly changed the misfit from near 50% to a fit near 40%. With enlargement of the designated P_{opt} from 13 h to 16 h the fit remained near 40%. Each longer designated P_{opt} between 16 and 20 h changed the fit toward being higher. All yet longer designated P_{opt} between 20 h and 24 h had essentially the same highest of all fits near 95%, thereby identifying 20 h to 24 h as the P_{opt} of the LDP. For this LDP, designated P_{opt} progressively shorter than 20 h more strongly reduced the fit. Therefore, the P_c of the LDP is 20 h.

The conclusions follow. First, the P_{opt} for sensitive LDP and SDP should be regarded as a range of P rather than an exact P. Second, the P_{opt} spans all P for the insensitive genotypes, spans about 20–24 h for the LDP and about 10–11.5 h for a SDP. Third, the P_c of the sensitive genotypes of both LDP and SDP is a precise P.

7.9.3 Easier determination of the optimal photoperiod

The GPTmodel was applied to the observed DTF data used for simulation of the DTF in sections 7.4.1–7.4.9. The range of designated P_{opt} that did not reduce the R^2 fit more than five percentage units below its highest percent was determined. Then, any P within this range that was common to all the compared genotypes of the species was identified. It was 11.3 h for the eight genotypes of bean discussed in section 7.4.1 (Tables 7.2, 7.3 and Figs 7.2 and 7.3). It was 12.7 h for the 18 maize genotypes discussed within section 7.4.5 (Table 7.7). The common P was 24 h for the four LD species modelled for which the DTF was simulated by the GPTmodel. The four include wheat (section 7.4.8 and Table 7.9); pea (section 7.4.1 and Fig. 7.3); *Poa pratensis* (section 7.5.2 and Fig. 7.4); and oat (section 7.5.3 and Table 7.10). The common P was then designated as the P_{opt} for every genotype of the species. Reapplication of the GPTmodel, with the photoperiod common to all the genotypes of each species designated as the P_{opt}, gave higher fits and more interpretable levels of the other five GPTmodel parameters than when the P_{opt} was directly estimated by the model for each genotype. This two step procedure enabled the GPTmodel to predict the phenology with even higher R^2 fit. And, it sharpened interpretation of the physiological-genetic controls over plant phenology by the genotype, the P, and the T.

7.9.4 Applied problems of the parameters of the linear rate and GPTmodel

The difficulty of interpreting levels of the parameters Pc and Pce of the additive linear rate model (including its three-plane version) is that the level of response to P changes with the T, and; the response to T changes with the P. This renders it inevitable that its parameters are an apparent Pc and apparent Pce of the genotype. Apparent parameters are all that can ever be observed, and directly interpreted, because the holistic system-established control over the DTF is by the simultaneous but opposite controls over the DTF through the gene activities that implement the *thermal response* and the *photoperiod response*. The delay of DTF by the photoperiod response is enlarged by a more non-optimal P and also enlarged by each higher T. On the contrary, the DTF is accelerated through the thermal response by each higher T. And, the thermal response is the integrated effect by T over the implementation of development by the majority of all the genes of the plant. For the linear rate model, the other problem is that the predictiveness of the DTF using these parameters depends on them being derived across a narrower range of T and P than the full ranges of where the crops are grown.

Researchers who apply the additive linear rate model are troubled by the P and/or T modulated changes of the parameters Pc, Pce and S_t that render them only apparent parameters as described in the prior paragraph. Researchers who have applied the GPTmodel have to deal with the apparent parameters also, since both the controls over DTF by P and by T change the expressed DTF. If the data set the GPT model was applied to has spanned enough P–T combinations to span the full range of both the P and the T, the true genotype-directed P_{opt} and T_{opt} and associated parameters will have been quantified by the GPTmodel. But, the GPTmodel-derived genotype-established S_p, P_{opt} and T_{bp}, and S_t, T_{opt} and D_b can never be observed directly. As described last in section 7.1.2, only the apparent optimum $T(T_{aopt})$ can ever be observed. And, proving that the T_{opt} has occurred requires experimental comparison of factorial P–T combinations that prove the P_{opt} and T_{opt} have occurred together in the same environment. The GPTmodel has incorporated the positive acceleration of the DTF through the *thermal response* and the retardation by the same higher T through the *photoperiod response* to quantify the genetic directions. The genetic control over the photoperiod response is due to the S_p, P_{opt} and T_{bp}, and the genetic control over the thermal response is due to the S_t, T_{opt} and D_b. The P_c of a genotype quantified using the GPTmodel will be the genotype established P_c. if it was derived using the full range of P–T combinations, and if the DTF was accurately measured across this very wide range of T and P.

Chapter 7 Summary

The R^2 fits by the GPTmodel (equation 7.4) with its assumption of both an optimum photoperiod and an optimum temperature were usually between 80 and 99%. These high descriptive and predictive fits validated paradigm shifts 8 to 13, these being the physiological-genetic bases used in Chapter 6 to derive the GPTmodel as equation 6.3. The equation incorporates: (i) that the DTF always results from a GPT interaction effect called the photoperiod response interacting with (ii) a genotype × temperature interaction effect (the thermal time response), and (iii) these two simultaneous controls interactively cause a curvilinear response of the DTF as summarized in Fig. 6.7.

The most published of all models for simulating the DTF uses 1/DTF, which is the rate of development to flowering, to (i) linearize the curvilinearity of the expressed DTF in response to the photoperiod and temperature, and (ii) to thereby avoid the photoperiod × temperature interaction effect on the DTF. The simulated 1/DTF (rate) must be converted back to the DTF. This is called the linear rate model.

A RATE version of the GPTmodel was developed (equation 7.5) for comparison with both the GPTmodel's DURATION version and the linear rate model. For both the linear rate model (equation 7.1) and the GPTmodel's RATE version, conversion from the simulated 1/DTF back to DTF usually caused small to large changes toward either better or poorer fit of the DTF than the fit of the directly simulated 1/DTF.

All three models predicted the DTF with high fits – if their requirements for model success were fulfilled. With its discretionary usages of equation 7.1 and/or 7.2 and/or 7.3, the linear rate model gave fits for all genotypes as high as those of the GPTmodel, if prediction of the DTF was never attempted for an environment having photoperiods or temperatures beyond the range for which conversion of the DTF to 1/DTF sufficiently linearized the curvilinear DTF responses. Such deviations are frequent. Predictiveness for these outlying photoperiods or temperatures was low, especially for the photoperiod sensitive genotypes, which express both the photoperiod and the thermal response.

The highest R^2 fits by both GPTmodel versions, and the highest accuracy of their model parameters, require the P–T combinations to span the extremes of both the P and T. This is the opposite of the requirement of the linear rate model that the data span a narrow range of P and especially T. The wide spans ensure for the GPT model that the optimum photoperiod (P_{opt}), and more essentially the optimum temperature (T_{opt}) are spanned. The more important of the six parameters of the GPTmodel are the sensitivity to photoperiod, the sensitivity to temperature, and the base (lowest) temperature that allows the photoperiod gene action. These parameters are established by the genotypes, because the model is based on the physiological-genetic interpretations of the control over the DTF derived in Chapters 4 and 5, extended in Chapter 6 and extended further in this chapter and Chapters 8 and 9.

The parameters for sensitivity to photoperiod of the linear model are not established by the genotype; they are altered by the temperature, i.e. they are apparent parameters as described in the summary of Chapter 6. Likewise, the linear model's parameters for sensitivity to temperature are altered by the photoperiod. The linear rate model is strictly a mathematical procedure that can predict the DTF. With use of only equation 7.1, the linear rate model simulated the DTF of the photoperiod insensitive genotypes with fits equal to the GPTmodel versions, because the DTF is responding to only the thermal response. But, for the photoperiod sensitive genotypes the linear equation usually gave much lower fits than the GPTmodel versions, because the effect by the temperature on the DTF through the photoperiod response is opposite to the effect through the thermal response, as illustrated best in Fig. 6.7.

Research with the GPTmodel completed after all the above conclusions were drawn, using the base-T of the photoperiod gene (T_{bp}), and the P_{opt} and T_{opt} as genotype-established model parameters, show the P_{opt} of SD plants to be a span of photoperiods near 11 h, while the P_{opt} of LD plants is a longer span of 20–24 h. In contrast, the critical photoperiod (P_c) is a definite photoperiod longer than about 11 h for the compared SD plants and shorter than about 20 h for the LD plants. Thus, the GPTmodel versions can quantify the critical photoperiod (P_c) as a genotype-established parameter. Doing so brings the GPTmodel into agreement with long-time usage of the P_c. This provides potential for simulating the DTF with higher R^2 fits than using the P_{opt}. Overall, the RATE version of the GPTmodel is now judged as superior to the DURATION version.

8 Interactive Control over Plant Development by Vernalization, Photoperiodism and Temperature

All the paradigm shifts 14 to 18 that involve vernalization-gene activity are derived in theoretical detail within this chapter from published research results. These shifts are then used to derive physiological–genetic equations for explaining the vernalization gene activity and its modulations by the temperature and its duration. Paradigm shifts 14 to 16, which were derived originally in Chapter 6, are most integral to deriving the physiological–genetic equations. Paradigm shift 15 is that the vernalization-gene activity occurs at high instead of low temperature. Paradigm shift 17 concerns control over the days to flowering (DTF) by the interaction between any vernalization and any photoperiod activity. Shift 18 calls for updating of the origin concept of phasic development to recognize that during vernalization an original short-day photoperiod response of wheat seedlings progressively changes to become a long-day response.

Two controls over crop adaptation and plant development are vernalization and photoperiodism. Both were discovered in the early 1900s. Vernalization is acceleration of plant reproductive development by a duration of low temperature (low-*T*). Photoperiodism is delay of plant development by the photoperiod (*P*). The third major control over plant development is the thermal-time effect, which Aitken (1974) called the universal effect of *T*. Many thorough reviews are available for each of the three controls (Murneek, 1948; Whyte, 1948; Salisbury, 1963; Lang, 1965; Vince-Prue, 1975; Napp-Zinn, 1984; Flood and Halloran, 1986; Hay and Kirby, 1991; Slafer and Rawson, 1994b, among others). However, the interaction controls over the DTF due to the simultaneous modulations of it by all three have been little studied. Understanding the physiological genetics of this holistic genotype × photoperiod × temperature (GPT) interaction control over the reproductive development is required for accurate modelling and prediction of plant phenology.

A mathematical model to describe the physiological genetics of the joint control over plant development due to just the photoperiod response and the thermal response is described in Chapters 6 and 7. That model indicates that any photoperiod response is actually always a GPT interaction control over the development, because the *T* modulates every gene-directed activity. The GPTmodel (Chapter 7) successfully predicted the DTF of 12 plant species. That model accounts for no effects by vernalization. Therefore, it is applicable only to plants that either have no vernalization requirement, or to only the post-vernalization period of the development.

This chapter reviews the control by vernalization and its interaction with the controls through the photoperiod response and the thermal response over the DTF, final leaf number (FLN), and the spikelet number (SN) of crops. Wheat has the most documented research. Other crops are mentioned as appropriate. The classical theory of phasic development is re-evaluated. Mathematical models that account for the vernalization by photoperiodism by thermal time interaction controls over the DTF, FLN, and SN of plants will account for the holistic $G \times E$ interaction controls over development. The relationships between low-*T* vernalization and both short-day (SD) and long-day (LD) vernalizations will be elucidated.

8.1 Reconsideration of the Classical Theory of Phasic Development of Plants

8.1.1 The classic concept of phasic development

The earliest suggestion for phasic development of plants was by Klebs. Whyte (1948) wrote:

> The research by Klebs and the postulation of a form of phasic development contain at least a reference to most of the problems at issue in the study of the developmental processes in plants, and of the reasons for transformation from a vegetative to a reproductive state. Klebs recognized the effect of decisive factors, temperature, light and/or darkness, alone or in combination, upon a series of phases or conditions, each one of which must be established before the next can begin; a hint of an after-effect of light treatment; evidence of reversal of development; the internal metabolic changes associated with vegetative growth or reproductive development.

Nevertheless, the classical theory of phasic development has usually been attributed to Lysenko (Napp-Zinn, 1984). Based on his research with winter wheat, Lysenko generalized that to flower, the plant must irreversibly pass first through a temperature-sensitive phase (a thermophase) and then through a light sensitive phase (a photophase). He suggested the photophase cannot begin before or during the thermophase; the plant passes to the photophase only after completing the thermophase, which is the vernalization (Napp-Zinn, 1984). Napp-Zinn wrote, 'Since then [since Lysenko proposed his theory] the thing's become more complicated, and even Lysenko himself seems to have lost confidence in his own theory.' The complications are from interaction between effects by the vernalization and by the photoperiod. Many plants do not respond to vernalization, i.e. they have no thermophase. For some but not all that respond to vernalization, its requirement can be replaced by a duration of an LD or an SD. This has been called LD vernalization, or SD vernalization. Most if not all plant species respond quantitatively rather than qualitatively to both vernalization and photoperiod.

8.1.2 Primary and secondary induction phases

Many perennial and biennial plants have a winter requirement for flower initiation. Its completion must be followed by the longer-*P* and higher-*T* of spring and summer before elongation of the reproductive stem will occur. The winter requirement of some long-day plants (LDP) can be fulfilled by a duration of either low-*T* or of SD. Based on these observations, Blondon (1972) introduced the terms primary induction and secondary induction. They replace Lysenko's concepts of the thermophase (acceleration of development by low-*T*) and photophase (control of development by the *P*). Blondon's primary induction was due to either a low-*T* or an SD, and his secondary induction was acceleration of development by an LD. Blondon's primary- and secondary-inductions were accepted by Heide (1980, 1984) who found some genotypes of *Bromus inermis* to be strict short-long-day plants (SLDP); their primary phase had to be induced by an SD. These SLDP species present the strongest support for the hypothesis of phasic development.

8.1.3 A three-phase framework of plant response to photoperiod

Roberts *et al.* (1988a) found for barley, and lentil (1986, 1988b), three phases of development in response to the *P*: a pre-inductive, post-inductive and a photoperiod-inductive phase positioned between the two non-inductive phases. The pre-inductive phase began at the onset of germination. The post-inductive phase began about 2 weeks before awn emergence. This three-phase framework of plant response to *P* was supported also by soybean data (Ellis *et al.*, 1992a). Soybean is an SDP, while barley and lentil are both LDP.

In Roberts *et al.* (1988a), ten spring and winter barley cultivars were grown at a mean diurnal *T* of 15°C, with reciprocal transfers – at 4, 8, 16 and 32 days after germination – between 16 h and either 8 h, 10 h or 13 h. All ten genotypes were initially insensitive to *P*, but differed in duration of the pre-inductive phase. The spring genotypes had a pre-inductive phase of 8–10 days, while this pre-induction for the winter genotypes was longer than the longest tested pre-transfer duration of 32 days. Low-*T* vernalization advanced DTF by reducing the duration of the pre-sensitive phase and causing earlier ear-initiation. LD accelerated the rate of development following the ear initiation.

The three-phase framework of Roberts *et al.* (1988a, b) for plant development differs little from the classical two-phase theory of development, except that, depending on the genotype, the pre-photoperiod sensitive phase may or may not respond to vernalization. It seems that the thermophase, i.e. the low-*T* vernalization of the classic theory of phasic development, is part of the pre-sensitive phase of the three-phase response to the *P*.

Discussion in section 8.4 will show that plant development does not involve a distinct vernalization phase that is completed before a photoperiod sensitive phase. Vernalization and photoperiodism will be shown to be separate but complementary, and to overlap across the early development. Across this time span, there is progression from the stronger control over development being through the vernalization-gene activity to stronger control through the photoperiod-gene activity.

8.1.4 Rejection of the concept of induction

8.1.4.1 Plant development is an autonomous process

Bernier (1988) stated: 'With time, many plants flower autonomously. All they require is that the environment be conducive to growth'. Koornneef *et al.* (1991) proposed a flowering model for *Arabidopsis thaliana* based on late flowering mutants, and understanding that their flowering occurs autonomously and cannot be stopped; it can only be accelerated or retarded by promoters or inhibitors. The term induction is not used in this book, except when referring to this concept as presented in the published literature. We rejected the term induction in section 6.4, based on the conclusion that plant development is not induced. Rather, rapid plant development is an autonomous capability of every plant, but its advance can be delayed (retarded) by the activity of the vernalization gene(s) and/or the activity of the photoperiod gene(s). Neither gene activity induces development (section 6.4.1; rather, both quantitatively delay it through a negative control over the autonomous capability for rapid development (Bernier, 1988; Koornneef *et al.*, 1991; Wallace *et al.*, 1993b; Yan and Wallace, 1995b, 1996).

8.1.4.2 The vernalization and photoperiod genes are inhibitor genes

We interpret genes for both sensitivity to vernalization and for sensitivity to photoperiod as being flowering inhibitors, because all genotypes that lack these genes flower intrinsically and early, without any vernalization and regardless of the *P*. With this interpretation, the so-called vernalization induction is a depression by low-*T* of the delay of development due to the vernalization (*vrn*) gene activity that occurs at higher-*T*. Likewise, the so-called photoperiod induction is depression by a non-delaying *P* of the photoperiod gene activity that would occur under a delaying *P* combined with a high-*T* (Chapter 6 and Murfet, 1985; Koornneef *et al.*, 1991; Summerfield *et al.*, 1991; Yan and Wallace, 1995b, 1996).

Photoperiod sensitive genotypes of pea produce a graft-transmissible flower inhibitor. Three genes (*Sn*, *Dne*, and *Ppd*) control the steps of the pathway leading to production of this inhibitor (Arumingtyas and Murfet, 1994). A genotype that lacks any of these genes fails to produce the flowering inhibitor, and therefore flowers early. For common bean, photoperiod-sensitivity versus insensitivity is due to a single photoperiod gene (Chapters 4, 5). The insensitive genotype (*ppd*) always flowers early and matures early regardless of the *P*. The photoperiod sensitive genotype (*Ppd*) expresses this same early phenotype under SD, but expresses late DTF and late maturity under LD. A second gene (*Hr*) in bean enlarges the delay of development caused by the *Ppd* allele, causing an extremely late phenotype (Mmopi *et al.*, 1995). A similar enhancer gene enlarges the photoperiod-caused delay of pea (Murfet, 1985).

That the vernalization and photoperiod gene activities inhibit development is supported additionally by the finding that photoperiod-insensitive and early-flowering genotypes of *Arabidopsis thaliana* do not contain phytochrome B, the molecule that implements the photoperiod response (Mozley and Thomas, 1995). This indicates that the effect of the *P* through the phytochrome B delays development of the photoperiod sensitive genotype; the *P* can neither induce nor accelerate the development of the insensitive genotype (Summerfield *et al.*, 1993; Yan and Wallace, 1995b, 1996; Chapters 6 and 7). Thus, vernalization is prevention by low-*T* of the activity of the vernalization gene(s) that would delay the development – if the seed had been planted at higher-*T*. Similarly, a near-optimal *P* prevents the photoperiod-gene activity that would delay the development – if the plant was growing under a non-optimal (delaying) *P*.

8.1.5 A new interpretation of the phasic development concept

We present the following hypotheses about phasic development of cereal plants (Fig. 8.1):

1. The autonomous process of developing to flowering consists of several steps. Completion of each step can be delayed by the activities of some inhibitor genes. In Fig. 8.1, A, B and C symbolize three physiological states of the tissues and/or cells of the meristem during progression of the autonomous development. Each state is expressed as a consequence of a sufficient quantitative level of a crucial substance or substances, the levels of which are modulated by the environmental factors.

2. Development from state A to state B can be quantitatively inhibited (delayed) by the vernalization-gene activity that will occur above a base-*T*, and will be quantitatively enlarged by each yet higher-*T*. Application of low-*T* until the vernalization requirement has been fulfilled (the autonomous development has sufficiently advanced) will reduce to completely prevent the vernalization-gene activity and its delay of development. This is how low-*T* vernalization accelerates development.

3. Similarly, the quantitative developmental advance from state B to state C is inhibited (delayed) by the photoperiod gene activity that for cereals will occur under short *P* and high-*T*. This gene activity and its delay are enlarged by each shorter *P* and/or higher-*T*. Therefore, both longer *P* and/or lower-*T* will reduce to completely prevent the photoperiod gene activity and its delay of development, a seeming acceleration of the development.

4. The responsiveness to LD is in proportion to the level of state B. As a result, the acceleration of development an LD will cause is enlarged progressively as the vernalization proceeds. The vernalization enlarges the responsiveness of the plant to LD. But, it is not essential to ultimate development of the response to LD.

These hypothesized concepts will be elucidated in the subsequent sections. They will be used as a basis for modelling the vernalization by photoperiod interaction control over plant development.

8.2 Control by Vernalization over the Days to Flowering

Vernalization is a primary (first occurring) control over reproductive development of plants that possess vernalization gene(s). Understanding vernalization is critical to discussion about the vernalization by photoperiodism by thermal response (the holistic GPT) interaction controls over the DTF.

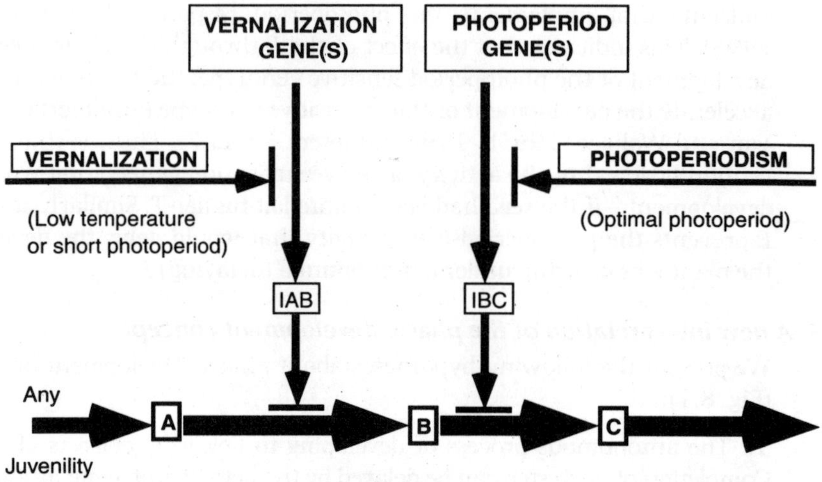

Fig. 8.1. A new hypothesis of phasic development in plants. The hypothesis integrates the following concepts. (i) Development from germination to flowering of winter cereal crops involves at least two phases, from state A to B, and B to C. A, B and C represent different physiological states of the meristem. (ii) Development from A to B to C occurs autonomously, i.e. the plant is inherently capable of rapid development to all its stages. Vernalization and photoperiod do not induce development. Rather, they quantitatively modulate the rate of the development. (iii) At high temperatures, the vernalization-gene activity inhibits (delays) the development of the first phase (from A to B), by producing the developmental inhibitor(s) (IAB). Low-temperature vernalization is prevention of the vernalization-gene activity, there by accelerating the development. (iv) Similarly, at high temperatures, the photoperiod-gene activity inhibits (delays) development of the second phase (from B to C), by producing some developmental inhibitors (IBC). An optimum photoperiod prevents the photoperiod-gene activity, thereby accelerating the development. (v) The sequence of the two phases cannot be reversed. This provides a basis for the vernalization × photoperiodism interaction effect on plant development.

8.2.1 Definitions of the vernalization requirement and vernalization response

Control over plant development by vernalization has been and will continue to be measured using the *vernalization requirement* and the *vernalization response.* Precise definitions follow.

The term 'vernalization requirement' began with the classic theory of phasic development. This theory considered vernalization to be required to induce the reproductive development of crops that respond to vernalization. The vernalization requirement was defined as the duration of the low-T needed to induce reproductive development of the genotype (Flood and Halloran, 1986). Evidence above in section 8.1.4 has indicated reproductive development occurs autonomously and rapidly – if it is not delayed by activity of the vernalization gene(s), or of photoperiod gene(s), or by low accumulated degree days. Delay but ultimate flowering does occur for vernalization-requiring genotypes when planted and grown under the warm T of spring followed by summer. The ultimate flowering indicates that vernalization does not induce the flowering. Rather, vernalization is required only for the most rapid development to flowering. Our definition of the vernalization requirement follows. *The vernalization requirement of a genotype is the duration in days at a low-T that is needed to largely eliminate the vernalization-gene activity and its delay of the development – that would occur had the planting been directly into the warm-T of spring or summer.* Exact fulfilment of the vernalization requirement establishes potential for the most rapid possible development of the genotype.

The definition of the vernalization response based on conventional understanding of plant development follows: *The vernalization response of a genotype is the reduction of its DTF by each longer duration of a low-T vernalization.* Our new definition is essentially the reverse. *The vernalization-gene controlled response is enlargement of the DTF due to lack of a sufficient low-T duration to prevent activity of the vernalization-gene and its delay of development.* The two definitions are quantitatively identical; they differ only in opposite assumptions about whether the gene activity occurs at high-*T* or at low-*T*.

8.2.2 Need for accurate quantification of the vernalization requirement

For applied comparisons among genotypes, qualitative rather than quantitative – measurement of the vernalization response based on the difference in DTF with versus without vernalization can be sufficient. But, this can never precisely quantify the vernalization requirement of the compared genotypes. Typically, comparisons of the vernalization responses of cultivars have used a single somewhat arbitrary duration of a low-*T* between 0°C and 8°C. The unvernalized control is simply planting of the dry seeds at a normal growing *T*. This procedure is less expensive, since only one low-*T* is needed. Advancing understanding of the physiological bases of vernalization, and accurate modelling of the phenology, require quantification of both the vernalization response and the vernalization requirement.

A few researchers have used multiple durations of vernalization at a given low-*T* (Purvis and Gregory, 1937; Gotoh, 1976; Berry *et al.*, 1980; Flood and Halloran, 1984; Kato and Yamagata, 1988; Miao *et al.*, 1990; Kato and Yamashita, 1991; Penrose *et al.*, 1991; Wang *et al.*, 1995). Unfortunately, all interpreted their data using the DTF measured from the termination of the vernalization. A few also measured the FLN on the main stem. Both the DTF and the FLN decrease gradually and level off progressively with each longer duration of vernalization. Due to the progressive levelling off, it has been impossible to quantify accurately the vernalization requirement. Only arbitrary criteria could be set to indicate completion of the vernalization requirement. For example, Gotoh (1976) defined the vernalization requirement as the days of low-*T* (1°C in the dark, or 8°C under illumination in another experiment) required for the given genotype to develop to flag-leaf unfolding within 34 days from the termination of the vernalization, i.e. after the vernalized plants were transferred to 20°C/24 h for post-vernalization growth. Similarly Miao *et al.* (1988) vernalized the germinating seeds of 40 genotypes for 10-day intervals of 0 to 70 days, and then planted the seeds into a warm greenhouse with 24 h. The average days to heading after these durations of low-*T* vernalization of a spring genotype was 33 days. Therefore, the vernalization requirement of each genotype was considered to be the duration of low-*T* that enabled the genotype to develop to heading within 33 days plus one standard deviation. Hunt (1979) used as the vernalization requirement the days of low-*T* required to develop to heading within 40 days (or within 100 days) from the termination of the vernalization treatment. For modelling the FLN of wheats, Kirby (1992) somewhat arbitrarily assumed 50 and 20 low-*T* days as the vernalization requirements, respectively, for the winter and semi-spring cultivars.

Therefore, the usual measurement of the DTF from the termination of the low-*T*, i.e. from the day the vernalized plant materials were planted into warm-*T*, like those of the spring and summer, will not quantify the vernalization requirement. In contrast, it will be shown below that a precise quantitative measurement of completion of the vernalization requirement results from measuring the DTF from the beginning of the vernalization, i.e. from the day the seeds are imbibed and subjected to the low-*T*. Hereafter, the DTF measured from the end (termination) of the vernalization treatment will be symbolized as the DTFe. The DTF measured from the beginning of the low-*T* vernalization will be abbreviated as DTFb.

8.2.3 Objective quantification of the vernalization requirement

8.2.3.1 The earliest published data indicative of a precise vernalization requirement

Purvis and Gregory (1937) studied vernalization using multiple durations of low-*T*. In 1935 seeds of Petkus winter rye were germinated in moist sand at 1–2°C for durations between 0 and 49 days. All durations terminated May 14, the day all germinating seeds or seedlings were planted into sand cultures. In 1936 the vernalization ranged from 5 to 20 weeks. Purvis and Gregory analysed both the DTFb and the DTFe (Fig. 8.2). These two measures of the DTF provided different conclusions as follows.

In 1935 (Fig. 8.2A) and for up to 14 weeks in 1936 (Fig. 8.2B), each longer duration of low-*T* progressively decreased the DTFe, indicating progressively smaller responses to continued vernalization. The slight enlargement of DTFe for the plants vernalized for 15–20 weeks can be ignored, based on the following explanation by Purvis and Gregory (1937): 'They were seedlings which had already received nine weeks chilling, and were planted out after further periods of five to 11 weeks at the low-*T*. These later-planted variants are not strictly comparable with the others, as the later stages in their development took place under a decreasing *P* – before the experiment was ended in October.' The conclusion is: the DTFe gave no exact quantification of the vernalization requirement.

In both 1935 (Fig. 8.2A) and 1936 (Fig. 8.2B) a clear indication of completion of the vernalization requirement was shown by curves of the DTFb. In 1935 the DTFb decreased gradually with increasing low-*T* duration to 35 days, each decrease being the vernalization response. But, this DTFb was slightly delayed by the longer (42 and 49 day) durations (Fig. 8.2A). The delays suggest completion of the vernalization requirement at 35 days (5 weeks). In 1936 (Fig. 8.2B), the DTFb was 110 days for the three vernalization durations of 5, 6 and 7 weeks. But, this DTFb was enlarged progressively rather than reduced with each longer

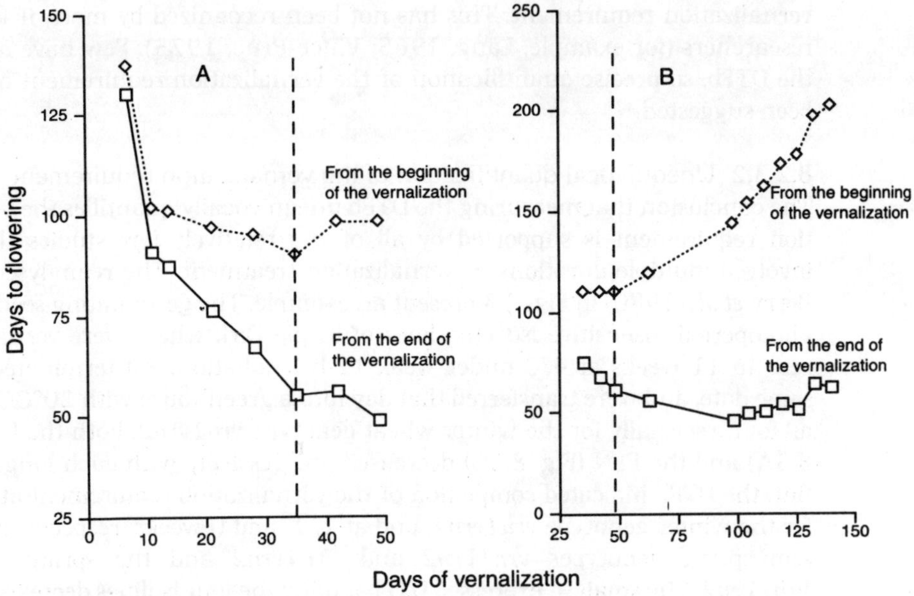

Fig. 8.2. The vernalization response of the winter rye cultivar Petkus measured as the days to flowering exclusive and inclusive of the duration of the vernalization. Measurement as the total time indicates an optimum duration of the vernalization. No comparable optimum is evident for measurement from planting at the end of the vernalization. (A) The results for a 1935 experiment with the seeds vernalized at 1–2°C for 0–49 days; (B) The results for a 1936 experiment with the seeds vernalized for 5–20 weeks. (The data are from Purvis and Gregory, 1937.)

low-*T*. That in 1935 the DTFb was delayed by each low-*T* longer than 5 weeks, and in 1936 also by longer than 7 weeks, indicates the vernalization requirement of Petkus winter rye to be no longer than 5–7 weeks. This re-analysis of these Purvis and Gregory (1937) data suggest measurement of the DTFb can give an unequivocal quantification of completion of the vernalization requirement of a genotype.

The winter rye cv. Petkus is actually an SLDP. The 5–7 week vernalization requirement indicated by the low-*T* vernalization experiments was reinforced by an SD-vernalization experiment of Purvis and Gregory (1937). Six weeks at 10 h clearly gave a smaller DTFb and a smaller FLN than both shorter and longer SD.

Purvis and Gregory (1937) recognized, as our re-analysis shows, that while the DTFe was a decrement curve that gradually approaches a minimum, the DTFb followed a U-shaped curve. The DTFb decreased progressively with the first several longer low-*T* durations, but increased after a sufficient duration. They stated: 'There appears to be little variation in this total time [the DTFb] with vernalization treatments of three to seven weeks duration, while beyond this range the total time increases.' They failed to interpret that the beginning of an increase of the DTFb with longer low-*T* indicates the completion of the vernalization requirement. This has not been recognized by most, if any, later researchers (for example, Lang, 1965; Vince-Prue, 1975). Few have measured the DTFb, so precise quantification of the vernalization requirement has never been suggested.

8.2.3.2 Unequivocal quantification of the vernalization requirement

The conclusion that measuring the DTFb unequivocally quantifies the vernalization requirement is supported by all of the relatively few studies that have involved multiple durations of vernalization treatment. The reanalysed data of Berry *et al.* (1980) in Fig. 8.3 present an example. The germinating seeds of four photoperiod insensitive isogenic lines of Triple Dirk wheat were vernalized for zero to 11 weeks at 4°C under weak-12-h irradiation. All terminated on the same date, and were transferred that day into a greenhouse with 20°C/12 h. For all four, especially for the winter wheat genotype *vrn1vrn2*, both the DTFe (Fig. 8.3A) and the FLN (Fig. 8.3C) decreased progressively with each longer low-*T*. But, the DTFb indicated completion of the vernalization requirement at 5 weeks for the winter genotype *vrn1vrn2*, and at 2, 1, and 0 weeks, respectively, for the semi-spring genotypes *vrn1Vrn2* and *Vrn1vrn2* and the spring genotype *Vrn1Vrn2*. The smallest expressed DTFb among the four isolines decreased in this same order. Our reanalysis reveals that the recessive vernalization genes (*vrn*) control the vernalization requirement; and simultaneously control the smallest DTFb expressible by the genotype. The vernalization requirement of each genotype is that duration of low-*T* that causes this smallest expressed DTFb. The latter is the intrinsic earliness (our basic duration, i.e. D_b) that is expressed when the autonomous development is not delayed by any vernalization-gene activity and/or by any photoperiod gene activity.

Miao *et al.* (1990) vernalized germinating seeds of 17 wheat cultivars with spring to extremely winter habits at 0–2°C for 0 to 70 days. The low-*T* all terminated on the same day and the vernalized seeds were planted in a greenhouse with

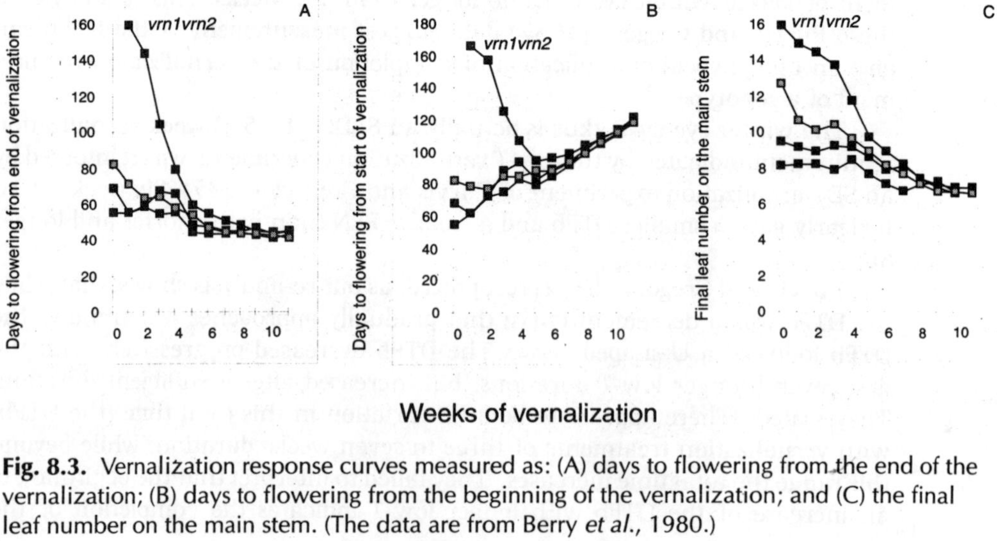

Fig. 8.3. Vernalization response curves measured as: (A) days to flowering from the end of the vernalization; (B) days to flowering from the beginning of the vernalization; and (C) the final leaf number on the main stem. (The data are from Berry *et al.*, 1980.)

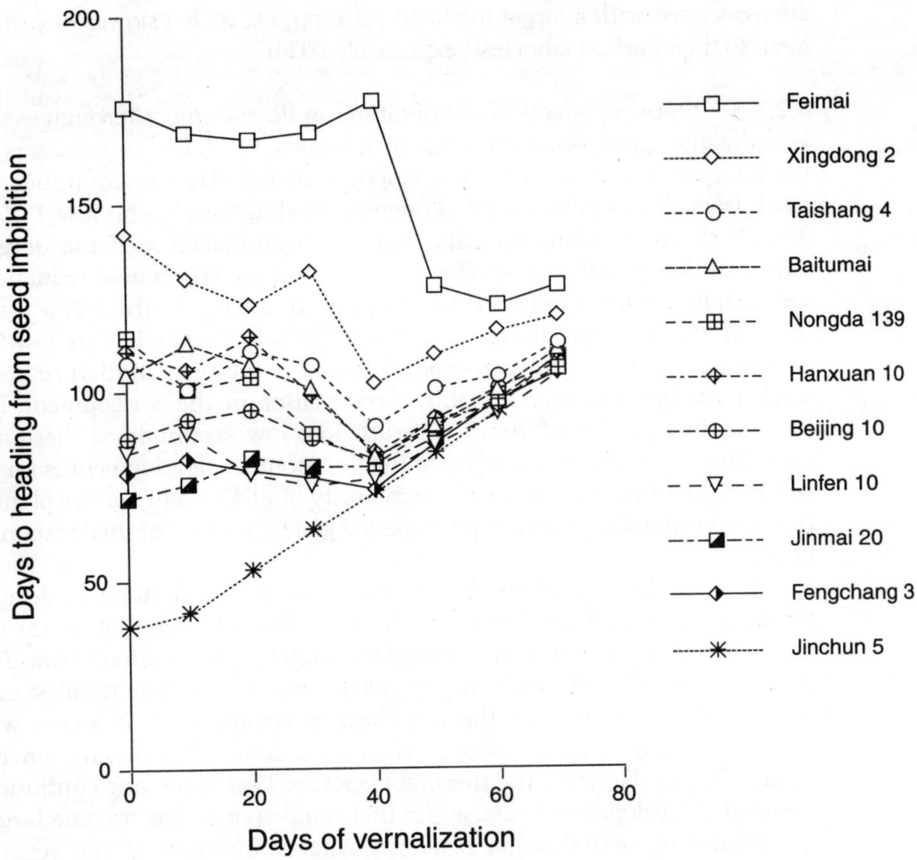

Fig. 8.4. Varietal differences in days to heading after different durations of vernalization, but measured from imbibition by the seed. The days to heading of genotypes having a vernalization requirement is reduced by longer durations of vernalization up to 40–60 days, but is delayed by yet longer vernalization. Thus, the beginning of delay by the low temperature indicates completion of the vernalization requirement. For the spring genotype Jinchun 5, which has no vernalization requirement, all vernalization delayed the days to heading. (The data are from Miao *et al.*, 1990.)

24 h and an average 22°C. As we did for Berry *et al.* (1980), we reanalysed the results (Fig. 8.4) using the DTFb obtained by adding the duration of the vernalization to the DTFe reported by the authors. All winter genotypes except Feimai showed a vernalization requirement of 40 days. The extremely late Feimai required 60 days. The spring genotypes such as Jinchun 5 had zero requirement for vernalization. A difference in the smallest expressed DTFb of the genotypes is again indicated (Fig. 8.4). It was about 33 days for all seven spring genotypes; the graph shows only two of the seven, because their DTFb responses and those of all five others overlapped precisely. In contrast this minimal DTFb varied from about 65 days to 128 days for the winter genotypes. Again, the stronger winter habits

are associated with a larger intrinsic duration, i.e. with a slower rate of development to their earliest (shortest) expressible DTFb.

8.2.3.3 Opposing effects of temperature on development through vernalization-gene versus other-gene activities

We interpret the U- or V-shaped response of the DTFb to continued low-*T* to result from two simultaneous but opposite modulations by this low-*T* (Figs 8.2A, 8.5). Both modulations indicate that for any imbibed seed the development advances slowly at the low-*T*. One low-*T* control is a Q_{10}-caused reduction of the vernalization-gene activity, which prevents the delay of the DTFb that would occur if the seed had been planted directly into higher-*T*. This low-*T*-caused acceleration of the reproductive development is the vernalization response. The second and opposite modulation is a retardation of the development. It occurs because the same low-*T* duration results in a low accumulated thermal degree days. This Q_{10} effect on growth and the autonomous development is implemented by the integrated activities of the majority of all the genes of the plant – other than the vernalization and/or photoperiod genes; it is the thermal response to the low-*T*.

At the beginning of the low-*T* vernalization, acceleration of development through the vernalization response is larger than the retardation through the low-*T* caused thermal response. Therefore, each longer duration of low-*T* reduces the expressed DTFb through the vernalization response. The smallest expressed DTFb marks completion of the vernalization requirement. It occurs when the decreasing vernalization response with each longer low-*T* duration becomes equal to or smaller than the thermal response. Thereafter, any continued low-*T* delays the development because the thermal response has become larger than any remaining vernalization response. After completion of the vernalization requirement but continuation of the low-*T*, retardation through the thermal response becomes the only control over the DTF (Fig. 8.5) until the *T* is raised sufficiently to allow any photoperiod response and/or acceleration of the DTF by the thermal response. It is probable that the indicated vernalization requirement of a given genotype interactively depends on the *T* during the vernalization, and on all of the post-vernalization *T* and *P*.

8.2.4 A mathematical description of the vernalization response of the days to flowering

Based on the verbal description above – as generalized in Fig. 8.5 – the response of the DTFb can be described using the following equation:

$$\mathrm{DTFb} = D_b + S_v(V_{req} - v)^2 \qquad (v \leq V_{req}) \qquad [8.1]$$

where:

DTFb is the days to flowering measured from the beginning of the vernalization.

v is the duration in days of the applied low-*T*.

V_{req} is the optimum duration of the low-*T* vernalization, that duration that causes the smallest expressed DTFb of the given genotype.

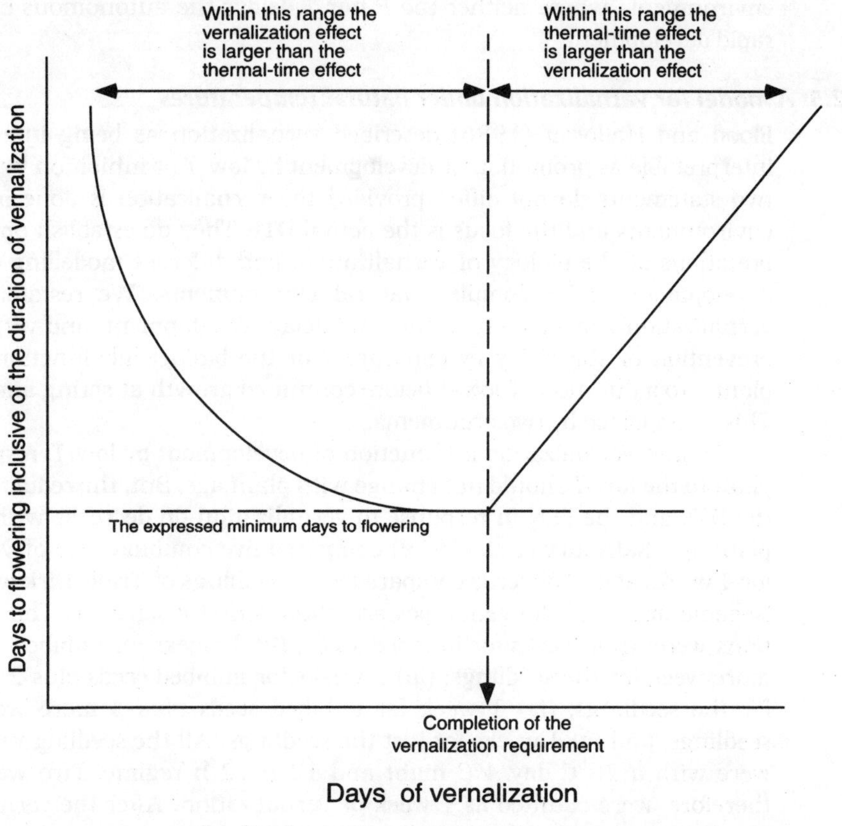

Fig. 8.5. Indication of the optimum duration of vernalization. When the response to vernalization is measured as the days to flowering (or to heading or other stage) from the beginning of the vernalization (DTFb), the vernalization requirement is indicated by the vernalization duration that causes the minimal days to flowering. After this requirement is completed, the days to flowering is delayed an equal time by each continued day of low-temperature vernalization. This delay is interpreted as due to the thermal control over the post-vernalization development. This thermal response caused delay by the low-temperature applied for the vernalization becomes observable as it becomes larger than any remaining vernalization response.

S_v is the sensitivity of the genotype to vernalization.
D_b is the smallest DTFb, which is expressed when $v = V_{req}$.

A quadratic function is used in equation 8.1 to account for the curvilinearity of the vernalization response. The restriction of $[v \leq V_{req}]$ indicates the vernalization will have been completed whenever $v > V_{req}$. After this completion, the thermal response rather than the vernalization response will become the predominant of these two opposite controls by the low-T over the plant development. This model applies when different durations of low-T are applied to germinating seeds or young seedlings, but only if the fully vernalized seeds or seedlings are placed immediately into and grown to flowering or heading in environments where neither the P nor T delays the autonomous capability for rapid development.

8.2.5 A model for vernalization under natural temperatures

Flood and Halloran (1986) described vernalization as being interchangeably interpretable as promotion of development by low-T or inhibition by high-T. The two statements do not differ, provided the vernalization is done in controlled environments and the focus is the actual DTF. They do establish opposite interpretations of the biology of vernalization, and different modelling of the plant development under complex natural environments. We restate: at high-T, vernalization-gene activity occurs and delays development, and vernalization is prevention of this delay by exposure – of the biologically-functioning seed or plant – to a duration of low-T before continued growth at spring and summer T. This is supported by two arguments.

First, if vernalization is induction of development by low-T, response of the plant to the low-T should not change with plant age. But, the reductions of both the DTF and the FLN in response to vernalization do decrease with advancing plant age. Salisbury *et al.* (1979) compared five combinations of vernalization for 4 weeks at 4°C of seeds compared with seedlings of Triple Dirk wheat and its isogenic lines. All the genotypes are photoperiod insensitive. The low-T durations were: (i) 4 weeks for imbibed seeds; (ii) 3 weeks for imbibed seeds plus 1 more week for the seedlings; (iii) 2 weeks for imbibed seeds plus 2 more weeks for the seedlings; (iv) 1 week for imbibed seeds plus 3 more weeks for the seedlings; and (v) 4 weeks for just the seedlings. All the seedling vernalizations were with a 20°C day/4°C night and 12 h/12 h regime. Two weeks of this, therefore, were counted as 1 week of vernalization. After the vernalization all plants were grown under 20°C/12 h. The DTFb of the respective five treatments were 80, 77, 93, 108 and more than 110 days, indicating seed vernalization is more effective than seedling vernalization, though there were opposite opinions (Gotoh, 1976). This same conclusion will be drawn from presentation below (section 8.3.2) of the data of Wang *et al.* (1995). Our interpretation of the data of Salisbury *et al.* (1979) is that the seedling vernalization resulted in larger DTFb than the seed vernalization, because the seedlings were exposed to longer high-T before becoming fully vernalized. The high-T the seedling plants were exposed to, before their vernalization requirement was completed, delayed their development. Vernalization of embryos or developing grains still on the maturing wheat ear is as effective or more so in reducing the DTFb and/or FLN than vernalization of the germinating seed (Riddle and Gries, 1958; Gotoh, 1976; Hoogendoorn, 1984; Evans, 1987; Sharma and Mascia, 1987; Whelan and Schaalje 1992).

Second, that low-T promote development is contrary to the universally recognized Q_{10} effect of T. Development is a gene-directed, enzyme-catalysed process. Biochemistry teaches that all enzyme-catalysed activities are enlarged by higher-T, up to T so high they destabilize the enzyme function. Based on the Q_{10} concept, the only rational interpretation is that the vernalization-gene activity inhibits development, and higher-T enlarge this inhibition (Fig. 8.1). The same interpretation was applied to the photoperiod response (Chapters 6 and 7).

Equation 8.2 describes that any high-*T* duration the germinating seeds or seedlings have been exposed to before being fully vernalized delays their reproductive development:

$$\text{DTFb} = D_b + S_v \sum \left(T - T_{bv}\right)^2 \left(v - V_{req}\right)^2 \qquad \left(v \le V_{req} \text{ and } T \ge T_{bv}\right) \qquad [8.2]$$

where:

DTFb, D_b, V_{req}, and S_v are as defined for equation 8.1.
v is the number of vernalizing days to which the plant has been exposed.
T_{bv} is the base-*T* above which the vernalization-gene activity begins and delays the development.
T is the daily temperature.
Σ is the summation of the daily temperature (T) that is above the T_{bv}, measured from the beginning of germination to the day the plant is fully vernalized, i.e. to when the DTFb is no longer reduced by continued low-*T* (see section 8.3.4).

Equation 8.2 shows that: (i) for a vernalization-sensitive genotype, the DTFb is enlarged by the summation of the daily non-vernalizing *T* the plant has been exposed to, measured from the day the seed germination began until the day the plant has been fully vernalized; and (ii) for a given duration of low-*T*, the DTFb will be larger if this vernalization begins later in the life cycle, since the accumulated non-vernalizing *T* before the vernalization will be larger. To illustrate, for multiple sowing-date experiments, an earlier fall sowing always results in a larger DTFb than a later fall sowing. On the contrary, a later spring sowing will result in a larger DTFb than an earlier spring sowing.

Two further decisions are needed before applying equation 8.2: (i) How to determine that the plant has been fully vernalized? and (ii) What is the base-*T* above which the vernalization-gene activity occurs and delays the development? Determination of the time when the plant is fully vernalized is a matter of both the plant age and duration of low-*T*. This will be detailed in section 8.3, along with the vernalization response of the FLN. Here, full vernalization is accepted as occurring the day the FLN is not reduced further by continuation of the low-*T* for vernalization. The temperature relation of the vernalization response will be detailed further in section 8.2.7.

8.2.6 Measuring the true vernalization effect on time to flowering

For almost all genotypes, any low-*T* duration will result in at least a small reduction of the post-low-*T* DTF, even if the genotype has no vernalization genes. It is necessary to determine, therefore, if a genotype has a true vernalization response. While investigating possible existence of a vernalization response in lentil, Roberts *et al.* (1988b) pointed out:

> The term vernalization is best restricted to the specific promotion of flower initiation by a previous cool temperature treatment given to imbibed seeds or young plants. . . . It does not seem to have been realized that even a preliminary cool temperature shortens the time taken to flower after transfer to a warmer environment, it does not automatically follow that a specific vernalization effect has been demonstrated. . . .

> The demonstration of a true vernalization effect, then, requires that any reduction in time to flower resulting from a vernalization treatment is greater than that which would be expected from the accumulation of thermal or photothermal time alone.

This criterion indicated no true vernalization response for two lentil genotypes (Roberts *et al.*, 1988b). The same was concluded for tested faba bean genotypes (Ellis *et al.*, 1988).

The same need exists in studies of vernalization of wheat. Most vernalization research is with imbibed seeds at 0–4°C in the dark. During this time little visible growth and morphological change are obvious. Therefore, researchers felt comfortable with measuring the vernalization response from the end of the vernalization, i.e. the DTFe. However, when the low-*T* applied to seeds involved some illumination, obvious growth occurred. Therefore, some researchers felt a necessity to account for the low-*T* duration while interpreting the results. Hoogendoorn (1984) pointed out: 'If differences in ear emergence are to be attributed solely to the effect of vernalization itself, allowance must be made for growth which occurs during vernalization'. He vernalized wheat at 5°C and 8 h for 2–9 weeks. For each genotype, he measured the increase in number of leaf primordia and plant height during the vernalization. Then, the vernalization duration was transformed into the equivalent days of the non-vernalizing 12°C/12 h that gave this increase. The 2-week and 9-week low-*T* were equal to 5.3 and 13.8 days of the 12°C/12 h. In a second experiment, Hoogendoorn (1985) vernalized diverse wheat genotypes under 5°C/8 h for 2–8 weeks before planting. He equated the 2–8 weeks of vernalization under 5°C/8 h to the 4.8 to 16.2 days at 12°C /12 h that gave equivalent increases of leaf primordia. These equivalent days were added to the measured DTFe, to quantify the earliness *per se* of the tested genotypes. Hoogendoorn (1985) defined the earliness *per se* (our basic duration symbolized as D_b) as the DTFe when the fully vernalized plants were grown under long days, including the equivalent days of the vernalization. True vernalization was accepted only if 2–9 days of 5°C/8h reduced DTFe more than the 4.8 to 16.2 days measured for the respective 2–9 week vernalization durations.

Kato and Yamagata (1988) adjusted also for growth during vernalization. Their experiments included several vernalization durations between 0 and 95 days at 5°C/24 h for 15 wheat genotypes. The vernalized plants were transferred to 20°C/24 h for continued growth to flag leaf unfolding. They counted the number of leaves that appeared during each low-*T* duration and transformed it into the equivalent days under 20°C/24 h, which were added to the days to flag-leaf unfolding measured from the end of each vernalization. They analysed the adjusted days.

Hoogendoorn (1984) and Kato and Yamagata (1988) failed to differentiate between development and growth. Vernalization is prevention by low-*T* of the retardation of reproductive development that would occur under higher-*T*. Initiation of leaf primordia and leaf appearance are growth processes. Both are accelerated by the accumulated thermal time, i.e. by higher-*T* (section 8.8). It is necessary to consider the argument of Roberts *et al.* (1988b), introduced in the first paragraph of this section, that any duration of low-*T* before high-*T* will at least slightly reduce the measured DTFe, even for genotypes with no vernalization gene(s). If the DTFb of a genotype is shortened by only the shortest duration of a low-*T*, as for the spring wheat genotypes in (Fig. 8.3B), the genotype does not have a true vernalization response (Fig. 8.3A).

We concluded in section 8.2.2 that measurement as DTFb indicates completion of the vernalization requirement, by delay rather than reduction of the DTFb by continuation of the low-*T*. Any delay will be through the thermal response. Delay by the low-*T* becomes observable only if the DTF is measured as the DTFb rather than the DTFe. Change from reduction to increase of DTFb with continued low-*T* is illustrated (Figs 8.2, 8.3, 8.4), and described theoretically in Fig. 8.5. The first observable delay of DTFb usually does not indicate there is no further vernalization response whatsoever. It does indicate all remaining vernalization response (change toward earlier DTFb) has become smaller than the simultaneous and opposite thermal response (delay of the DTFb) due to the same low-*T*. With the first observable delay of the DTFb by low-*T*, the vernalization requirement may be fully completed, but more likely, for another day to several days there will be a small but unobservable continued acceleration of the development through prevention by the low-*T* of activity by the vernalization genes.

The reasoning of the last paragraph reinforces conclusion that the DTFb is the superior measure for research directed toward improving understanding of the biology of vernalization. Measurement as the DTFb incorporates the assumption that the duration of the low-*T* becomes an integral part of the time used to develop to flowering. Our first introduction of this concept met strong negative arguments. In contrast, it is taken for granted by all plant scientists that the duration of a photoperiod constitutes part of the total time used to develop to flowering. We suggest the duration of low-*T* vernalization should be interpreted in the same way.

8.2.7 About the optimum temperature for vernalization

The terms chilling, cold treatment, low-*T* vernalization, optimum temperature for vernalization, etc. occur often in the published literature but none have had precise definitions. The only quantified optimum-*T* found for vernalization of wheat and barley is in Trione and Metzger (1970). They vernalized a winter wheat cultivar for 15–25 days at *T* between 1°C and 14°C. All plants were then kept for 1 week under 15°C/16 h, then grown for 5 weeks at 25°C/16 h. Each plant was then dissected to measure the distance between the crown and the apical meristem. Elongation of this meristem did not begin until the seedlings had at least 15 days at 5–8°C. The maximum elongation was at 7°C, and it decreased if the *T* was raised to 9°C or lowered to 3°C. Trione and Metzger concluded a sharp optimum-*T* for vernalization was indicated. Atherton *et al.* (1990) also reported a sharp optimum-*T* for vernalization of carrot. They subjected non-juvenile carrots to 3–12°C for 9, 12 or 15 weeks before transfer to 16–20°C. Bolting and flowering increased linearly with vernalization at −1–5°C, but declined linearly between 7°C and 16°C. Craigon *et al.* (1995) derived a similar conclusion from re-analysis of data of Chujo (1966), who vernalized young wheat seedlings of cv. Norin 27 at 1°C, 4°C, 8°C, 11°C, 15°C and 18°C for 20, 30, 40, 50 or 60 days under varying illumination. The vernalized seedlings were transferred to a higher-*T* and 20 h. The days to final leaf emergence was smallest at a sharp optimum-*T* at 11°C and 8°C, when the vernalization duration was 20 and 30 days, respectively. Sharpness of this optimum-*T* was reduced as the vernalization duration was longer than 30 days.

On the contrary, winter cv. Petkus rye had a broad optimum-T between −1°C and 10°C, when vernalized at different *T* for 6 weeks (Purvis and Gregory, 1935 cited in Salisbury, 1963). Salisbury (1963) interpreted this broad optimum-*T* as follows. Flowering depends on a balance between the concentrations of an inhibitor and a promoter; both being enlarged by higher-*T*, with the inhibitor having the larger Q_{10} coefficient. Our interpretation (Fig. 8.1) is similar to Salisbury's. His promoter can be interpreted as arising through the majority of all the genes and their implementation of the autonomous development, with this development being delayed (inhibited) by activity of the vernalization and/or photoperiod genes. The autonomous development and the integrated Q_{10} effect on it constitute the long recognized thermal-time response. We call this simply the thermal effect, to recognize that acceleration of the DTF with each higher-*T* is reversed to delay of the DTF if the applied *T* becomes higher than the T_{opt} established by the genotype. Salisbury's inhibitor is due to the activity of the vernalization gene(s). The 'vernalization response' is due to their inhibiting (delaying) of the development under high-*T* but not under *T* lower than the base-*T* that permits the vernalization gene activity. The Q_{10} effect of each higher-*T* than the base-*T* further enlarges this gene activity and delay. We interpret the optimum-*T* for vernalization indicated in the few cited studies to result from the simultaneous and opposing acceleration versus delay of plant development by the same *T* through the vernalization response versus the thermal response. Through the vernalization response the development is accelerated by a sufficient duration of low-*T* near 0–10°C, because they reduce to prevent the vernalization-gene activity. But, simultaneously, these low-*T* retard development through the thermal response, by slowing the many gene actions that collectively implement the plant's autonomous capability for rapid development. All *T* the seeds and seedlings are exposed to that are progressively higher than the 'vernalizing temperatures' progressively delay development more through the vernalization-gene activity, while simultaneously accelerating it more through the thermal-time response.

Several wheat phenology models have endorsed a broad optimum-*T* for vernalization. For example, Porter *et al.* (1987) and Kirby (1992) assumed 3–10°C were equally optimal for vernalization of winter wheats, while either lower than 3°C or higher than 10°C were gradually less effective. The highest-*T* causal of a vernalization response was assumed to be 17°C. This contrasts with 0–3°C being successfully used to vernalize germinating seeds in many studies.

Equation 8.2 includes no optimum-*T* for vernalization. Only a base-*T* (T_{bv}) is involved, below which no vernalization-gene activity occurs. All *T* lower than the T_{bv} (but higher than the biological zero temperature) are assumed to be equally effective for vernalization, because all prevent any vernalization-gene activity. All *T* higher than this base-*T* allow vernalization-gene activity, and the level of this activity is enlarged by each yet higher-*T*. With this understanding, no assumption is needed about an upper *T* causal of the vernalization response. Because of delay of development by the thermal control due to the low-*T*, it seems appropriate to regard T_{bv} in equation 8.2 as the low and optimum-*T* for vernalization.

8.3 Vernalization Response of the Final Leaf Number

8.3.1 Use of the final leaf number as a measure of the vernalization response

The FLN on the main stem is the most often used alternative to the DTF for measurement of the vernalization response. Vernalization is believed to facilitate transformation from vegetative growth (initiation of leaves) to reproductive growth (initiation of flowers). Based on this belief, Hay and Kirby (1991) preferred the FLN as their measure of both the vernalization and photoperiod responses. They concluded the FLN is an absolute measure of both, whereas the DTF is modulated additionally by the thermal time. Wang *et al.* (1995) used the FLN as their measure of the vernalization response.

It is often concluded in the literature that the FLN is determined solely by vernalization and the photoperiod (Miglietta, 1989, 1991a, b; Slafer and Rawson, 1994b). Cutforth *et al.* (1992) compared the effects of vernalization, the post-vernalization *T*, and water stress on the FLN of spring wheat cvs HY320 and Neepawa. They used five *T* with and without vernalization. Vernalization reduced the FLN for all but the lowest day/night of 7/2°C. After vernalization, the FLN was independent of the growth *T* under 16 h and optimum moisture. Slafer *et al.* (1994b) used spring wheat cvs Condor and Thatcher and found that the FLN was modulated by the *P* but not by the rate of change of the *P*.

That the FLN is determined solely by low-*T* vernalization and the *P* suggested the FLN is a better measure than the DTF for vernalization and photoperiod research. But, there are negative opinions. Berry *et al.* (1980), Pugsley (1966) and Levy and Peterson (1972) all considered the FLN to be a less sensitive index of the vernalization response than the DTF. We conclude that the superior measure of the vernalization response depends on the research goal. In section 8.2 we concluded that DTF measured from the beginning of the vernalization, the DTFb, will best assist phenological modelling, because it quantifies a precise measure of the 'vernalization requirement' (Figs 8.3B and 7.4). Completion of the vernalization requirement is not indicated by either the DTFe (Fig. 8.3A) or by the FLN (Fig. 8.3C). The FLN and the DTFb are complementary measures of the vernalization response; each provides information that is essential to understanding the mechanisms of vernalization, and to accurately modelling the plant phenology.

8.3.2 The vernalization response curve of the final leaf number

For wheat genotypes with the strongest spring habit, the FLN is independent of vernalization (Miao *et al.*, 1992), because they possess no vernalization genes and vernalization requirement. For all winter genotypes the FLN decreases as the vernalization duration is longer (Berry *et al.*, 1980; Miao *et al.*, 1992; Wang *et al.*, 1995, among many others). The response of the FLN to the vernalization durations follows a hyperbolic or logistic curve – it decreases and then levels off to a minimum FLN, after a sufficient low-*T* duration (Figs 8.3C, 8.6). This mirrors the response of the DTFe (Fig. 8.3A).

Conclusions that can be drawn from the hyperbolic response curve of the FLN follow: (i) Vernalization is not a mandatory requirement for development.

With no vernalization the plant develops more leaves, but will eventually initiate floral organs. Vernalization quantitatively accelerates the reproductive development by reducing the FLN. (ii) The response of the FLN to vernalization diminishes gradually as the low-*T* duration is extended, suggesting vernalization saturation will result after a sufficiently long duration of the low-*T*; and (iii) For a given wheat genotype a minimum FLN is expressed when the plants are fully vernalized. And, a maximum FLN is expressed with no vernalization – if the post-vernalization *P* does not delay the DTF, as Purvis and Gregory (1937) concluded for rye.

8.3.3 Responsiveness to vernalization decreases with plant age

Decrease of the vernalization response with advancing plant age has been long recognized (Vince-Prue, 1975; Flood and Halloran, 1986). Salisbury *et al.* (1979) indicated that seed vernalization for 4 weeks at 4°C reduced the FLN more than vernalization of seedlings. Low-*T* during ripening of the embryos was even more promotive of development (Sharma and Mascia, 1987).

Wang *et al.* (1995) measured the FLN in response to the durations of low-*T* across a wide range of plant ages. They vernalized seedlings of two winter wheat cultivars for 0–70 days at 5°C day/2°C night and 8 h at the 0, 1, 2, 3, 4, 5, 6, 7, or 8 leaf age. Respectively, these were at 0, 5, 8, 13, 17, 23, 30, 39, and 46 days after sowing. Before and after vernalization, the plants were in a greenhouse with 20 ± 5° C and 20 h. The results for cv. Pioneer 2548 are presented (Fig. 8.6). The conclusions follow. (i) Response of the FLN to the low-*T* durations followed a logistic curve for each of the nine plant ages. (ii) Except for a visually obviously-different curve for leaf age zero, the steepness of the response curve decreased with increase of the plant age at which the low-*T* was applied. (iii) A maximum FLN occurred with no vernalization, and for each leaf age a sufficiently long low-*T* duration established a minimum FLN. This minimum became larger as the vernalization was delayed to a later leaf age.

8.3.4 A model for response of the final leaf number to different durations of vernalization as modified by the plant age

8.3.4.1 The model

Based on the three conclusions in the previous paragraph, a logistic equation for response of the FLN to different durations of low-*T* vernalization – as modified by the plant age – is proposed as:

$$L = L_0 + B\,e^{-(b - cAge)v} \qquad [8.3]$$

where:

L is the FLN;
L_0 is the genetically-established minimum FLN of the genotype;
Age is the plant age at which vernalization is applied; it may be expressed as either the days after germination, or the number of leaves on the plant;
v is the duration of the low-*T* vernalization;
B is the potential number of leaves that can be eliminated by vernalization;
b is the sensitivity to vernalization of the FLN when germinating seeds are

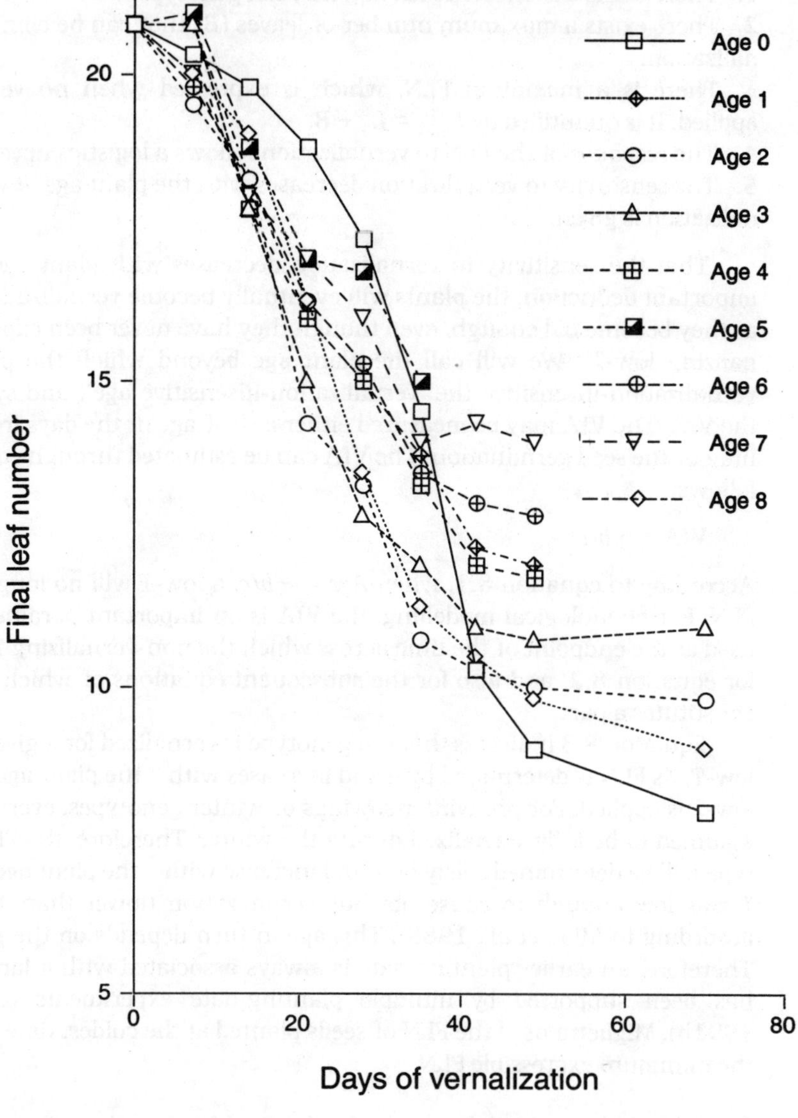

Fig. 8.6. Final leaf number in response to the duration of vernalization applied at nine leaf ages of the winter wheat cv. Pioneer 2648. (The data are from Wang *et al.*, 1995.)

vernalized, i.e. when $Age = 0$;

c is the modification by the plant age of the vernalization sensitivity.

When the germinating seeds are vernalized for different durations (with $Age = 0$), equation 8.3 reduces to:

$$L = L_0 + B\,e^{-bv} \qquad [8.3a]$$

Five concepts are integrated into equation 8.3:

1. There exists a minimum FLN (L_0) for each genotype.
2. There exists a maximum number of leaves (B) that can be eliminated by vernalization.
3. There is a maximum FLN, which is expressed when no vernalization is applied. It is quantified by $L_{max} = L_0 + B$.
4. The response of the FLN to vernalization follows a logistic curve.
5. The sensitivity to vernalization decreases with the plant age at which the vernalization is given.

That the sensitivity to vernalization decreases with plant age provides an important deduction: the plants will eventually become vernalization insensitive as they become old enough, even though they have never been exposed to a 'vernalizing low-*T*'. We will call the plant age beyond which the plants become vernalization-insensitive the 'vernalization-insensitive age', and symbolize it as the VIA. The VIA may be measured either as leaf age or the days from the beginning of the seed germination. The VIA can be estimated through equation 8.3 as follows:

$$\text{VIA} = -b/c$$

According to equation 8.3, when $Age = -b/c$, a low-*T* will no longer reduce the FLN. For phenological modelling, the VIA is an important parameter. It can be used as the endpoint of the time across which the non-vernalizing *T* are summed for equation 8.2, and also for the subsequent equations of which equation 8.2 constitutes a part.

Equation 8.3 indicates that if a genotype is vernalized for a given duration of low-*T*, its FLN is determined by – and increases with – the plant age at which the low-*T* is applied. For pre-winter-sowings of winter genotypes, every plant can be assumed to be fully vernalized during the winter. Therefore, the FLN of a genotype will be determined solely by – and increase with – the plant age at which the *T* was low enough to cause obvious vernalization (lower than 10°C or 12°C according to Miao *et al.*, 1988). This age in turn depends on the planting date. Therefore, an earlier planting date is always associated with a larger FLN. This has been supported by multiple planting-date experiments (e.g. Miglietta, 1991b). Miglietta used the FLN of seeds planted at the coldest time of the year as the minimum expressible FLN.

8.3.4.2 Application of the model for the final leaf number

Application of equation 8.3 to the data (Fig. 8.6) of Wang *et al.* (1995) gives the following equation:

$$L = 6.0 + 15.6\, e^{-(0.026935 - 0.000289\, Age)v}$$

The fitted R^2 between the estimated and observed FLN is 91.5% ($n = 74$). Several observations for leaf age 0 (for vernalization beginning with the just-imbibed seeds) are most distant from the fitted line (Fig. 8.7). Excluding them raised the R^2 to 95%. This implies that the vernalization response of seeds differs from the responses of seedlings. The vernalization-responses of all nine leaf ages

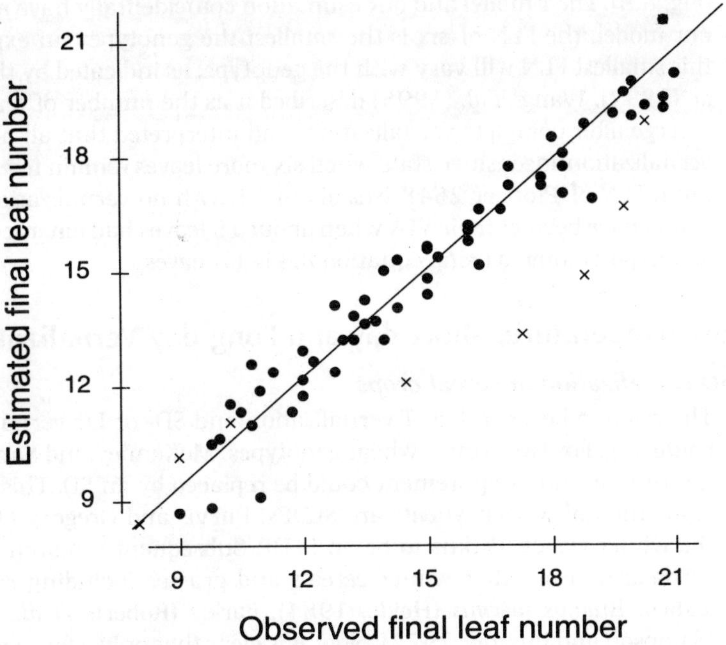

Fig. 8.7. Observed final leaf number of wheat cv. Pioneer 2548 plotted against that simulated by application of equation 8.3. Plants of leaf age 0 to leaf age 8 were vernalized for 0–70 days. The solid points are plants with vernalization applied at leaf ages 1–8. X represents plants with vernalization applied at leaf age zero, i.e. to the germinating seeds. The R^2 fit across all nine leaf ages is 91.5%, and across all but leaf ages zero it is 95%. (The analysed data are shown in Fig. 8.6 and are from Wang *et al.*, 1995).

are typical exponential curves. In contrast, the response curve of the FLN for the germinating seeds is an S-shaped curve, indicating a lag-time before response to the low-*T* vernalization begins (Fig. 8.6). This 'lag-period', also apparent in data of Jedel *et al.* (1986) and Miao *et al.* (1992), largely disappeared when seedlings were vernalized (Fig. 8.6). Gotoh (1976) concluded this also. This lag period may provide a clue for studying the mechanism of vernalization. Accurate modelling requires accounting for this lag-period.

Applying equation 8.3 to the data of Fig. 8.6 provided the following information about the cv. Pioneer 2548. (i) Its minimum FLN is six; its maximum number of leaves that can be eliminated by vernalization is 15.6, and; its maximum FLN is 21.6 (the observed maximum was 20.8). (ii) Pioneer 2548 has a VIA of 93 days from seed germination, which is equivalent to the 17-leaf stage. Chujo (1966) concluded for wheat cv. Norin 27 that at age 90 days the plants no longer respond to vernalization.

This model interprets the data of Wang *et al.* (1995) differently than they did. First, they ignored that the response to vernalization was reduced as the plants aged (Fig. 8.6). Second, they used a single linear function for all leaf ages while, the data show the vernalization response for each age to be curvilinear

(Fig. 8.6). Their model and our estimation coincidentally have a constant of 6. In our model, the FLN of six is the smallest the genotype can express. We suggest this smallest FLN will vary with the genotype, as indicated by the data of Miao *et al.* (1992). Wang *et al.* (1995) described it as the number of leaves remaining to emerge after complete vernalization, and interpreted that all plants will reach a vernalization-insensitive state when six more leaves remain to emerge. The maximum FLN of Pioneer 2548 was about 21 with no vernalization – so the plants would have been at their VIA when about 15 leaves had emerged. The VIA of this genotype estimated using equation 8.3 is 17 leaves.

8.4 Low-temperature, Short-day and Long-day Vernalizations

8.4.1 SD vernalization of cereal crops

The relation between low-*T* vernalization and SD- or LD-vernalization has been confusing. For two winter wheat genotypes, McKenney and Sando (1935) found the vernalization requirement could be replaced by an SD. This led them to conclude that all winter wheats are SLDPs. Purvis and Gregory (1937) also found the winter rye cv. Petkus to be an SLDP. Subsequent research has revealed SD-vernalization of other winter cereals and grasses including rye grass (Cooper, 1960), *Bromus inermis* (Heide, 1984), barley (Roberts *et al.*, 1988a) and oat (Sampson and Burrow, 1972). Now, it is clear that only a few genotypes of wheat that respond to low-*T* vernalization also flower earlier in response to an SD (Austin and Ford, 1976; Davidson *et al.*, 1985; Evans, 1987; Miao *et al.*, 1990, 1992; Manupeerapan, 1992).

For genotypes that seemingly are vernalized by both low-*T* and SD, the following generalizations can be made.

1. The SD vernalization is actually a photoperiod response since this SD response can be eliminated by a night break (Napp-Zinn, 1984).
2. When low-*T* and SD are applied at the same time, the low-*T* effect dominates and makes the SD ineffective (Cooper, 1960).
3. For a given genotype, there is an optimum duration of SD vernalization, just as there is an optimum duration of low-*T* vernalization. Either will give the shortest expressible DTFb. For rye cultivar Petkus these were about 6 weeks (Purvis and Gregory, 1937, as detailed in section 8.2).
4. The responses to low-*T* and SD applied separately are additive, until the vernalization requirement has been completed (Hartman, 1964, 1966).
5. If the low-*T* before an SD is long enough, it pre-empts the response to the SD (Roberts *et al.*, 1988a for barley; Evans, 1987 and Miao *et al.*, 1990, 1992 for wheat).

Based on these five observations, Yan and Wallace (1995b) suggested for the SLDP, their first (primary) phase (Fig. 8.1), which is equivalent to the primary induction of Blondon (1972), is a photoperiod-gene-established SD requirement, instead of a vernalization-gene-established low-*T* requirement. But, this SD requirement can be replaced by an equivalent duration of a low-*T*. The hypothesized mechanism follows. (i) There is a photoperiod gene(s) the activity of which occurs under an LD to delay the development of the primary phase. (ii) The level of this photoperiod-gene activity is modulated by the Q_{10} effect, so it is reduced to prevented completely by a sufficiently low-*T*. This mechanism is a central concept of our GPT interaction model (Chapters 6 and 7), which assumes that any photoperiod-gene-controlled response is due to a GPT interaction effect on the development. That model, equation 6.1 of Chapter 6, when modified, describes the primary phase of an SLDP as follows:

$$L = L_0 + S_{p1}\sum\left(P - P_{opt1}\right)^2\left(T - T_{bp1}\right)^2\left(V_{req} - v - s\right)^2 \qquad [8.4]$$

where:

L is the observed FLN of cereals, or the number of nodes (or leaves) to the first flower (NTF) of other plant species. This L can also represent the duration in days of this primary phase.
L_0 is the minimum FLN or NTF. It is expressed by the genotype when there is no delay by the photoperiod-gene activity.
P_{opt1} is the genotype-established optimum photoperiod for development of this primary phase; it is an SD for an SLDP, while the optimum photoperiod for the secondary phase is an LD. For an SLDP wheat, the P_{opt1} can be at least as short as 6 h (Miao *et al.*, 1990, 1992).
S_{p1} is the sensitivity to delaying photoperiods established by the photoperiod gene(s).
T_{p1} is the base (lowest) temperature that allows activity of the photoperiod gene(s). It is equivalent to the T_{bv} of equation 8.2.
P is the applied photoperiod.
T is the applied temperature. When $T \leq T_{bp}$, i.e. under 'low-*T*', there is no photoperiod gene activity, and consequently no delay of development.
V_{req} is as defined in equations 8.1 and 8.2.
v is the number days of low-*T* the plant has been exposed to.
s is the number of SD the plant has been exposed to.
Σ is the summation from the beginning of the seed germination to the day (time) that the plants are fully 'vernalized'.

Only a few of the genotypes that respond to low-*T* vernalization respond also to an SD. Therefore, the statement of earlier researchers that low-*T* and SD can each replace the other should be corrected. Low-*T* vernalization can replace SD vernalization but, on the contrary, the so-called SD-vernalization cannot replace a true low-*T* vernalization requirement, because the latter is established by a vernalization gene(s) rather than by a photoperiod gene(s).

Most physiological-genetic details of the control over SD vernalization remain to be elucidated. One fact is that the SLDP genotypes are generally both later and more responsive to low-*T* than the typical winter genotypes. Do the SLDP produce a different inhibitor through their photoperiod gene than is produced by the vernalization gene(s)? Or, alternatively, is the SD response established by a specific combination of vernalization-genes? This can be tested by crossing an SLDP and a spring genotype to determine if they give any progenies that respond to low-*T* vernalization but do not respond to an SD. To our knowledge, this has never been tested.

8.4.2 LD vernalization of other plant species

Many LDP in addition to the cereals and grasses respond to low-*T* vernalization. It has been interpreted that the vernalization requirement of these plants can be replaced by an LD (Vince-Prue, 1975; Napp-Zinn, 1984). The most documented examples are *Pisum* and *Arabidopsis thaliana*.

In *Pisum*, there are three complementary photoperiod genes, *Sn*, *Dne* and *Ppd* (Arumingtyas and Murfet, 1994). These genes cause both the responsiveness to *P*, and a response to low-*T*. As commonly interpreted in the past, genotypes with these three photoperiod genes can be vernalized by either low-*T* or an LD. Based on our GPT interaction model (Chapters 6 and 7) this should be interpreted as follows. The activity of these photoperiod genes inhibits development under an SD, but not under LD. The photoperiod-gene activity is modulated by the Q_{10} effect, and through it a low-*T* reduces to completely prevents this photoperiod-gene activity and its delay of development. Therefore, a low-*T* can replace the LD requirement, but not vice versa. It has been reported that there exists a low-*T* vernalization response in pea that cannot be replaced by an LD. This true low-*T* vernalization response is conferred by the gene *Lf* (Murfet and Reid, 1974).

It is well established for *Arabidopsis thaliana*, that the response to light quality (to the R:FR ratio) of the DTF and NTF of some ecotypes can be eliminated by a prior low-*T* (Martinez-Zapater and Somerville 1990; Bagnall, 1993). We interpret this as indicating that both responses are controlled by photoperiod-gene activity. The following data from Karlsson *et al.* (1993) show variable interaction between the vernalization and photoperiod responses that depends on the genotype. Karlsson *et al.* studied the photoperiod and vernalization controls over the number of leaves to flowering (FLN) of 32 *Arabidopsis* ecotypes. Vernalization of the imbibed seeds at 4°C under low illumination for 24 days was compared with an unvernalized control, after being transferred to 8 or 20 h and 22°C. All 32 ecotypes responded with smaller FLN under the LD. Two genotypes, CAN-0 and LE-0, were so sensitive to vernalization that, if not vernalized, they did not flower within the time span of the experiment under either the SD or the LD. Of the remaining 30 ecotypes, all except Co-4 had a smaller LD response after being vernalized (Fig. 8.8A). Thus, we interpret that the low-*T* partially replaced to fully filled the requirement for an LD of most of the ecotypes. All but five ecotypes showed a vernalization response (reduction of the FLN) under the 8 h post-vernalization *P* (Fig. 8.8B). However, only eight of the 30 flowered with smaller FLN after vernalization when grown under the 20 h. This change of the vernalization response under post-vernalization SD versus LD has been interpreted as replacement of the vernalization by the LD. The following alternative interpretation is based on the GPTmodel detailed in Chapters 6 and 7. The eight ecotypes that showed a vernalization response under the post-vernalization LD are the only ones that might have a true vernalization-gene established response. The 17 ecotypes that responded to the vernalization, i.e. flowered with a reduced FLN only under the post-vernalization SD, actually have no vernalization-gene established response. Their response to low-*T* is established by a photoperiod gene(s), but the photoperiod-gene activity and consequent delay of development are eliminated – by the prior low-*T*. The ecotype Co-4 differed from all the others in that it responded to LD slightly more strongly after being vernalized, as compared with not being vernalized.

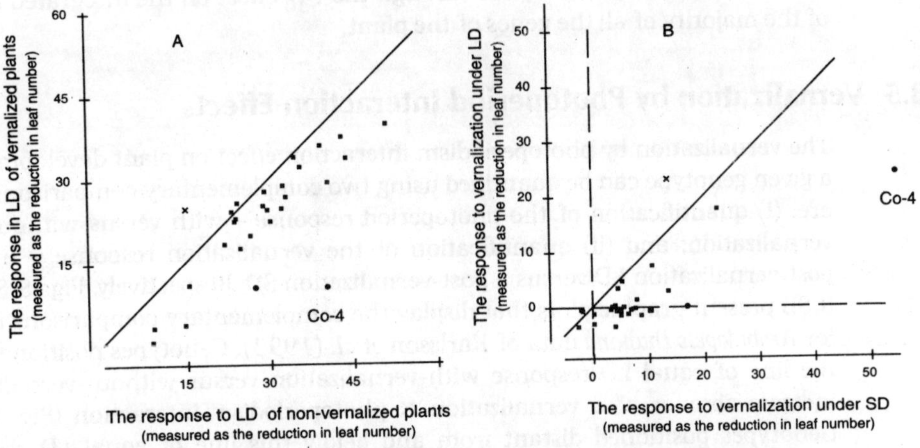

Fig. 8.8. The vernalization by photoperiod interaction effect on leaf number to flowering of 30 ecotypes of *Arabidopsis thaliana.* (A) The reduction in number of leaves in response to LD of vernalized versus non-vernalized plants. (B) Reduction in number of leaves in response to vernalization measured under LD versus SD. (The data are from Karlsson *et al.*, 1993.)

Diverse responses to vernalization and LD were revealed also by research of Koornneef *et al.* (1991). They identified many gene mutations that alter the maturity of *Arabidopsis.* Quantitative trait loci (QTL) research has identified at least 11 QTLs that influence the DTF of *Arabidopsis thaliana,* some of which showed $G \times E$ interaction while others did not (Jansen *et al.*, 1995).

It must be recognized that what seems to be a photoperiod response may be a response to the photoperiod *per se*, a response to the light quality (wave length ratio), a response to the total light energy, or a combination of these (Martinez-Zapater and Somerville, 1990). If a response to LD is actually a response to the total light energy, rather than a true photoperiod response, low-*T* cannot be expected to replace or reduce this response.

In summary, based on the GPTmodel of Chapters 6 and 7, no true vernalization (a vernalization-gene established acceleration of development by low-*T*) can be replaced by any photoperiod. On the contrary, replacement of an SD- or an LD-requirement by a low-*T* is real; it is due to the ever-present Q_{10}-implemented modulation by *T* of the photoperiod-gene activity. There are but limited data about the quantitative relations between the SD or LD requirements and their replacement by a low-*T*. Carefully planned experiments are required to achieve detailed understanding of these quantitative GPT interaction effects. The GPT interaction control over development described in Chapters 6 and 7 involves only a photoperiod-gene activity. Any vernalization × photoperiod interaction involves both vernalization- and photoperiod-gene activities, as explained in the following sections. Not mentioned in the last two sentences is a further interaction with both any photoperiod and with any vernalization responses. Both responses are always interacting with the simultaneous and always functioning thermal response, which arises through the Q_{10} effect on the integrated activity of the majority of all the genes of the plant.

8.5 Vernalization by Photoperiod Interaction Effects

The vernalization by photoperiodism interaction effect on plant development for a given genotype can be quantified using two complementary comparisons. They are: (i) quantification of the photoperiod response – with versus without prior vernalization; and (ii) quantification of the vernalization response – under a post-vernalization LD versus a post-vernalization SD. Respectively, Figs 8.8A and 8.8B present graphic plots that display the complementary comparisons i and ii for *Arabidopsis thaliana* data of Karlsson *et al.* (1993). Genotypes positioned near the line of equal LD response with-vernalization versus without-vernalization indicate absence of a vernalization × photoperiodism interaction (Fig. 8.8A). Genotypes positioned distant from and below this line of equal LD response indicate partial to complete replacement of the LD response by the low-*T* vernalization. Responses positioned distant from and above the equality line indicate enlargement of the LD response by the low-*T* vernalization. This is the true vernalization by photoperiodism interaction. The vernalization by photoperiodism interaction detected using the effect of the post-vernalization *P* on the response to the low-*T* vernalization indicates either no interaction for some genotypes, or partial replacement of the LD response by the low-*T* vernalization (Fig. 8.8B). Only ecotype Co-4 showed a slight complementation effect on the FLN through a vernalization by photoperiodism interaction effect.

8.5.1 Vernalization by photoperiod interaction patterns among wheat genotypes

Comparison across a broad range of genotypes is required to reveal the range of possible vernalization by photoperiodism interaction patterns. Davidson *et al.* (1985) measured the days to heading from planting (DTFe) of 117 wheat genotypes in response to a two-vernalization-treatment by two post-vernalization-photoperiod factorial. The germinating seeds were vernalized at 1–2°C in the dark for 6 weeks. The vernalized seeds and unvernalized controls were then grown in a glasshouse with an average of 20°C and both the natural *P* of less than 11 h (until 55 days after planting) or a constant 16 h. For the late genotypes, flowering occurred as late as 200 days after sowing the seed. The LD response was quantified as the reduction of the DTFe under the 16 h compared with the DTFe under the natural SD. The vernalization response was quantified as the reduction of the DTFe by the 42 days of low-*T* vernalization as compared with the unvernalized control.

A continuum of near zero to large vernalization by photoperiodism interaction is indicated (Fig. 8.9A and B). Near zero interaction is shown by the genotypes positioned adjacent to the line of equal response. Large vernalization by photoperiodism interaction is indicated for the genotypes positioned far above this line. No genotype was far below the equality lines, indicating that the vernalization response and the LD response in wheat cannot replace each other, which does occur in *Arabidopsis thaliana* (Fig. 8.8). Instead, in wheat and the biennial cereal crops, vernalization and the LD response are complementary. From Fig.

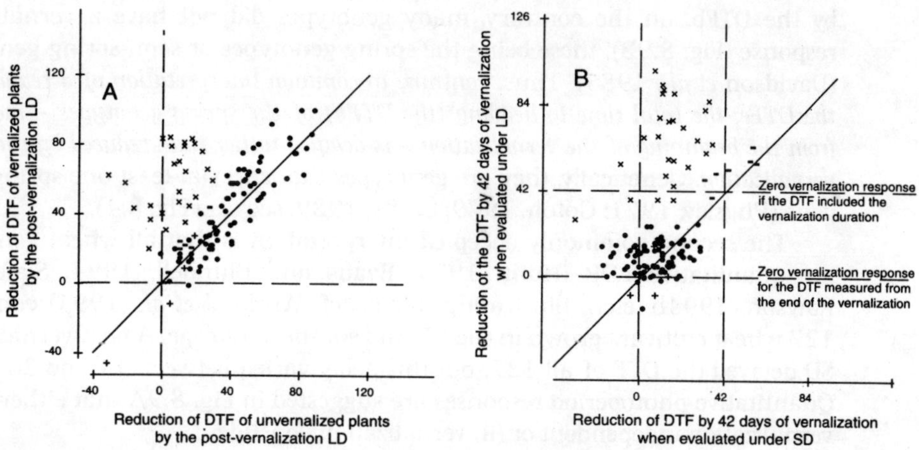

Fig. 8.9. The vernalization by photoperiod interaction effect on the days to heading (DTF) of 117 wheat cultivars. This interaction is indicated by: (A) The effect by vernalization on the response to LD, which is quantified as the reduction in days to heading measured from the end of the vernalization (DTF) in 16 h versus in the natural 11-15 h. (B) The effect by photoperiod on the vernalization response measured as the reduction in DTF by the vernalization of 42 days at 1–2°C compared with no vernalization. X represents wheat genotypes that are photoperiod sensitive; they show significant vernalization by photoperiod interaction. Genotypes positioned adjacent to the equality response lines show little vernalization by photoperiod interaction, and they include spring wheats (circles), winter wheats that are photoperiod-insensitive (–), and late cultivars with no response to vernalization (+). (The data are from Davidson *et al.*, 1985.)

8.9B, the genotypes showing no vernalization by photoperiodism interaction include three groups. The majority are 'spring' genotypes; seven are photoperiod-insensitive-winter genotypes; four are late flowering genotypes that showed no vernalization response, the latest flowering of all (French cv. Cote d'Or) being one. Cote d'Or behaved like a short-day plant with no obvious response to the 42 days of low-*T* (Fig. 8.8A). A second experiment by Davidson *et al.* (1985) indicated that for some winter genotypes 4 weeks of low-*T* did not affect the DTFe while 6 weeks caused a small response, so it may be the 6 weeks of low-*T* was too short to reduce the DTFe of Cote d'Or. The genotypes showing the vernalization by photoperiodism interaction are all winter cultivars that are also photoperiod sensitive. Two, Templar and Huntsman, are well-known SLDP (Fig. 8.9A) studied by Evans (1987). They are characterized by an SD response (delay by the LD) when unvernalized. But after being vernalized, they have an LD response (acceleration by the LD). The conclusion is for a cultivar to show a vernalization by photoperiodism interaction effect, it must be a 'winter' genotype that is also photoperiod-sensitive.

The conclusion stated in the last sentence seems contradictory to some accepted interpretations. First, that almost all genotypes respond to vernalization, as indicated by both the FLN (not shown), and the DTFe (Fig. 8.9B). Judged by the DTFb, on the contrary, many genotypes did not have a vernalization response (Fig. 8.9B), these being the spring genotypes or semi-spring genotypes (Davidson *et al.*, 1985). Thus: *contrary to common interpretation of a reduction of the DTFe, the total time to heading (the DTFb) of the spring genotypes – measured from the beginning of the vernalization – is delayed rather than reduced by the low-T vernalization.* Genetically, they are genotypes that carry at least one spring (*Vrn*) gene (Pugsley, 1971; Gotoh, 1979; Crofts, 1989; see also Fig. 8.4).

The second commonly accepted interpretation is that all wheat genotypes are quantitative LDP (Hunt, 1979; Evans and Blundell, 1994; Slafer and Rawson, 1994b, etc.). Illustrating this belief, Worland *et al.* (1994) compared 127 wheat cultivars grown in the UK and southern Europe. A post-vernalization SD delayed the DTF of all 127, but this delay varied between 14 and 200 days. Quantitative photoperiod responses are suggested in Fig. 8.9A that either are (i) vernalization independent or (ii) vernalization dependent.

The vernalization-independent photoperiod response is a universal LD response expressed by all wheat genotypes, including the commonly recognized photoperiod-insensitive genotypes. This indicates: *the often-called day-neutral genotypes are simply less responsive to the photoperiod than those classified as sensitive or highly sensitive genotypes.*

The vernalization-dependent photoperiod response is an LD response that is expressed by most but not all winter genotypes. It arises through the vernalization by photoperiodism interaction effect on the plant development, and occurs only for genotypes that both have the 'winter habit' and are 'photoperiod-sensitive'. The vernalization-dependency of the photoperiod-response of these winter genotypes is indicated by their being positioned far above the line of equal LD response (Fig. 8.9A). No vernalization by photoperiodism interaction is indicated for the genotypes that are either 'spring' and/or 'photoperiod-insensitive'; they are all positioned adjacent to the equality line. Recognition and differentiation of the vernalization-dependent and vernalization-independent photoperiod responses is integral to full understanding of the mechanisms of both the photoperiod- and the vernalization-response. It will be needed for modelling of the phenology.

Griffiths *et al.* (1985) reported for the chromosome-substitution line Chinese Spring (Hope 5D) that no vernalization response of the FLN occurs if the vernalized plants were grown under SD but it did occur if grown under an LD. If after vernalization, the plants were grown under SD, their FLN was restored to the level of the unvernalized plants. Double-ridges formed eventually even in SD, but it required the same time after planting, and the same number of leaves were formed whether or not the apex was vernalized. This shows a post-vernalization LD is required for the vernalization to reduce the DTF and FLN. Griffiths *et al.* concluded: 'There are clearly two stages in the progress to flowering in Chinese Spring (Hope 5D) wheat, since for vernalization to be effective it must be followed by LD, and without vernalization Chinese Spring (Hope 5D) produces the same number of leaves in LD and so LD is not effective alone either'. Thus, the vernalization by photoperiodism interaction is a complementation between the vernalization and the post-vernalization LD. If preceded by a duration of low-*T*, an LD accelerates the plant development, this acceleration being a vernalization-dependent LD response. The dependency on vernalization for the subsequent LD response (the vernalization by photoperiodism interaction) is responsible for the fact that – *although all wheat genotypes are LDP, many winter genotypes do not express an LD response or even express an SD response if they have not been vernalized* (Fig. 8.9A, and Miao *et al.*, 1990, 1992).

Halloran (1975) studied the vernalization by photoperiodism interaction effect on the FLN of eight wheat genotypes. Vernalized and unvernalized seedlings were grown to the 0, 3, 5, 7 and 9 leaf stages under 8 h, and then transferred to 18 h and natural *T* of summer plantings. The FLN was plotted against the leaf number at the time of the transfer. All genotypes responded to the 18 h. The FLN increased as transfer to the LD was delayed to later-leaf ages. But, there was no vernalization by photoperiodism interaction for six of the eight genotypes, and only one genotype showed a significant vernalization by photoperiodism interaction effect on the FLN, i.e. showed vernalized plants to be more responsive to the photoperiod. Davidson *et al.* (1985) pointed out that the eight cultivars Halloran (1975) studied are mostly *spring* genotypes for which no vernalization by photoperiodism interaction effect should be expected.

Rawson and Zajac (1993) studied the development of the isogenic wheat lines of Hartog using three low-*T*-vernalizations and four-post-vernalization *P* (9 h, 11 h, 13 h and 15 h) combined with two *T* (25/15°C and 17/7°C). The results indicated the following. (i) There was little vernalization by photoperiod by temperature interaction for the DTF of Hartog, but interaction was obvious for Late Hartog, which is more responsive to both vernalization and photoperiod. (ii) This vernalization by photoperiodism interaction was evident under the higher- but not under the lower-*T* of the post-vernalization period.

8.5.2 *The photoperiod response after different vernalization durations*

Miao *et al.* (1988, 1990, 1991, 1992) studied the photothermal responses of 17 wheat cultivars and kindly provided their unpublished data which will be analysed next. The 17 genotypes spanned the range of ecological adaptations across China. Germinating seeds were vernalized at 0–2°C for 0, 10, 20, 30, 40, 50, 60 or 70 days. The vernalized seeds were planted and the seedlings exposed for 30 days to 6 h, 10 h, 14 h, or 24 h and 20–25°C in four growth chambers. All were then grown in a greenhouse with continuous light (24 h) and an average of 22°C (range of 15–30°). The days to heading (DTF) from vernalization termination (DTFe) and the FLN were recorded.

8.5.2.1 Response of days to heading to photoperiod after different durations of vernalization

Responses of the DTF to the four post-vernalization *P* are presented for two spring and ten winter cultivars (Fig. 8.10). The photoperiod responses are indicated by the slope of the linear regression of the DTF on the *P*.

All seven spring cultivars responded to the vernalization durations combined with four post-vernalization *P* like the responses for Liaochun 6 and Jinchun 5 (No. 1 and 2 in Fig. 8.10 and 8.11). All had no response to vernalization but lower DTF in response to longer *P*, with no obvious effect on the slope of the photoperiod response due to the eight low-*T* vernalization durations.

Cultivars 3 to 7 are semi-winter to typical winter cultivars. With zero

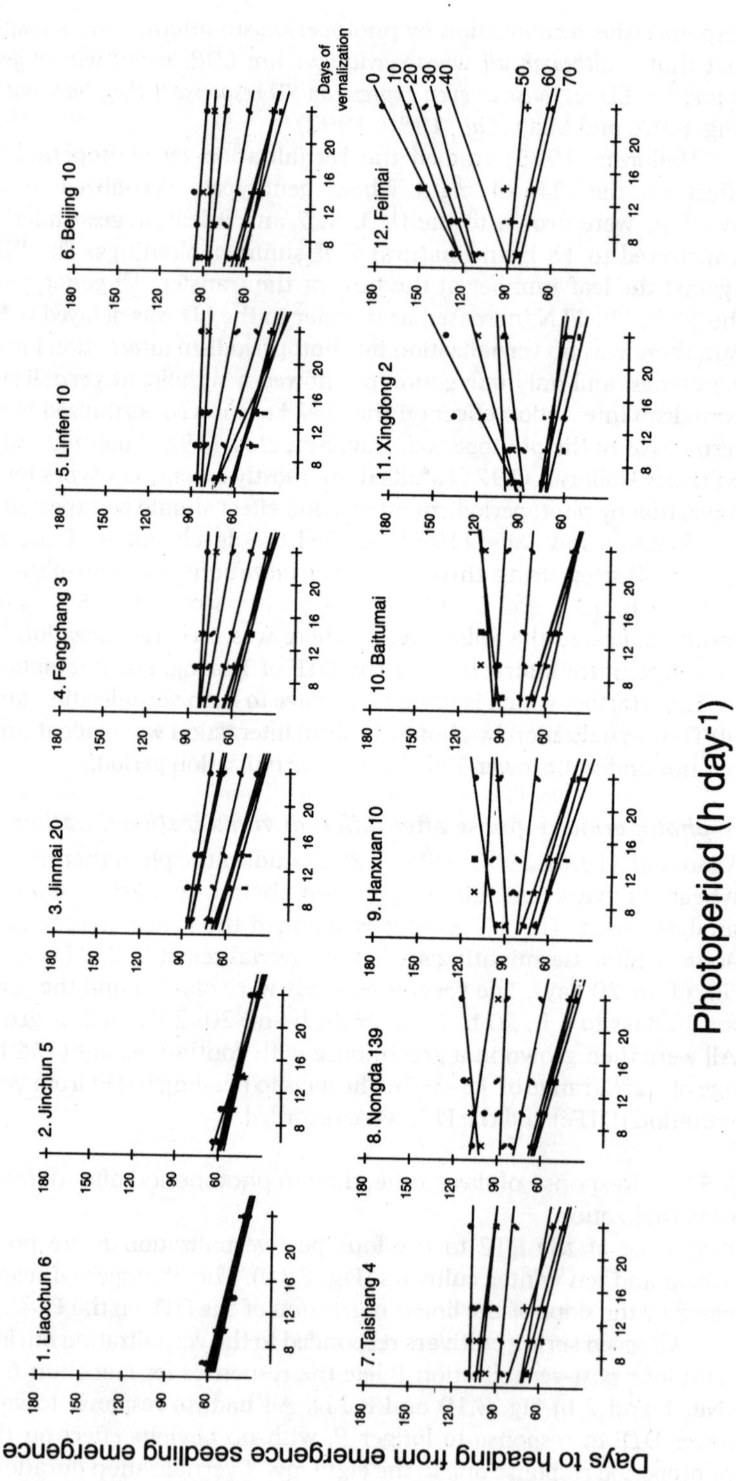

Days to heading from emergence seedling emergence
Photoperiod (h day-1)
1. Liaochun 6
2. Jinchun 5
3. Jinmai 20
4. Fengchang 3
5. Linfen 10
6. Beijing 10
7. Taishang 4
8. Nongda 139
9. Hanxuan 10
10. Baitumai
11. Xingdong 2
12. Feimai
Days of vernalization
0
10
20
30
40
50
60
70
180
150
120
90
60
8
12
16
20

Fig. 8.10.

vernalization they responded either not at all or only slightly to LD (Fig. 8.10). As the low-*T* vernalization was longer, the post-vernalization LD response enlarged progressively, and its slope stabilized as the low-*T* duration was 30–40 days or longer. Cultivars 8 to 10 are typical winter cultivars. With no vernalization, they developed to heading slightly earlier under SD, i.e. had slight SD response (Fig. 8.10). Ten to 30 days of the low-*T* eliminated the SD response, and replaced it with an LD response that tended to stabilize as the vernalization was 40 days or longer. Cultivars 11 and 12 are SLDPs. Cultivar 11 expressed the SD response while the vernalization was 30 days or shorter, but changed to an LD response when it was 40 days or longer (Figs 8.10, 8.11). Cultivar 12 showed the SD response up to 40 days of vernalization, but the LD response after 50 days or longer (Figs 8.10, 8.11). In summary for the genotypes that require vernalization, as the low-*T* was longer, the SD response expressed before any vernalization changed gradually to become an LD response.

8.5.2.2 Response of final leaf number to photoperiod after different durations of vernalization

The response of the FLN was similar to that of the DTF, with minor differences (Fig. 8.11 compared with 8.10). For genotypes 5, 6, 10, and 12, there seemed to be a threshold response, an abrupt reversal from an SD to an LD response occurred after a sufficient vernalization duration (Fig. 8.11). A threshold response to vernalization was reported also for four isogenic wheat lines differing in vernalization genes by Salisbury *et al.* (1979), and shown also by Berry *et al.* (1980).

8.5.2.3 Quantitative variations of the photoperiod response among genotypes after different vernalization durations

It has long been believed that vernalization makes the meristems capable of responding to the *P* (Murfet, 1985). Devay (1967) reported that an active phytochrome causal of the LD response was present only in vernalized winter wheat

---

Fig. 8.10. (*opposite*) The days to heading (DTH) in response to photoperiod after 0 to 70 days of vernalization for 12 selected wheat cultivars. Germinating seeds were vernalized at 0–2°C before planting. The vernalization durations are labelled only for cultivar 12, but usually they are similarly positioned vertically for all the winter cultivars. The seedlings emerged into the initial photoperiods of 6 h, 10 h, 14 h or 24 h for 30 days, and were then transferred to a greenhouse with 24 h and an average 22°C. Cultivars 1 and 2 have spring habit; cultivars 3 to 12 have increasingly stronger winter habit. The photoperiod response of the spring cultivars is not influenced by the vernalization duration, but responses of the winter cultivars are dramatically influenced by it. With no vernalization, the winter cultivars showed only marginal reduction of DTH by the initial longer photoperiod, or had an SD response as evidenced by increase in DTH under longer photoperiod. Increase of the DTH in response to longer photoperiod of cultivars having the strongest winter habit is attended by gradual reduction of the SD response as the vernalization requirement progressed toward completion. After this completion, all cultivars expressed an LD response. (The analysed DTF are from unpublished data of Miao *et al.*)

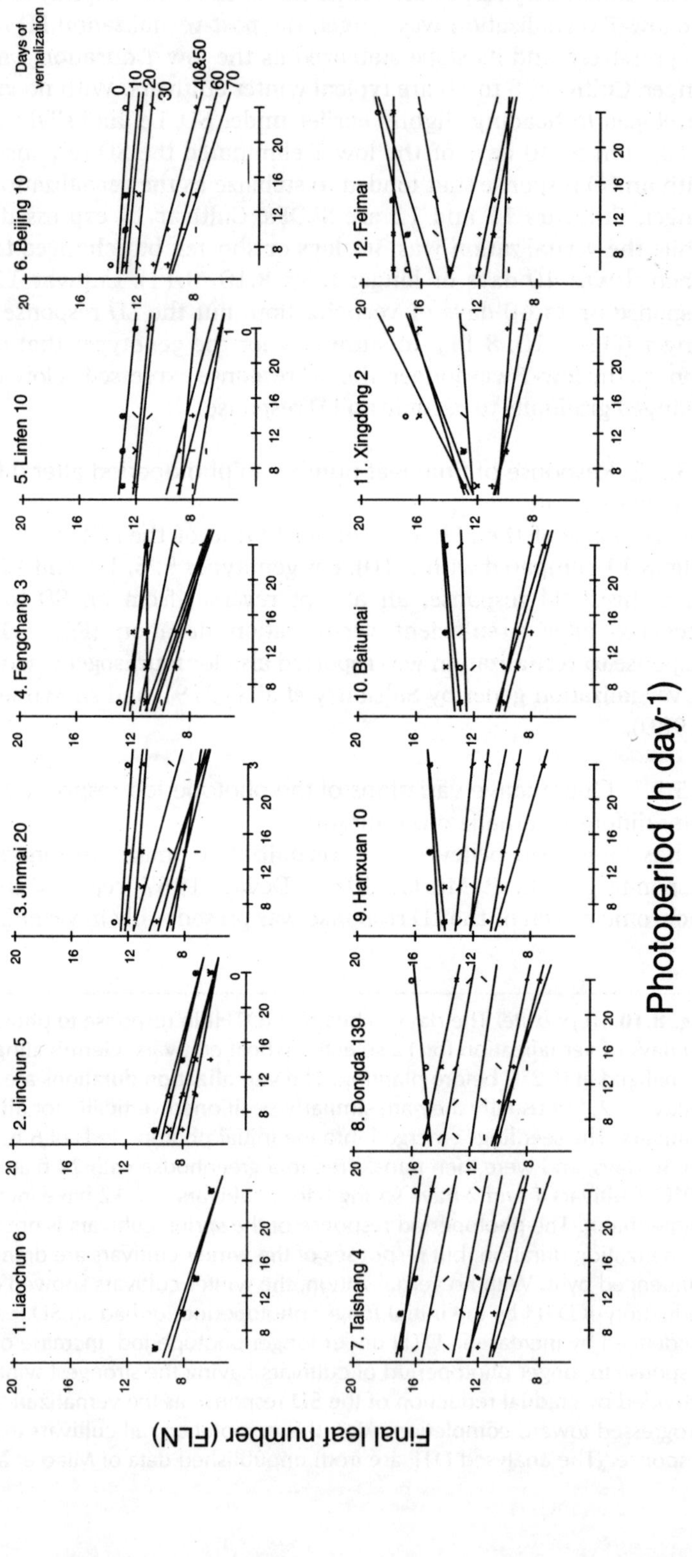

Final leaf number (FLN)
Photoperiod (h day-1)
Days of vernalization
0
10
20
30
40&50
60
70
1. Liaochun 6
2. Jinchun 5
3. Jinmai 20
4. Fengchang 3
5. Linfen 10
6. Beijing 10
7. Taishang 4
8. Dongda 139
9. Hanxuan 10
10. Baitumai
11. Xingdong 2
12. Feimai

Fig. 8.11.

plants. Accumulation of this phytochrome followed an S-shaped response in relation to the low-*T* duration. The data of Miao (Figs 8.10 and 8.11) confirm that winter wheats gradually gain response to LD as they are vernalized. Interpretation of Miao's data is summarized as follows.

1. With no vernalization, the spring genotypes of wheat express an LD response. The common winter genotypes initially have responses that vary between a slight SD- to a slight LD-response. The extreme winter genotypes are actually SLDP. They initially express a strong SD response which varies continuously and quantitatively, across the genotypes – from a strong SD response and a zero LD response to a strong LD response.
2. Progressively longer vernalization durations quantitatively eliminate any initial SD response, while progressively enlarging the LD response to the stable level characteristic of each genotype.

Through the vernalization by photoperiodism interaction, a photoperiod sensitive plant can be interpreted as being fully vernalized when it has become maximally responsive to an LD. The minimum days of low-*T* vernalization required to develop to this maximum LD responsiveness is the vernalization requirement. For SLDP genotypes, which have the strongest winter habit, vernalization is required to: (i) eliminate their SD response, and (ii) enlarge and then maximize their LD response. For the typical winter genotypes, the vernalization only enlarges the LD response. No vernalization is required for the spring genotypes; the photoperiod sensitive ones fully express the LD response without any vernalization. There is a genotype-established continuum from spring habit through a winter habit to an SLDP habit, and their respective vernalization requirements vary quantitatively from zero to as long as 60 days. This is illustrated theoretically in Fig. 8.12.

Fig. 8.11. (*opposite*) The final leaf number (FLN) in response to photoperiods of 6 h, 10 h, 14 h and 24 h after 0–70 days of vernalization for 12 selected wheat cultivars. Germinating seeds were vernalized at 0–2°C before planting. The vernalization durations are labelled only for cultivar 6, but usually they are similarly positioned vertically for all the winter cultivars. The seedlings emerged into the initial photoperiods of 6 h, 10 h, 14 h or 24 h for 30 days, and were then transferred to a greenhouse with 24 h and an average 22°C. Cultivars 1 and 2 have spring habit; cultivars 3 to 12 have increasingly stronger winter habit. The photoperiod response of the spring cultivars is not influenced by the vernalization duration, but responses of the winter cultivars are dramatically influenced by the vernalization duration. With no vernalization, these cultivars had only marginal reduction of the FLN by an initially longer photoperiod, or had an SD response as evidenced by increase in FLN under longer photoperiod. Increase of the FLN in response to longer initial photoperiod of cultivars having the strongest winter habit indicated gradual reduction of the SD response as the vernalization requirement progressed toward completion. After completion, all cultivars expressed an LD response. (The analysed FLN are from unpublished data of Miao *et al.*) Compare with the response of the days to flowering in Fig. 8.10.

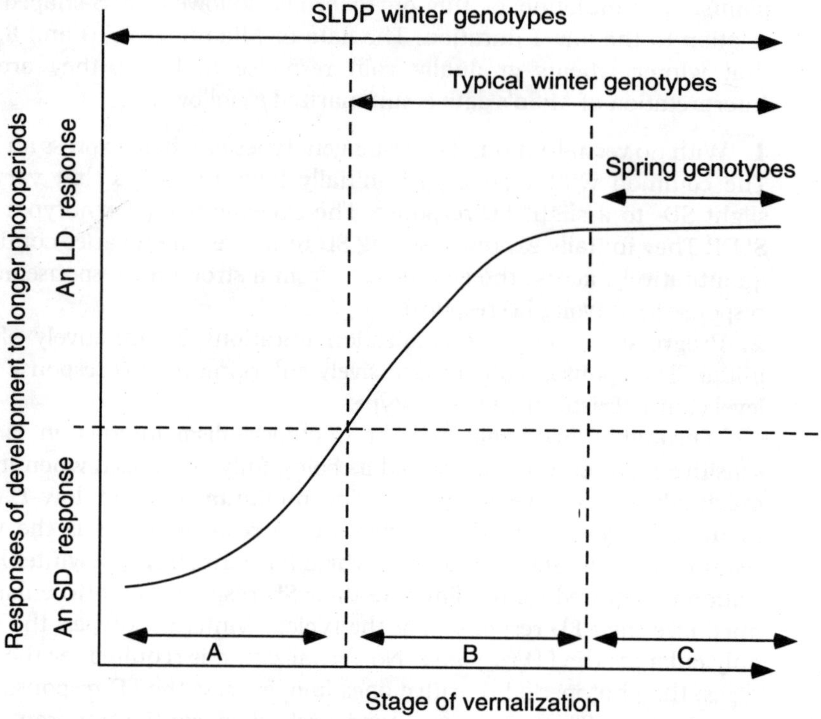

Fig. 8.12. The relationships between vernalization and the short-day versus long-day responses to longer photoperiod of wheat cultivars possessing the short-long day, typical winter and spring habits. The vernalization of an SLDP genotype can be divided among three advancing stages of the developmental response (A, B and C), which are described additionally in Fig. 8.1. Within developmental stage A, a longer vernalization reduces the initial SD response and the LD response begins; within stage B continued vernalization increases the change toward stronger LD response; within segment C further vernalization no longer enlarges the LD response. SLDP genotypes pass progressively through all three response stages. Spring genotypes express only stage C. These gradations advance understanding of the different vernalization requirements of the three habits of wheat cultivars as illustrated further in Fig. 8.13.

3. All wheats become LDP after being vernalized. And, the quantitative responsiveness to the LD is a function of both the genotype and the vernalization. This relation can be described by a simple linear function:

$$S_p = \alpha + \beta(v/V_{req}) \qquad (v \le V_{req}) \qquad [8.5]$$

where:

S_p = the sensitivity of the genotype to the LD.
α = the original sensitivity to LD before any vernalization; it can range from positive for spring genotypes through zero for typical winter genotypes to negative for the SLDP.
β = the increase in responsiveness to LD per unit increase in degree of the

vernalization. This is a quantitative measure of the vernalization by photoperiodism interaction.

V_{req} = the shortest duration of vernalization needed to result in the full LD response of the given genotype. The limited data (compare Fig. 8.4 with Fig. 8.9) indicate the V_{req} of equation 8.5 is the same as the V_{req} of the previous equations.

v = the duration of low-*T* the plants have been exposed to, and; when $v \geq V_{req}$, the plant has been fully vernalized.

The ratio v/V_{req} in equation 8.5 is a measurement of the degree of vernalization. There can be better ways to measure this.

For simplicity, the response to *P* after progressively longer durations of low-*T* vernalization was presented as linear regression lines (Figs 8.10 and 8.11), in order to present a broad interpretation of the control by vernalization over the photoperiod responsiveness. Except for the spring genotypes, longer low-*T* quantitatively enlarged the genotype's responsiveness to LD. However, this response to *P*, measurable as either the DTF or the FLN, is actually curvilinear as shown by the scattering of the observed responses (Figs 8.10 and 8.11). This is why a quadratic function was used in the GPTmodel of Chapters 6 and 7 to more accurately model the GPT interaction effect on the DTF.

Recognition of the general pattern that longer vernalization quantitatively enlarges the responsiveness to LD of the winter genotypes (shown in Figs 8.10 and 8.11) is due to the intelligent experimental design of Miao *et al.* (1990). They (i) compared a wide range of genotypes after (ii) eight different low-*T* vernalization durations, followed by (iii) four initial *P* of 30 days, followed by (iv) a common 'growth environment' of continuous illumination (24 h) and the same summer-time *T*. This was the appropriate environment for testing the vernalization by photoperiod interaction, because its 24 h and 22°C average *T* are close to the optimal post-vernalization *P* and *T* for wheat development. More such systematic and insightfully-designed experiments are required, with yet more accurately controlled environments, to acquire full understanding of the vernalization by photoperiodism interaction effects on plant development,

Roberts *et al.* (1988a) concluded barley and other cereals have a photoperiod insensitive stage before becoming photoperiod sensitive. They indicated this insensitive stage is short for spring genotypes and longer for winter genotypes. Vernalization shortens this insensitive stage, which they called the pre-photoperiod inductive phase. The implication for winter cereals is that the plants will respond to an LD eventually, whether vernalized or not. That vernalization only shortens this insensitive stage is generally consistent with the interpretations from Figs 8.10 and 8.11. However, those figures show no stage that is absolutely photoperiod-insensitive for all genotypes. Although commonly regarded as LDP, the winter wheat genotypes (and probably those of other winter cereals) gradually gain their responsiveness to LD, as they are progressively vernalized. Alternatively, they gain it more slowly as the plant age advances.

The insights derived from Figs 8.10 and 8.11 (compare Fig. 8.1) also expand understanding about phasic development. There are primary and secondary 'phases' of the development of winter cereals. But, there is no absolute differentiation between them in terms of either time or any morphological stage. The two

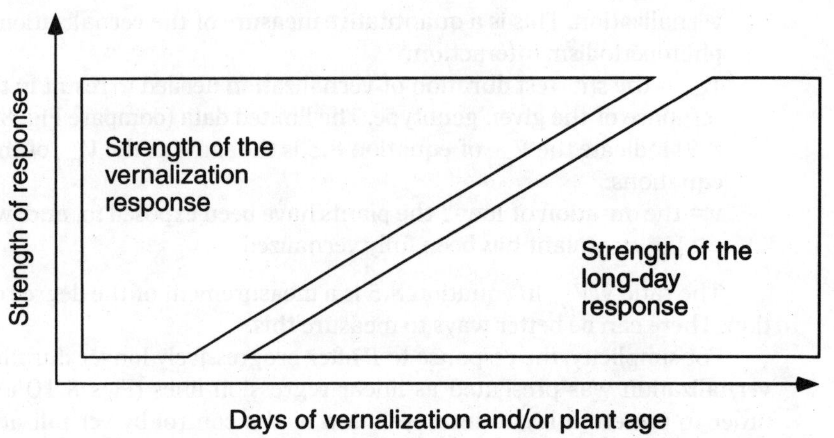

Fig. 8.13. A new overview of phasic development in plants. Vernalization and photoperiodism are separate controls over development. Responsiveness of the plant to vernalization is reduced as the plant is exposed to longer durations of low temperature and also as the plant ages. And, responsiveness of the plant to long photoperiod arises and is increased as the vernalization duration is longer and/or as the plant ages. There is no definite differentiation in time between the vernalization responsiveness and the photoperiod responsiveness; they overlap across a segment of time.

'phases', or more precisely the two physiological (or biochemical) processes overlap across time. And, magnitudes of the vernalization- and photoperiod-responses change quantitatively and reciprocally (Fig. 8.13). As the vernalization duration is extended, responsiveness to the vernalization decreases, and responsiveness to the LD increases. As the plant ages it eventually will lose all responsiveness to vernalization, while at the same time becoming LD-responsive, even with no vernalization. Whether the plant can eventually become maximally responsive to an LD without any vernalization is a non-investigated question.

8.6 Modulation of Days to Flowering by Vernalization, Photoperiod and Temperature

8.6.1 General considerations

DTF is a function of the genotype, and of strong modulations of its genetic instructions by the (i) low-T vernalization, (ii) post-vernalization T and P, and modulations due to the variations of other environmental factors:

$$\text{DTF} = f(G, V, T, P, M)$$

where G symbolizes the genotype, V the low-T vernalization, T the post-vernalization temperature, P the photoperiod, and M the miscellaneous factors such as soil fertility and moisture (Marcellos and Single, 1971). The V, T, and P explain most of the variation of the DTF of each genotype. For simplicity the M is excluded from the following thinking about modelling of the DTF. When V, P, and T are all optimum for the given genotype, the DTF is the smallest that can be expressed by that genotype. This minimum DTF has been called the basic vegetative period (Slafer and Rawson, 1995b); narrow-sense earliness (Yasuda and Shimoyama, 1965; Kato and Yamagata, 1988); flowering tendency (Aitken, 1974; Berry and Aitken, 1979; Wallace, 1985); earliness *per se* (Hunt, 1979; Hoogendoorn, 1985; Malse *et al.*, 1989); intrinsic earliness (Penrose *et al.*, 1991), and innate earliness (Sampson and Burrows, 1972). For a given genotype, the above expression can be modified as:

$$\text{DTF} = D_b + f(V, T, P) \qquad \text{or}$$

$$1/\text{DTF} = 1/D_b - f'(V, T, P)$$

where D_b symbolizes the basic (smallest expressible) DTF of the genotype, and this D_b is expressed when all environmental factors are at their optimal level. This provides the framework for constructing a DTF model.

8.6.2 A daily rate model

From discussion above in this chapter:

1. The rate of development toward flowering (or to heading for small-grain cereals) of a genotype is determined primarily by the duration of the vernalization, the post-vernalization T, and the post-vernalization P.
2. Vernalization is prevention by low-T of delay of development by the vernalization-gene activity expressed at high-T.
3. Response to vernalization decreases as the duration of the low-T is extended, and/or as the plant age advances.
4. The photoperiod response becomes larger as the plants are progressively vernalized, and the always-present thermal response becomes obvious upon completion of the vernalization requirement.

These four concepts are integrated into the following equation.

$$r = r_b - S_v\left(T - T_{bv}\right)^2\left(V_{req} - v\right)^2 - \left(\alpha + \beta\frac{v}{V_{req}}\right)\text{POSTV} \qquad [8.6]$$

in which POSTV is the post-vernalization response to the T and P:

$$\text{POSTV} = S_p\left(P_{opt} - P\right)^2\left(T - T_{bp}\right)^2 + S_t\left(T - T_{opt}\right)^2$$

where

r = the daily rate of development towards flowering, or heading, or any other stages that indicate the transition from vegetative growth to the commitment of the apex to reproductive development.

r_b = the maximum daily rate of development, which occurs autonomously, if there is no delay by the vernalization-gene activity and/or by photoperiod-gene activity and/or by non-optimal T. Any delay by T other than through its modulations of the vernalization- and photoperiod-gene activities is through the average effect on the development due to all other gene

activities that also control it, i.e. is through the thermal time response.
T = the daily temperature.
P = the daily photoperiod.
V_{req} = the days of vernalization required to completely vernalize the genotype and result in full responsiveness to an LD.
v = the days of vernalization the plant has been exposed to.
S_v = the sensitivity of the genotype to the vernalization.
S_p = the sensitivity to photoperiod when the plant has been fully vernalized.
S_t = the sensitivity to non-optimal temperatures when the plant has been fully vernalized.
T_{bv} = the temperature above which the vernalization gene activity is expressed.
T_{bp} = the temperature above which the photoperiod gene activity is expressed.
T_{opt} = the optimum temperature of the post-vernalization development, which is a fairly high-T.
P_{opt} = the shortest optimum photoperiod for development.
α = the sensitivity to photoperiod when there has been no vernalization. This value can be negative, zero, or positive, depending on the genotype.
β = the vernalization by photoperiod interaction coefficient. The relation between α and β is so defined that $\alpha + \beta = 1$.

The quadratic functions are again used to account for any curvilinearity of the responses.

8.6.3 A days to flowering model based on the daily temperature and photoperiod

The transition from vegetative development to reproductive development will occur when:

$$\Sigma r = 1.$$

Therefore we have:

$$1 = \sum \left\{ r_0 - S_v\left(T - T_{bv}\right)^2\left(V_{req} - v\right)^2 - \left(\alpha + \beta\frac{v}{V_{req}}\right)\text{POSTV} \right\}$$

or

$$\sum r_0 = 1 + \sum \left\{ S_v\left(T - T_{bv}\right)^2\left(V_{req} - v\right)^2 - \left(\alpha + \beta\frac{v}{V_{req}}\right)\text{POSTV} \right\}$$

Dividing both sides of the equation by r_b, which is a constant for a given genotype, results in something like:

$$1/\text{DTFb} = 1/D_b - \sum \left\{ S_v\left(T - T_v\right)^2\left(V_{req} - v\right)^2 + \left[\alpha + \beta\left(v/V_{req}\right)\right]\left[\text{POSTV}\right] \right\} \quad [8.7]$$

because $r = 1/\text{DTFb}$, and $r_b = 1/D_b$.

Where: the summation is from the day of planting to the day of anthesis;

DTFb is the number of days to flowering (or to other specified stages) from sowing or imbibition of the seeds, and D_b is the basic (shortest expressible) duration to flowering. Verbally, the expressed DTFb of a genotype is determined by its basic duration to flowering, plus any delay due to vernalization-gene activity, and/or due to photoperiod-gene activity, and/or due to non-optimal *T*. The days of delay by the (i) vernalization-gene activity, (ii) photoperiod-gene activity, and (iii) non-optimal *T* are all influenced by the duration of the vernalization. If the genotype establishes a V_{req} but has not been vernalized, the delay of development is predominantly due to the vernalization-gene activity. After the plant has been vernalized beyond that measurable as $[v/V_{req} \geq 1]$, the predominant cause of any delay of the development becomes due to the photoperiod-gene activity and/or any non-optimal *T* for growth and development, i.e. the thermal response.

8.6.4 Environment-specific forms of the days to flowering model

Equation 8.7 is a comprehensive model including all vernalization, photoperiod and temperature controls over the DTFb. For specified environments it can be simplified to special forms.

When the LDP are grown under continuous illumination (P = 24), *or any photoperiod long enough to prevent delay of development, that part of the equation representing delay by photoperiod-gene activity disappears.* This reduces equation 8.7 to equation 8.8:

$$1/\text{DTFb} = 1/Db - \sum\left\{S_v\left(T - T_v\right)^2\left(V_{req} - v\right)^2 + \left[\alpha + \beta\left(v/V_{req}\right)\right]\left[S_v\left(t - T_{opt}\right)^2\right]\right\} \quad [8.8]$$

Thus, under an optimal *P*, the DTFb is modulated by the *T* to give two simultaneous but opposite responses: (i) the vernalization response, and (ii) the thermal response. Because of interaction between these two genotype-established controls over the DTF, the response will be indicated by an optimum *T* for a given duration of the vernalization, or alternatively, by an optimum duration of vernalization for a given *T*. Both optima will be indicated by the most rapid expressible development of the genotype, as described in section 8.2. Both the indicated optimum *T* for vernalization and the indicated optimum duration of the low-*T* vernalization result jointly from the vernalization temperature and the post-vernalization temperature.

*When the seeds have been vernalized under a constant low-*T *for different durations,* the method used in most vernalization research, equation 8.7, can be reduced to equation 8.9:

$$1/\text{DTFb} = 1/D_b - \left\{S_v\left(T - T_v\right)^2\left(V_{req} - v\right)^2 + \sum\left[\alpha + \beta\left(v/V_{req}\right)\right]\left[\text{POSTV}\right]\right\} \quad [8.9]$$

The summation is from the end of the low-*T* vernalization.

When $v \geq V_{req}$, equation 8.9 is reduced further to:

$$1/\text{DTFb} = 1/D_b - \sum[\text{POSTV}] \quad [8.10]$$

Equation 8.10 models the response to only the post-vernalization *P* and *T*. For all pre-winter plantings in most temperate regions, vernalization is not a limiting factor for development, because the duration of the low-*T* of winter is always long enough for genotypes with even the strongest winter habit. Therefore, equation 8.10 applies to the late fall or winter plantings or even early spring plantings. For these, it is the SD and/or low-*T* of the spring that limits (delays) the development. This conclusion is supported by most if not all multiple planting-date studies. It is evidenced most strongly by extensive research by Chinese wheat ecologists (summarized in Jing, 1991, 1993).

If the average temperature and photoperiod after the vernalization are used, equation 8.10 is reduced further to:

$$1/\text{DTFb} = 1/D_b - S_p(24-P)^2(T-T_{bp}) - S_t(T-T_{opt})^2 \quad [8.11]$$

where respectively, *T* and *P* symbolize the average *T* and average *P* measured from completion of the vernalization requirement until flowering. Equation 8.11 is the GPTmodel (equations 6.3 and 7.4 detailed in Chapters 6 and 7).

Data indicating GPT interaction effects on the DTF of many crops are abundant (Chapter 6). Rawson and Richards (1993) provided a wheat example. They compared seven isolines of Triple Dirk that differed for both photoperiod- and vernalization-genes. The plants were grown under 9 h, 11 h, 13 h and 15 h in all combinations with both 15°C and 25°C. The single winter type (TD-C) never headed in any of the eight environments. The conclusions from the spring genotypes are: (i) Lines TD, TD-B, TD-D, and TD-E had similar DTF at 25°C and 15°C under SD (9 h), but under 15 h their DTF averaged 16 days earlier at 25°C than at 15°C. The low-*T* shortened the DTF only when combined with a non-delaying LD. The genotypes TD-A and TD-F had DTF 77 days longer under 9 h than 15 h at 15°C but 50 days longer under 25°C, indicating the *P* × *T* interaction was small for these two lines. (ii) The days to the double-ridge stage provided more insight than the days to heading, as concluded also by Evans (1987). In general, under SD, low-*T* delayed the days to appearance of the double ridges. Conversely, low-*T* reduced the days to double ridges under LD. Thus, the Q_{10} effect by *T* modulated the photoperiod gene action that functioned under the SD, and simultaneously modulated the acceleration of development through the majority of all the other genes and their control over the autonomous development. Similar *P* × *T* interaction in wheat was reported by Rahman and Wilson (1978). For eight wheat genotypes the days to floral initiation responded to the range of *T* with a U-shaped curve, and the optimum *T* was higher under the LD than the SD.

If the plants have been fully vernalized and then grown under continuous illumination, or a long enough photoperiod, equation 8.11 is reduced further to:

$$1/\text{DTFb} = 1/D_b - S_t(T-T_{opt})^2 \quad [8.12]$$

All extension of the DTFb beyond the D_b is modulated entirely by the *T*, i.e. by the thermal response. The curvilinearity of the DTFb in response to *T* in the absence of vernalization and photoperiod gene activities is well documented for many crops as reviewed in Chapters 6 and 7. Additional evidence for this U-shaped response comes from recent carefully planned phytotron studies of wheat by Slafer and Rawson (1994b). Four diversely adapted cultivars were compared. The seedlings were vernalized for 50 days, and then grown under a near-optimal 18 h in combination with six *T* between 10°C and 25°C. The 50 days of vernalization and near-optimal *P* reduced confounding of the responses to the vernalization, and to the post-vernalization *P* with the thermal response. The thermal response reduced the days to terminal spikelet initiation as the *T* rose from 10°C to 19°C, but at all progressively higher-*T* than 19°C these days were progressively enlarged. A second experiment repeated this result (Slafer and Rawson, 1995a, b). Other special forms of equation 8.7 are also possible for additional experimental designs based on optimal vernalization, or photoperiod, or temperature.

8.7 Vernalization and Photoperiodism in Relation to Morphogenesis of Wheat

It is thoroughly accepted that vernalization and photoperiodism modulate the days to plant maturity, and thereby strongly determine the expressed plant yield. The relations between phasic development and morphogenesis can be an important cause of this phenomenon. However, these relations remain poorly understood, and many questions are unanswered.

8.7.1 The morphogenesis of wheat

Development of the wheat inflorescence has been described and illustrated in detail by many workers, as cited by Pinthus (1985) and McMaster (1997). Development toward a completed inflorescence of a small-grain cereal crop involves the successive morphological stages of an unelongated shoot apex, an elongated shoot apex, single-ridges on the apex, double-ridges on the apex, terminal spikelet initiations, followed by stem elongation, heading and flowering. Normally, elongation of the shoot apex precedes onset of development of the inflorescence. At the base of the elongated apices, several additional primordia are visible in the form of single ridges. Further development of these ridges is restrained, while the buds in their axils grow rapidly. These buds, along with their subtending leaf initials, appear in the form of double-ridges, which are the first morphological evidence of floral initiation, i.e. of the spikelet primordia. As the spikelet primordia at the centre of the embryonic spike swell and extrude beyond their respective lower ridges (which are the leaf initials), simultaneously, additional single ridges appear acropetally, before the apical dome is differentiated to become the terminal spikelet. An important fact is that, while the single-ridge initiation (the leaf primordia) on the elongated apex (the embryonic spike) proceeds *acropetally*, the double-ridge formation (the spikelet initiation) begins at about the middle of the embryonic spike and proceeds *both acropetally and basipetally*. As a result, the lowest (and oldest) spikelet primordium may not become committed, and consequently; the FLN is not determined, until the

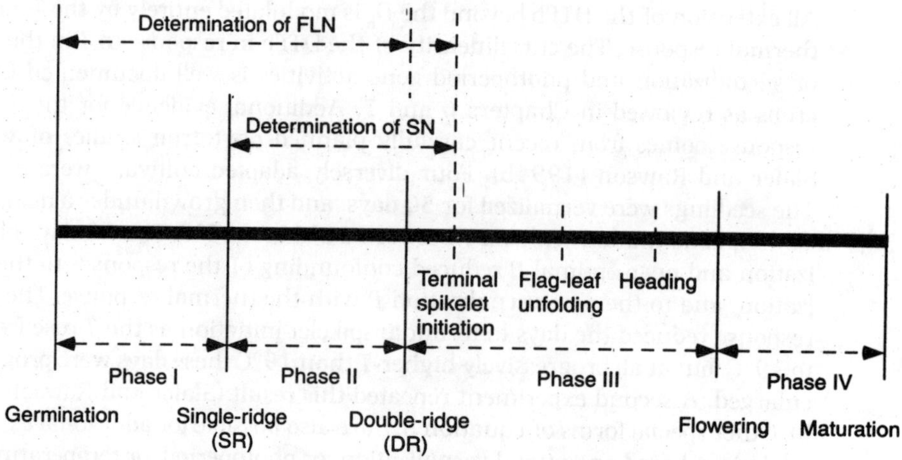

Fig. 8.14. Morphological stages and developmental phases and their relation to the final leaf number (FLN) and spikelet number (SN) of wheat.

terminal spikelet initiation stage. *Therefore, both the FLN and the number of spikelets are determined at about the terminal spikelet initiation stage.* Nevertheless, several researchers have reported that the FLN has already been determined at the double-ridge stage, and cannot be altered by any later environment (Griffiths *et al.*, 1985; Huang *et al.*, 1988). The time from the single-ridge stage to the double-ridge stage is related closely to the spikelet number (Yan *et al.*, 1993).

To repeat, the period of leaf formation spans from seed germination to the terminal spikelet initiation or to the double-ridge stage, and the period of spikelet formation spans from the single-ridge stage (shortly after the shoot apex elongation) to the terminal spikelet initiation. Thus, the leaf formation and the spikelet formation time spans overlap – from the single-ridge stage to the double ridge stage or to the terminal spikelet initiation. Therefore any environmental factor that modulates the time-span of this overlap will influence both the FLN and the spikelet number, as well as the DTF. For example, unpublished data of Miao (personal communication) indicate that vernalization reduced both the FLN and the spikelet number of the winter genotypes. Likewise, post-vernalization LD reduced both the FLN and the spikelet number as compared with SD.

The sequence of inflorescence development and its relation to the FLN and spikelet number of wheat and other small grain cereals is illustrated in Fig. 8.14. Phase I is from seed germination to shoot apex elongation or to first single-ridge appearance; phase II is from the single ridge to the first double-ridge appearance, and; phase III is from double-ridge appearance to anthesis (Yan *et al.*, 1993).

8.7.2 Effect of vernalization on the morphological stages

Lysenko (from Whyte, 1948) indicated vernalization is an invisible process in that there is no specific stage that indicates its completion. Invisibility arises also from the great flexibility of the timing of applying vernalization. Plants can be vernalized as developing immature embryos (Purvis and Gregory, 1937 for rye; Riddle and Gries, 1958; Gotoh, 1976; Hoogendoorn, 1984; Sharma and Mascia, 1987; Evans, 1987, Whelan and Schaalje, 1992 and McMaster, 1997 for wheat); or as germinating seeds, which is the most common experimental procedure; or as seedlings of different age (Gotoh, 1976; Wang *et al.*, 1995). This is the reason no morphological stage indicates either the beginning or completion of the vernalization. After plant materials have been vernalized at a low-*T* for a sufficient time, however, it can be confidently said that they are vernalized.

We cannot judge whether a plant has or has not been vernalized by observing its morphological stage. This can be judged only by examining its rate of development, i.e. by examining which morphological stage(s) are delayed by lack of vernalization under a post-vernalization *P–T* combination that *per se* does not delay the development.

The literature consistently shows the double-ridge stage to be the first morphological stage that can be delayed by lack of vernalization. Hu and Luo (1988) vernalized germinating seeds of 15 winter and 16 spring wheats at 1–3°C for 0–60 days, before planting at the end of March in Hefei, China (32°N). All genotypes after all vernalization durations developed to the single-ridge stage. Most plants that failed to develop to heading failed also to develop to the double-ridge stage, indicating the latter is the first morphological stage that can be delayed by lack of vernalization. Huang *et al.* (1988) grew spring and winter wheat genotypes planted in the spring (March 5, 20 and April 5) at Yingchuan, China (38°N). The winter genotypes failed to develop to heading for the latest April 5 planting, because of the higher-*T*. Nevertheless, as compared with both March plantings, their days from planting to apex elongation (phase I of Fig. 8.14) were shortened rather than delayed. Huang *et al.* concluded that the developmental rate of phase I of Fig. 8.14 was promoted by higher-*T*, rather than by vernalization, although all winter genotypes have an intrinsically longer phase I than spring genotypes. It was the development from single ridge to the double ridge stage (phase II), that was delayed by lack of vernalization, as concluded by Yan *et al.* (1993a). Additional confirmation came from He *et al.* (1988); they reported the most vernalization-sensitive stage of wheat to be from the single ridge to the double ridge stage. For barley it was from single ridge to the triple-mound stage, which corresponds with the double ridge stage of wheat.

There has been argument as to whether there is any effect of vernalization beyond floral initiation indicated by the double ridge stage. Slafer and Rawson (1994b) and Flood and Halloran (1984, 1986) concluded there was no vernalization effect beyond this stage. On the contrary, their data indicated a strong vernalization effect after the double ridge stage. In Flood and Halloran (1984), development of the only winter genotype Triple Dirk C (*vrn1vrn2*) was delayed significantly by inadequate vernalization, both for the days to the double ridge stage and for the days from the double ridge stage to heading. Other studies (Rahman, 1980 and Manupeerapan *et al.*, 1992, for example) have reported that vernalization did not delay development to the double ridge stage, but did delay subsequent development to the terminal spikelet initiation and therefore to heading. A question arises. Now that it is known development toward apex elongation and the appearance of single-ridges on it are not modulated by vernalization, is there any factor, in addition to the thermal-time, that evoke this stage? Thus far, there is no evidence.

In summary, lack of vernalization delays development after but not before the single-ridge stage. As a consequence, for winter genotypes, both the FLN and the spikelet number are enlarged by inadequate vernalization.

8.7.3 *Effect of photoperiodism on the morphological stages*

The stage at which the photoperiod response begins has also been much debated. Brooking *et al.* (1995) cited conclusion of Cooper (1956) that photoperiod sensitivity of wheat begins at emergence of the first leaf, and sensitivity to photoperiod can continue until terminal spikelet initiation. To demonstrate their hypothesis that the FLN, like the leaf appearance rate, is determined by a direct and continuously updated environmental response, Brooking *et al.*, indicated that the stage at which the FLN is determined by the *P* for *spring* wheat depends on the genotype. It ranged from about the 1.5-leaf stage for the less photoperiod sensitive genotype to the final-leaf-primordium plus four-spikelet-primordia stage for the most photoperiod-sensitive genotypes. Roberts *et al.* (1988a) reported for barley that the photoperiod sensitive stage began from about the double-ridge stage for both spring and winter genotypes.

That photoperiod sensitivity commences at the first leaf stage cannot possibly be true for winter wheats, since they require some duration of vernalization for any LD response. And, fully-vernalized seeds cannot respond immediately to an LD, because the *P* is perceived by the leaves. It is not likely that the LD response commences until the single-ridge stage, the reason being, as indicated in Figs 8.10 and 8.11, that 10 days of low-*T* vernalization provides the winter genotypes with limited ability to respond to *P*. It is doubtful that by this time these winter genotypes could have already developed to the single ridge stage. Therefore, it may be most rational to say that commencement of the response to *P* depends on both the vernalization and the leaf (plant) age, the latter requiring a minimum number of leaves.

Unlike for vernalization, the effect of the *P* is expressed immediately as it is perceived by the plants (Brooking *et al.*, 1995). And, its effect continues until heading and flowering (Slafer and Rawson, 1995b). As a consequence, both the FLN and the spikelet number, and the rate of stem elongation, can all be strongly modulated by the *P*, if the plant has been adequately vernalized. Generally, for LDP, longer *P* results in smaller FLN, smaller spikelet number, and smaller DTF.

8.8 Determination of Final Leaf Number and Rate of Leaf Appearance

There are two obvious morphological stages just before flowering of wheat and other small-grain cereals: the flag (final) leaf unfolding and heading. Many studies have used days to heading instead of the actual DTF, because it is more easily observed. Others used the days to flag leaf unfolding to represent the DTF (Chujo, 1966; Gotoh, 1976). The days to flag-leaf unfolding (DTFL) is determined by the FLN and the rate of leaf appearance (RLA) since: DTFL = FLN × RLA. This was used by Miglietta (1991a, b) to model and predict the DTF.

8.8.1 Final leaf number is determined by the vernalization- and photoperiod-gene activities

The FLN is determined by the vernalization response – and the plant age at which the vernalization is applied – plus by the photoperiod response. Slafer and Rawson (1994b) found that when wheat plants are fully vernalized and then grown under LD, their FLN was modulated only marginally by the growth *T*. Slafer *et al.* (1994b) reported that both the genotype for sensitivity to *P* and the applied *P* significantly altered the FLN, but the FLN was not altered by the rate of change of the *P*. Most of the variation of the FLN that resulted from differences in the *P* resulted from change of the total time-span of the initiation of leaves. Brooking *et al.* (1995) proved for several *spring* wheat genotypes that the FLN and RLA were both determined by a direct, continually-updated response to the *P*. The FLN was linearly or exponentially related to the *P* at a given genotype-specific leaf-age. By comparing multiple planting dates, for example, they showed the FLN of cv. Otane to be determined by the *P* at the 1.5-leaf age. The FLN of cv. Rongotea was determined by the *P* at the flag leaf primordia initiation plus two initiated-spikelets (FLPI+2) stage. And the FLN of the cv. CRSW6 was determined by the *P* at the FLPI+4 stage. Thus, cultivars clearly differ in the morphological timing of their response to the *P*, as well as in magnitude of the response. The more photoperiod sensitive genotypes had a longer duration of response to the *P*, and their FLN was determined later than that of the less sensitive genotypes. A model to predict the FLN based on these findings does not require a response to factors other than those of the immediate environment, i.e. does not require the plant to 'remember' prior levels of the environmental factors. Nevertheless, we think it safer to say that the FLN of spring genotypes is determined by the *P* before the 1.5 leaf stage to a later genotype-specific stage.

The influences of (i) vernalization and (ii) of the plant age at vernalization on the FLN were detailed in section 8.3. And, the vernalization by photoperiodism interaction effect was described in section 8.6. This vernalization by photoperiodism interaction can be modelled using equation 8.7, which does ignore the thermal response during all the post-vernalization growth:

$$\mathrm{FLN} = L_0 + \sum \left\{ S'_{\mathrm{v}}\left(T - T_{\mathrm{bv}}\right)^2 \left(V_{\mathrm{req}} - v\right)^2 - \left(\alpha + \beta \frac{v}{V_{\mathrm{req}}}\right) \left[S'_{\mathrm{p}}\left(P_{\mathrm{opt}} - P\right)^2 \left(T - T_{\mathrm{opt}}\right)^2 \right] \right\} \quad [8.13]$$

Equation 8.13 can be simplified for applied use, depending on the specific experimental *P* and *T*, as described above in section 8.6.4 for equation 8.7. Kirby (1992) modelled the FLN of 12 wheat genotypes using just (i) the accumulated thermal-time for a base-*T* of 0°C measured from sowing until full vernalization and, (ii) the *P* at full vernalization. His procedure was similar to equation 8.13, but differed in its theoretical interpretation of the vernalization effect.

8.8.2 Rate of leaf appearance is modulated by only temperature and the leaf age of the plant

8.8.2.1 Rate of leaf appearance is not influenced by the photoperiod

Baker *et al.* (1980) planted the winter wheat cv. Maris Huntsman in the field from October 4 to June 8. They found the RLA to be a linear function of the thermal-time from sowing, with a base-*T* of 0°C. Kirby *et al.* (1987) reported little effect on the RLA by the *P*. Kirby (1990) confirmed the conclusion of Baker *et al.* (1980) that, expressed as thermal time, the leaves emerged at an approximately constant rate. Additional supports are in Kirby *et al.* (1985); Kirby and Perry (1987); Malse *et al.* (1989); Slafer *et al.* (1994b) and Evans and Blundell (1994). The strong linear relation between leaf number and the cumulative thermal time establishes that other factors cause relatively small if any effects on the RLA.

Miglietta (1989) re-analysed data from the literature and concluded the RLPI is independent of the *P*, and is a linear function of only the *T*. Jamieson *et al.* (1995) verified this and indicated that, in the absence of severe stress, the RLA depends on only the *T* and the already-existing leaf number; it is independent of both the *P* and its rate of change. Jamieson *et al.* (1995) confirmed the model of Miglietta (1991a, b), but using the tissue- rather than air-*T*, and an assumed base-*T* of 0°C. Jamieson *et al.* wrote:

> The link between leaf appearance and tissue temperature is causal. Our model does not require that the plant has a 'memory' of something that happened previously, or to know whether it is spring or autumn; the phyllochron does not depend on planting date. The only preconditioning the plant goes through is obvious – leaf appearance follows initiation after some delay. The mechanisms are clear and simple.

8.8.2.2. Rate of leaf appearance is a quadratic function of the temperature

Most above reviewed studies reported a linear response of the RLA to the growth *T*. Based on the Q_{10} effect by *T*, however, if *T* spans a wide range, biochemically, it is rational to assume that the RLA is a quadratic function of the *T*. Kirby *et al.* (1987) presented data indicating the response to *T* of the rate of leaf initiation to be curvilinear. This was supported by a carefully controlled study of four winter wheats and four spring barleys grown in seven constant *T* between 7.5°C and 25°C (Cao and Moss, 1989). For all genotypes, the RLA increased parabolically with increasing *T* until an optimum *T* (maximum RLA), above which the rate declined with higher-*T*. Cao and Moss reported these responses were described by quadratic polynomials with R^2 greater than 0.96. Therefore, the average RLA can be determined by:

$$\mathrm{RLA} = R_0 + S_t'\left(T - T_{\mathrm{opt}}\right)^2 \qquad [8.14]$$

where R_0 is the maximum RLA when the *T* is optimum. Equation 8.14 has the same mathematical form as equation 8.12 for the DTF. This similarity implies that when the plants have been fully vernalized, and then grown under optimum *P* (24 h), the DTF is primarily controlled by the *T* through the RLA.

8.8.2.3 Rate of leaf appearance is reduced by advance of the plant age

Miglietta (1991a, b) proposed that the RLA is slightly slower than the RLPI, and their relation can be described by:

$$\text{RLA} = \text{RLPI}(1 - cL),$$

or $$\text{RLA} = \text{RLPI}\, e^{-cL}$$

where *L* symbolizes the number of leaves that have emerged and c is a fitted constant.

This equation involves the hypothesis that the RLA is reduced slightly as the plant ages, i.e. as the number of existing leaves becomes larger, because the later leaves have longer leaf sheaths, and both the length and width increase with the leaf age (McMaster *et al.*, 1991). Miglietta (1991a, b) proposed that the already existing number of leaves (*L*) and the existing number of initiated leaf primordia (which he symbolized as *P*) have the relation: $L = b\,(P - 4)$. That is, four leaf primordia existed before germination, while at germination the leaf number was zero. Therefore, Miglietta argued, if the FLN is known, these relations will facilitate prediction of the phenology of wheat.

There has been little research about the effect of vernalization on the RLA. Chun (1993) found for three winter barley genotypes that longer vernalization duration accelerated the RLA. This implies an interaction between vernalization and response of the RLA to the thermal time, as suggested in equations 8.6 to 8.9. It is also possible that the vernalization effect on the RLA reflects the control by low-*T* vernalization over the FLN. Lack of vernalization enlarges the FLN and the RLA is reduced with advance of the leaf age, so the average RLA may be reduced by lack of vernalization. Clarification requires more research.

8.9 Determination of the Spikelet Number

The SN is a physical component of the yield of wheat and other small-grain cereals. Rawson (1971) and Rahman and Wilson (1977) pointed out: The SN is the product of the duration of spikelet initiation (DSI) and the rate of spikelet initiation (RSI):

$$\text{SN} = \text{DSI} \times \text{RSI}.$$

8.9.1 Duration of spikelet initiation is determined by vernalization, photoperiod and temperature

Rawson (1971) reported that the SN of wheat increases as the *P* is shortened (more delaying), even though this decreases the RSI. Photoperiod insensitive genotypes maintained a high RSI under all *P*. Wall and Cartwright (1974) compared 14 wheat cultivars that spanned spring to moderate winter growth habit after vernalization (28 days) versus no vernalization, followed by 25/20 and 23/10°C combined with 14 h or 10 h. The results follow. (i) The SN was closely related to the days to heading (DTF). (ii) For spring genotypes, there were more spikelets under low-*T* and SD, and both tended to lengthen the DSI. (iii) For the three winter cultivars of the 14, the SN seemed to be independent of *T* during growth, but to be enlarged by SD. Rawson (1971) and Wall and Cartwright

(1974) indicate that usually the DSI is the primary determinant of the SN, as supported also by Pugsley (1966, 1971).

The DSI is the time in days between the single-ridge stage and the terminal spikelet initiation stage (Fig. 8.14). In general, the DSI is shortened by longer low-*T* vernalizations (of winter genotypes), longer *P*, and higher growing *T*. Thus, the DSI can be modelled using equation 8.7, or its simplified forms in section 8.6.4.

Scarth *et al.* (1985) studied the effects by the photoperiod genes *Ppd1* and *Ppd2* on the growth and development of the shoot apex in wheat under 8 h at 18 ± 1°C. The photoperiod-sensitive genes (*ppd1* and *ppd2*) increased the DSI and the duration of later inflorescence development, compared with the insensitive photoperiod genes (*Ppd1* and *Ppd2*). These photoperiod genes had little effect on the RSI. The increased DSI and later phases by the photoperiod-gene activity under the SD, can be interpreted in terms of photosynthate partitioning as described in Chapters 4 and 5 for bean. For wheat, under the delaying SD and high-*T*, the stronger photoperiod-gene activity would negatively control (reduce) partitioning of the available assimilate toward immediate growth and continued development of the already initiated spikelet primordia. This reduction would leave more photosynthate to be partitioned to continued initiation of new spikelets. As a consequence, the DSI would be longer for the photoperiod sensitive genotypes, so more spikelets would be initiated for them under SD than for the photoperiod insensitive genotypes.

8.9.2 Rate of spikelet initiation is determined by temperature and assimilate availability

Halse and Weir (1970) reported: (i) The enlarged DSI due to SD was associated with an enlarged SN. (ii) A higher SN also attended enlargement of the DSI due to lack of vernalization. (iii) On the contrary, an enlarged DSI due to a post-vernalization low-*T* was not accompanied by a significant change of the SN.

Contrary to the increased rate of leaf initiation with higher *T*, which was not affected by the *P*, the RSI was modulated by both *T* and *P* (Kirby *et al.*, 1987). Evans and Blundell (1994) also concluded that, whereas the leaf initiation and its rate of appearance were modulated little by the *P*; the RSI responded to *P* from the beginning of the floral initiation, i.e. about the single-ridge stage.

The response of the RSI to the *P* may be a response to the total light irradiation and the availability of assimilate. Rahman and Wilson (1978) studied eight wheat genotypes grown under 8/16 h thermoperiods of 16/9°C, 23/16°C and 30/23°C combined with the natural 12.5–14.0 h compared against 24 h. The RSI had an optimum *T* which varied with the *P*. Rawson and Richards (1993), as described in section 8.6, grew six isolines of Triple Dirk – that were isogenic for both vernalization and photoperiod genes – under five *T* combined with two *P*. Short *P* combined with moderate *T* (15°C) delayed appearance of the double ridges and facilitated initiation of a large SN. At 25°C, in contrast, large delay to double ridge appearance by the short *P* increased the SN little if any. RSI at the shoot apex slowed and eventually ceased for the more photoperiod sensitive isolines. At the higher-*T*, the leaves appeared faster, but the RSI was reduced. Rawson (1993) considered the effect of the radiant energy on the rate of development of wheat cv. Late Hartog grown under different *P–T*. He concluded that development of wheat is generally not source-limited (photosynthate-limited) in environments where wheat is traditionally cultivated. But, under high-*T* combined with SD, the RSI is slowed by a limiting source : sink ratio. This interpretation agrees with earlier reports of reduction of RSI by low light intensities (Friend, 1965; Lucas, 1971; Rahman *et al.*, 1977). The conclusion is that the RSI will be limited by: (i) a deficiency of assimilate supply due to low-daily radiant energy, (ii) a non-optimal *T*, (iii) a non-optimal *P*.

8.9.3 *Overall effects of photoperiod and temperature on spikelet number*

Simultaneous but somewhat opposite modulations by *P* are described above for the two components of the SN: the DSI and the RSI. The *T* also causes opposite effects, which at a given time may be either the same or opposite to the effects modulated by the *P*. Consequently, interpretation of a net effect on the SN due to *P–T* combinations is difficult; it requires the direction of all the effects to be determined (McMaster, 1997; McMaster *et al.*, 1992). When the opposite effects of *P* on the DSI and RSI components of the SN are equal, no effect due to the *P* upon the SN will be obvious. The same applies for the modulations by the *T*. Consequently, the DSI and RSI are so buffered that the SN of a genotype can be relatively constant across a wide range of environments (Stern and Kirby, 1979; Kirby *et al.*, 1987; Hay and Kirby, 1991). Kirby *et al.* (1987) reported that in fields in England, sowing dates between mid September and mid November did not alter SN of the winter wheat cv. Avalon, even though later sowings resulted in a smaller FLN. Halse and Weir (1970) reported also that the SN was not influenced by a relatively wide range of *T*.

8.10 Convergent Versus Divergent Development

8.10.1 *The phenomena of convergent and divergent development*

Hay and Kirby (1991) introduced the concept of convergent development in wheat with the following remark: 'When a wheat crop reaches maturity and is ready for harvest, all of the ears of the crop are mature and, within each ear, all of the grains are ripe. This synchrony of development between and within the ears, which is taken as granted by farmers who can harvest the entire crop in a single operation, is, in reality, a remarkable feat of coordination by the plants.' They continued: 'The fact that temperate wheat crops sown over a wide range of sowing dates can be ready for harvest within a period of 1 or 2 weeks, is an equally striking case of convergence of development and (near) synchrony of maturity.' Hay and Kirby identified two mechanisms for this remarkable convergence toward synchrony of plant development. One is the effect of *T* including vernalization, and the associated *P*. The second is the mortality of organs due to intra- and inter-plant competition. Mortality of the later-developed tillers and florets of wheat begins after the terminal spikelet initiation stage on the main stem (Hay and Kirby, 1991). At terminal spikelet initiation the main stem unequivocally becomes programmed toward reproductive growth and stem elongation. Because development to terminal spikelet initiation is modulated by both the *T* (including vernalization) and the *P* (section 8.7), timing of mortality of smaller tillers and florets is also modulated by the *P* and *T*, and; consequently, the two controls become one. After the terminal spikelet initiation stage, i.e. initiation of the reproductive organs, the tillers that continue rapid reproductive development and stem elongation become stronger competitive sinks than the later tillers that are still undergoing vegetative growth. Consequently, the latter are deprived of assimilate required for their survival. This mortality of the disadvantaged tillers and florets caused by intra-plant competition will be intensified, by stress factors such as high plant density and drought.

Convergent development among the tillers and florets of a single plant, and among different planting dates of a genotype, which were extensively reviewed by Hay and Kirby (1991), are interpretable using the vernalization by photoperiodism by thermal response interaction models. Equations 8.6 and 8.7 provide the basic forms of these models.

8.10.1.1 Convergent development among tillers of a single plant

Suppose: tillers can arise from the time the main stem has three leaves until this stem begins to elongate; that the tillers individually perceive the *P* and *T* (Hay and Kirby, 1991), and; all tillers initiated before *T* becomes high in the spring are adequately vernalized. Therefore, development of these tillers will be promoted by longer *P* and also higher *T*. In the northern hemisphere, the *P* increases steadily with time after December, and the *T* does so after a lag period. Therefore the earlier tillers will have been exposed to relatively shorter average *P* and lower average *T*, than the later tillers. As a consequence, the earlier tillers develop relatively slowly, as if waiting for the later tillers, resulting in synchrony among all the tillers developed in early spring or before. The tillers arising in late spring will not be fully vernalized, so the longer *P* will accelerate their development minimally, while the higher-*T* will delay rather than promote their development because of the lack of vernalization. As a result of their lack of reproductive organs, these slowly developing tillers will be disadvantaged in competition for the available photosynthate. The later tillers will die when the early tillers begin to elongate and initiate spikelets (reproductive organs), because of the stronger competitiveness of the latter for the available assimilate. Spring genotypes do not respond to vernalization. Their early versus late tillers respond to the longer *P* and higher *T*, however, in the same way described for the winter genotypes.

8.10.1.2 Convergent development among planting dates of a genotype

Convergence of development among different sowings of a given genotype can be explained using the same concept as convergence among the tillers of a single plant. All sowings in early spring or before will be adequately vernalized, so vernalization gene activity will not cause differences in the rate of development. Therefore, the post-vernalization low but rising *T* and the short but lengthening *P* will become the major controls over development and causal of the convergent development. Temperature will be the major control until it becomes high enough to allow the *P* to become an also strong control. Plants of an earlier sowing will have been exposed for longer time to low *T* (thus a lower average *T*) than a later sowing, which will slow the development as compared with the later sowings.

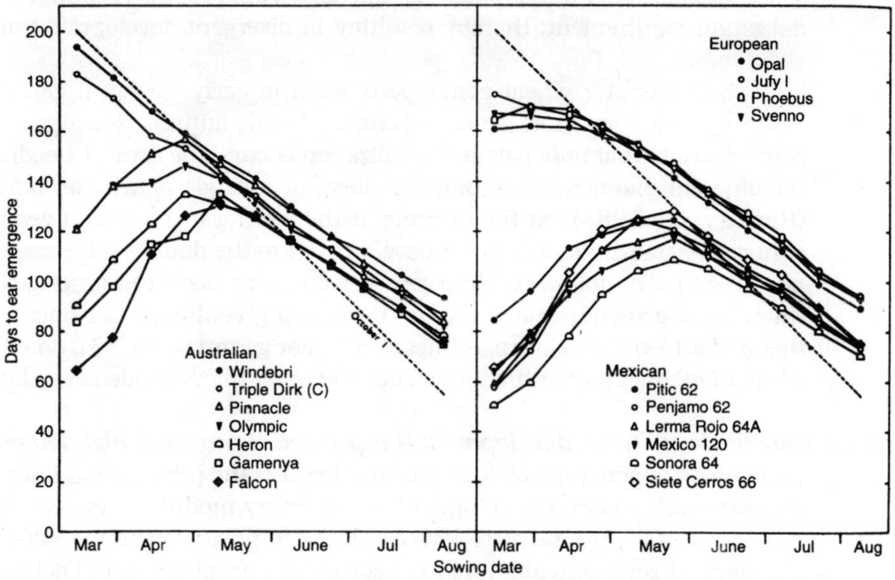

Fig. 8.15. The influence of time of sowing on days to ear emergence of Australian, European and Mexican wheats. October 1 indicates the approximate optimum time of ear emergence at Wagga Wagga, Australia, where these data were collected. (Reprinted from Syme, 1968.)

Consequently, the difference in heading or flowering date among the sowings will be smaller than suggested by the difference in sowing dates (Fig. 8.15).

8.10.1.3 Convergent versus divergent development of early and late genotypes

Hay and Kirby (1991) discussed only convergent development. But, it is only half of the modulation of plant development by *P–T* combinations. Divergent development is the other half. It is remarkable that when planted in the autumn or winter at temperate locations all genotypes and all plantings develop to heading and flowering within a short span of dates. Divergent development is equally common and remarkable. For example, when a winter genotype is planted from early to late spring and more so for summer plantings, the difference in maturity can be much greater than expected from the difference in planting dates. Similarly, for different wheat genotypes, the range of maturity among them will change from being very small for an early sowing to very large for a late sowing. Some genotypes will never develop to heading when planted in late spring or summer. This divergent development results also from differential modulations by the *T* and *P* of the development of the different genotypes. The low *T* of early spring cause the early genotypes to develop more slowly while causing the late genotypes to develop more rapidly, thereby causing convergent development among different genotypes. For late spring and summer plantings, on the contrary, the high-*T* modulate more rapid development of the early (spring) genotypes, but later development of the late (winter) genotypes that have a vernalization requirement, thereby resulting in divergent development among the genotypes.

When a winter wheat genotype is sown in early spring, it can develop to flowering in a very short time, as compared with autumn sowings. However, if planted so late that only partial vernalization occurs, the time of heading among the different plants, and among the tillers of a single plant, can differ greatly (Huang *et al.*, 1988). At the extreme, if the sowing is too late, a genotype that requires vernalization will never develop even to the double-ridge stage (Hu and Luo, 1988). The following divergent development occurred among tillers of a single plant grown in the winter of 1994 in a greenhouse at Ithaca, New York that had a 12 h *P* and average *T* near 25°C. For genotype 89(33), one tiller headed about 60 days earlier than a second, and other tillers headed even later.

8.10.2 Genetic diversity for development is expressed strongest at high temperatures

From above, a general conclusion is that low *T* is the primary modulator of convergent development, while high-*T* is the primary modulator of divergent development. This occurs because low-*T* reduces to prevents the vernalization- and photoperiod-gene activities and their consequent inhibitions and delays of development, while high-*T* enlarges these activities and delays the development.

Murfet (1977b) described the variation in DTF among a natural population of *Themeda australis* under two *T*. A population collected from a frost hollow at altitude 1500 m expressed low variation when grown at 15/10°C. Growth at 27/22°C enlarged the variation, and the population split into a group with early DTF and one with late DTF. Murfet wrote:

> This suggests the existence of genetic diversity for a quantitative cold requirement which is normally covered up at the natural site by the frequent occurrence of low temperatures. It also illustrates a fundamental principle – the chance of detecting genetic differences is increased by growing the plants under conditions which maximize the difference in response between the genotypes.

He continued, 'Biometrical genetics is like using a computer to unscramble a picture seen through a frosted glass window. It might be practical in some cases to open the window and get a direct Mendelian view by the simple device of choosing environmental conditions which maximize the effect of particular genetic differences'. Murfet's statements emphasize the crucial role of *T* and *P* in research of genetic resources.

We now ask: What is the environment that will maximize expression of the genetic diversity for plant maturity? The vernalization by photoperiodism by thermal response interaction models provide a generalization, equations 8.6 and 8.7 being their basic forms. *The phenotypic expression of genetic diversity for the DTF of an SDP species will be maximized by high* T *combined with a delaying LD. For an LDP species, expression of the genetic diversity will be maximized by high* T, *alone or in combination with a delaying SD.* Data of Wallace and colleagues summarized in Chapters 4 and 5 (Figs 4.1, 4.2, 5.3, 9.2, 9.3) and in Mmopi *et al.* (1995) for bean, an SDP, agree with this conclusion. Temperature is the primary factor that influences the plant development because high *T* modulates large activity of the

(i) vernalization-genes, (ii) photoperiod-genes, and (iii) all other genes and their control through the thermal-time response (equations 8.6 and 8.7).

Likewise, higher *T* maximizes phenotypic expression of the genetic diversity of LDP, by enlarging the vernalization by photoperiodism by thermal-time interaction controls over the DTF. Figure 8.15 from Syme (1968), illustrates this. Syme planted seven Australian, four European and six Mexican wheats in New South Wales at 2-week intervals between March 15 and August 7. The mean *T* changed from 13.2°C to 10.3°C, 5.9°C, 5.7°C, 2.9°C, 4.8°C, to 5.1°C. Corresponding mean *P* were 12.6 h, 11.5 h, 10.8 h, 10.3 h, 10.3 h, 11.1 h and 12.3 h. The difference in days to heading (DTF) among the genotypes was greatest for the earliest planting date which had the highest *T*. This difference decreased progressively with lower mean *T*. Difference was minimal and more stable for May to August plantings with low mean monthly *T* between 5.9°C and 2.9°C. Syme (1968) did not include late spring-time sowings, but Miao *et al.* (1988) did for 40 diverse wheat cultivars representing all ecological adaptation types grown throughout China. The China plantings were at 37.7°N and 800 m elevation between September 20, 1984 and May 20, 1985 at 10-day intervals. The long winter killed the spring genotypes if planted earlier than November 20. Difference in DTF measured from emergence across all surviving genotypes was small and stable for all sowings earlier than December – but consider also the longer time to emerge under the low-*T*. Difference in DTF became progressively larger as the *T* became higher for the sowings after December. The difference became infinite for the later spring sowings, since many winter genotypes failed to develop to heading (Fig. 8.16).

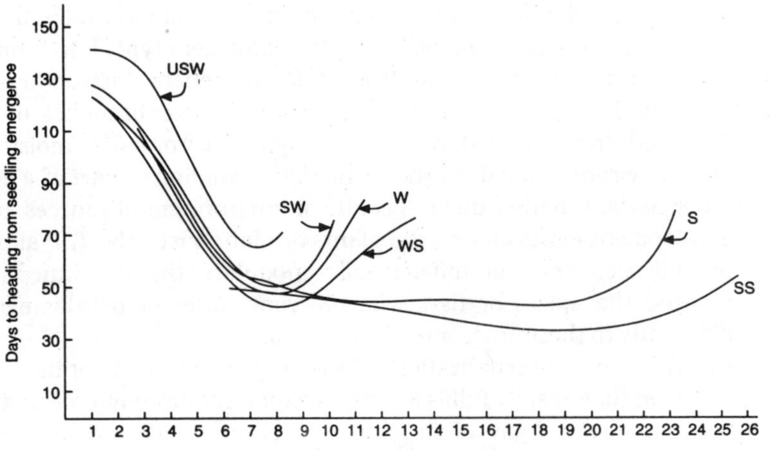

Fig. 8.16. Days to heading from seedling emergence of different types of wheats in response to 24 planting dates in 10-day intervals starting from September 20. USW = ultra strong winter type; SW = strong winter habit; W = typical winter habit; WS = winter–spring facultative habit; S = spring habit; SS = strong spring habit. (Reprinted from Miao *et al.*, 1988.)

Controlled-environment experiments (Fig. 8.4) illustrate further the strong control by *T* over variability of the phenotypic variation among diverse genotypes. Miao *et al.* (1990) vernalized germinating seeds of 17 wheat cultivars at 0–2°C in the dark for 0–70 days at 10-day intervals. Then, all the vernalized seeds were planted on the same day in a greenhouse with 24 h and a mean 22°C. The difference in DTF measured from seed imbibition (DTFb) was largest for seeds planted directly into the high-*T* of the greenhouse with no vernalization (Fig. 8.4). The difference in DTF among the genotypes decreased progressively to a minimum as the seeds had been vernalized for 40 to 60 days.

The principle that high *T* – especially when combined with delaying *P* – maximizes expression of the existing genetic variability can facilitate research on genetic resources, genetic inheritance and plant breeding. Planting segregating wheats in early autumn facilitates differentiation between spring and winter segregates, since the relatively high *T* cause the spring genotype to develop rapidly during the autumn. On the contrary, the winter genotypes develop slowly because they have not been vernalized. The same differentiation occurs for sowings in late spring. For selection of only winter genotypes, the segregates should be sown in early autumn. Late spring plantings facilitate effective selection of only the spring genotypes.

Making crosses between extremely-early versus extremely-late genotypes can be facilitated by vernalizing the late genotype at low *T* long enough to fulfil their vernalization requirement. Alternatively, both genotypes can be grown during late-autumn or early-winter to minimize their difference in DTF.

8.10.3 Acquired diversity of development is also enlarged by high temperature

Another common but poorly understood phenomenon is that high *T* in combination with a delaying *P* (short for LDP and long for SDP) enlarges any differences in plant development that originate through environmental differences met across time by individual plants of the same genotype. Environmentally caused variation does not originate from different genetic instructions. It is acquired diversity. It can occur even among the tillers or branches of a single plant. Acquired diversity of development originates from soil, moisture, nutrient or other environmental differences, including shading of part of a plant by leaves of other parts. Whether due to genetic or environmental sources, once diversity of development exists among the plants or plant parts, the *T*, *P* and other environmental factor(s) will differentially modulate the continued development – because the apices or tissues at different states of development will respond differently to these environmental factors.

Whenever a vernalization-requiring genotype has not been vernalized for a duration that exactly fulfils its vernalization requirement, i.e. not been optimally vernalized, the delay of development becomes associated with an increased variation of the developmental phenotype among the individual plants. This is illustrated by different wheat genotypes which Huang *et al.* (1988) planted on March 5, 20 and April 20 at Ningchuan, China. For all three dates, the spring genotypes expressed earlier heading (a smaller DTF) for the later sowings. For the first two sowings, all plants of the semi-spring genotypes headed, but for the third sowing only part headed, ranging from 52% to 8% across 3 successive

Table 8.1. Days to heading and percentage of headed plants of wheat as influenced by planting date.

		Planting date					
		March 5		March 20		April 5	
Habit	Year	DTF	%	DTF	%	DTF	%
Spring	1983	63	100	61	100	56	100
	1984	56	100	56	100	52	100
	1985	59	100	57	100	53	100
Semi-spring	1983	68	100	77	100	77	52
	1984	62	100	71	100	81	37
	1985	61	100	66	100	69	8
Winter	1983	78	36	80	33	–	0
	1984	75	40	84	35	–	0
	1985	66	46	76	22	–	0
Strong winter	1983	–	0	–	0	–	0
	1984	–	0	–	0	–	0
	1985	–	0	–	0	–	0

The modelled days to heading are from Huang *et al.* (1988).

Table 8.2. Effect of the duration of vernalization at 6°C under 12 h on the days to heading and its variation among plants of cv. Norstar grown in a post-vernalization environment of 20°C and 18 h.

Vernalization duration (weeks)	Average days to heading (DTFe)*	Average total days to heading (DTFb)*	Range in days to heading among the plants	Coefficient of variance of days to heading
4	105	133	48	112.8
6	62.7	105	28	37.9
8	41.5	98	10	3.9
10	36.5	107	6	2.2
12	33.2	117	11	4.7

*Respectively, DTFe and DTFb are the days to flowering measured from the end and the beginning of the vernalization. The original data are from Whelan and Schaalje (1992).

years. For some winter genotypes no plants headed for any sowing. For others, the proportion ranged between 36% and 46% for the first sowing, and 22% to 33% for the second sowing (Table 8.1). This acquired diversity among spring sowings may have originated from different depths of soil through which the emergence occurred, or from soil-moisture differences, etc. High *T* will enlarge such acquired diversity.

A second example of acquired diversity is presented in Table 8.2, based on data of Whelan and Schaalje (1992). They vernalized winter wheat cv. Norstar for 4–12 weeks before transfer to 20°C/12 h. The optimal vernalization was 8 weeks. Incomplete vernalization resulted in larger average and variation of the DTF among the individual plants. Extensive recent data (Worland *et al.*, 1994) indicated that both the genetically-established diversity among 127 European wheats, and acquired-diversity were enlarged by a delaying SD.

8.10.4 Two forms of environmentally-modulated divergent development

Wallace *et al.* (Figs 4.3, 4.4 and Chapters 4, 5) reported that a delaying LD combined with a high *T* resulted in only a 5-day-span of DTF among 129 plants shown to be homozygous for the photoperiod-insensitive genotype (*ppdppd*) of bean. This contrasted with a 28-day span among 132 plants shown to be homozygous for the photoperiod-sensitive allele (*PpdPpd*) of the same photoperiod-gene (Fig. 4.3). Under near-optimum SD and high *T*, both parental genotypes, and also both the homozygous photoperiod-insensitive (*ppdppd*) and the photoperiod-sensitive (*PpdPpd*) F_8 segregates flowered across a short time-span (Fig. 4.4). Therefore, it was the delaying *P* in combination with the high *T* that caused the divergence of development (sixfold larger range of DTF) among the 132 plants that inherited allele *PpdPpd*, than among the half of all the plants that inherited the allele *ppdppd* (Fig. 4.3). Delay of development by the *P* combined with a high *T* is the well recognized photoperiod response, which is expressed only by a photoperiod sensitive genotype. This delay caused two forms of divergent development. One was divergence of development between the insensitive and the photoperiod sensitive genotype, as evidenced by their respective early versus late (delayed) average DTF (Fig. 4.3). The second was divergent development among the individual plants of the sensitive genotype, indicated by the sixfold larger span of their DTF than that of the insensitive genotype. The photoperiod sensitivity of all plants was due to the dominant gene *Ppd*, so implementation of the divergence in DTF among these F_8 generation randomly derived inbred lines would be due in part to differences in genetic instruction through the variation in their background genes (Mmopi *et al.*, 1995). From Fig. 9.8, such background variation caused a small range in the average DTF of the insensitive genotype, and a much larger variation in the average DTF of the sensitive genotype.

Chapter 8 Summary

Plant development toward flowering is interactively controlled by: (i) any vernalization response which may or may not be associated with a simultaneous photoperiod response; (ii) any post-vernalization photoperiod response; and (iii) will be associated with the always-functioning thermal response. This three-component interaction was interpreted, based on three concepts: (i) The phasic development of plants that require vernalization changes a pre-vernalization SD photoperiod-response that shortens the DTF of some genotypes to the far-better recognized post-vernalization LD photoperiod-response and its shortening of the DTF. (ii) Separate Q_{10} effects by the applied temperature (*T*) modulate the level of activities of all three subsets of genes that jointly control the DTF. These subsets are (a) any vernalization- and/or (b) any photoperiod-genes, both of which delay development at higher *T*, and (c) the majority of all the other genes and their implementation of development, including the Q_{10} effect on their joint activity known as the thermal time response. (iii) Temperature becomes the primary factor that influences plant development, because a higher average *T* of the growing season enlarges the activity of both any vernalization-genes and any photoperiod-genes, with both activities delaying the autonomous development of the plant, while simultaneously this same higher *T* enlarges the activities of all the other genes as they collectively implement and advance the autonomous development. More detailed conclusions follow about the photothermal controls over plant development.

1. Plant development toward flowering occurs autonomously. No induction of development is needed. Both vernalization- and photoperiod-gene activities inhibit the rate of the autonomous development at all *T* above the base *T* that allows any activity of these genes. The Q_{10} effect due to each *T* higher than the vernalization gene's base-*T* enlarges the vernalization-gene activity, and the Q_{10} effect due to the same *T* will enlarge the photoperiod-gene activity and its delay of the development.

2. Vernalization is prevention by low-*T* of the vernalization-gene activity that will occur under higher *T* and will inhibit the autonomous development. Similarly, the so-called photoperiod induction is prevention by appropriate *P* and/or low *T* of the photoperiod-gene activity that will occur under a delaying *P* – if the *T* is higher than the base-*T* that allows the photoperiod-gene activity. Through its Q_{10} effect, a low-*T* reduces any vernalization-gene activity and any photoperiod-gene activity, so a sufficiently low-*T* will prevent all of each gene activity.

3. Vernalization and photoperiodism are physiological and complementary processes established by distinct subsets of genes. Their controls overlap across the time-span of the earliest growth and development. This overlap is such that with increased duration of low-*T*, the shortening of the DTF due to the continued vernalization becomes progressively smaller as the potential shortening of the DTF by a long photoperiod (LD) becomes progressively larger.

4. Neither low-*T* vernalization nor an appropriate *P* is required to induce development. For genotypes that have vernalization gene(s), however, an initial low-*T* is required for the most rapid development toward flowering. Likewise, for genotypes that have photoperiod gene(s), a non-delaying *P* is required for the most rapid development.

5. The vernalization requirement can be precisely quantified – if and only if – the DTF is measured from the beginning of the low-*T* vernalization, rather than from the end of the vernalization.

6. Both the FLN and the SN of wheat are determined at about the stage of the terminal spikelet initiation. The initiations of both the final leaves and the terminal spikelets overlap across at least the single-ridge stage to the double-ridge stage. This time-span is strongly determined by both the vernalization- and the photoperiod-responses. This explains why the FLN, SN and DTF tend to respond similarly to low-*T* vernalization and to the *P*. In addition, winter wheats have a longer time-span from seed germination to the single-ridge stage, which is associated with a larger FLN and greater winter hardiness. A consequence is that greater winter hardiness and/or larger spikes are always associated with greater lateness (a larger DTF).

7. The so-called SD-vernalization and LD-vernalization responses are directed by photoperiod-genes rather than by vernalization-genes. These photoperiod responses can be replaced by a duration of low-*T*, because the photoperiod-gene activity is reduced to prevented by progressively lower *T*. This has been misinterpreted as low-*T* vernalization. However, any vernalization-gene directed low-*T* acceleration of development is not modulated by the *P*. Low-*T* can replace the so-called SD- and LD-vernalization responses. But, neither an LD nor an SD can replace any vernalization-gene implemented response. Only a few plant species and genotypes that respond to low-*T* vernalization respond also to the mislabelled SD- or LD-vernalizations.

8. The above statements about the vernalization by photoperiod by thermal response interaction and consequent differentials of the partitioning of the available photosynthate fully interpret the remarkable phenomena of convergent and divergent development among genotypes grown across a range of environments. Low-*T* is the major cause of convergent development, while high-*T* combined with delaying *P* is the major cause of divergent development. Models are presented that summarize present understanding of the interactive control over plant development by vernalization, photoperiodism and the thermal response. These models can facilitate further elucidation of the vernalization response by photoperiod response by thermal time response interaction controls over plant development. The models have not been tested, because DTF data accompanied by recording of all pre- and post-vernalization *P*, *T* and their durations were not found in published literature.

9

Yield System Analysis: An Adjunct to Yield Trials

This chapter involves paradigm shifts 19 to 27. Shifts 19 and 20 have been progressively derived through all prior chapters. Shift 19 is that a major control over crop adaptation and yield arises through the phenology. This control results from the complex three-way interaction among: (i) any photoperiod-response caused $G \times E$ interaction; (ii) the always-functioning thermal-time response caused $G \times E$ interaction, and (iii) any vernalization-response caused $G \times E$. Shift 20 is derived in part in this chapter, by showing that the interacting photoperiod and thermal time responses can be quantified holistically. In Chapter 7, the GPTmodel (equation 7.4) quantified the two interaction effects simultaneously but separately, and therefore also holistically, to simulate and predict with high fit the days to flowering (DTF) for previously untested *P–T* combinations. Shift 21 is that the accumulated growing degree days (thermal-time) accurately indicates the DTF regardless of whether each higher temperature modulates change toward earlier DTF as always occurs for photoperiod insensitive genotypes, or modulates later DTF as occurs across the higher temperatures for the photoperiod sensitive genotypes. This calls for acceptance that the accumulated thermal degrees indicates the DTF even if it is not linearly related to the temperature. Paradigm shift 22 is that using the multilocation (or year or season) yield trials already being conducted to measure the yield to also measure the aerial biomass at harvest maturity, and the DTF and the days to harvest maturity, is the cheapest possible way to advance understanding of the system controls over crop yield. Shift 23 enlarges the physiological–genetic explanation of all quantitative traits. Shift 24 indicates need for more application of whole-system research for which a philosophical and procedural basis is defined in paradigm shift 25. Shifts 26 and 27 present additional procedural details, which contribute to whole-system research and its interpretation.

9.1 Procedures of Yield System Analysis

9.1.1 Biomass-based yield system analysis

Donald (1968) interpreted the total accumulated biomass and harvest index (HI) to be two components that account in full for the crop yield. Arithmetically, they do so. He did not use the term yield system analysis (YSA). Donald's procedure failed to measure the third major physiological–genetic component of the process of accumulating yield: the genotype–environment implemented (system-established) time each genotype requires (uses) within the referenced environment to develop to harvest maturity.

9.1.2 Thermal-time based yield system analysis

Williams and Saxena (1991) and Williams (1992), at the International Crops Research Institute for the Semi Arid Tropics (ICRISAT) in India described a thermal-time based YSA, also without calling it a YSA. They measured the durations and rates of biomass and yield per accumulated growing degree day (GDD) to maturity, rather than per calendar day. The calendar days to flowering (DTF) and days to harvest maturity (DTHM) were measured as for a calendar-based YSA. Then, however, the daily accumulations of GDDs were added across all the days from planting to the DTF and also to the DTHM. For each day, this accumulated thermal time is the average of: (i) the lowest degrees C (base *T*) that allows advance of plant development, and of (ii) that higher degrees C.

The concept of a thermal-time based YSA arose through much evidence that the accumulated GDDs each genotype requires to develop to its DTF is the same for every season and location (Squire, 1990; Hodges, 1991; Jordan, 1993). This is validated for both photoperiod sensitive and insensitive genotypes below in section 9.3.3.2.

9.1.3 Calendar-time based yield system analysis

Like CO_2, water and mineral nutrients, time is an external physical input that is integral to plant system functioning. A time duration of biological activity must occur for any biomass to be accumulated beyond that of the seed, and for any of this net accumulated biomass to be partitioned to the yield. The holistic genotype and holistic environment jointly determine (Fig. 1.1) how much input of CO_2, water and mineral nutrients can be used to give the expressed plant growth and yield. Likewise, the genotype and environment jointly determine how much time will be needed to develop to the DTF and to the DTHM. Through its variations of *T* and/or moisture, the environment alone determines for each growing season whether more or less than the actually needed time is available for the genotype to develop to harvest maturity.

Some genotypes have genetic direction for early flowering and maturity, others for intermediate time, and others require longer time to complete growth and development. Measurement of the: (i) days each genotype required and/or was able to use to develop to harvest maturity is as essential to complete YSA as measurement of (ii) the net accumulated biomass; and (iii) the partitioning of this biomass as quantified by the HI (Wallace, 1991). Each of these three major physiological–genetic components of accumulating yield is genetically directed

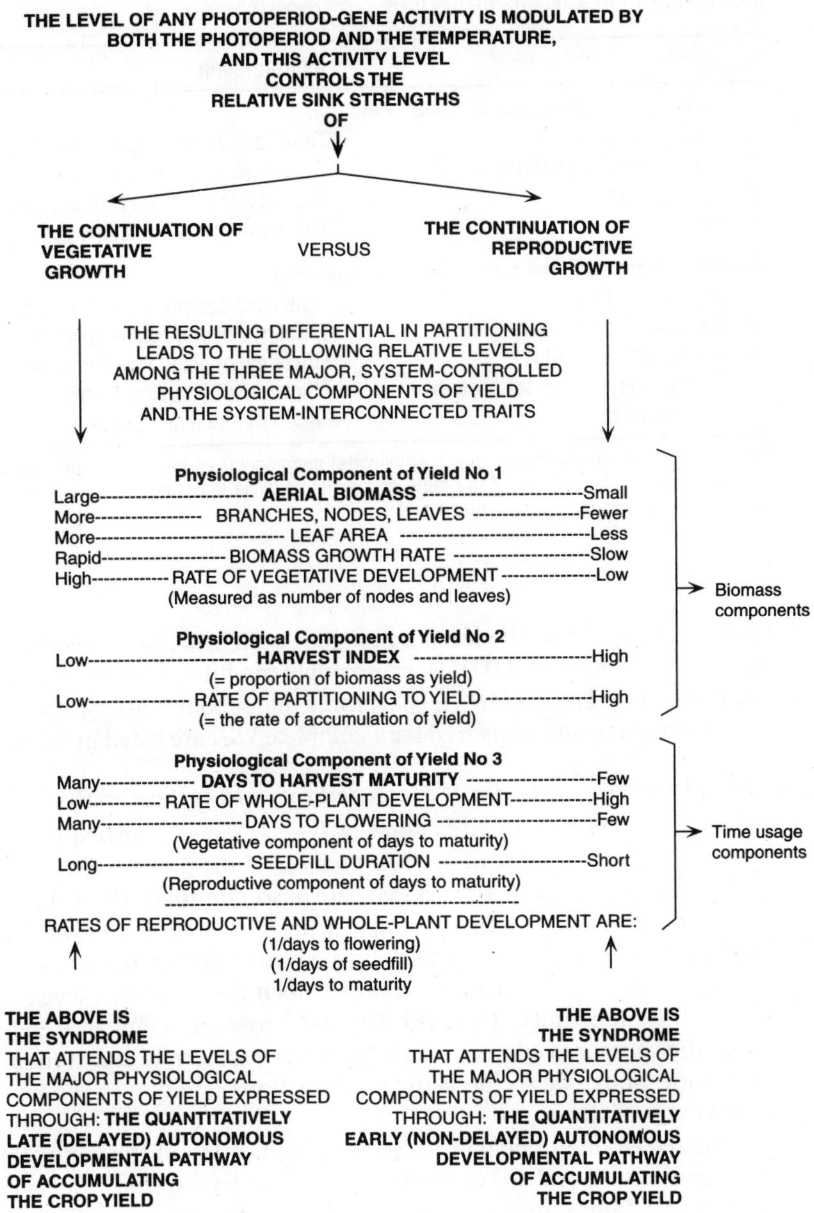

Fig. 9.1. Effects by competitive partitioning of photosynthate between reproductive versus vegetative growth on levels of the three major, physiological components of yield and levels of traits with system-established interconnections. (Modified from Wallace *et al.*, 1993a.)

Table 9.1. The nine[1] outputs from the yield system of a cultivar that are measured by the yield system analysis procedure.

Output	Interpretation	
Four direct measurements within each yield trial		
1. Days to flowering	Time used to develop to flowering	
2. Days to harvest maturity	Time used to develop to harvest maturity	
3. Aerial biomass	The net accumulated photosynthate	
4. Yield	The economically valuable system output	
Five calculations from the four direct measurements		
5. Days of seed fill	Time used for actual yield accumulation	
6. Yield day^{-1} to maturity	Efficiency of yield accumulation	*The rate of partitioning*
7. Yield day^{-1} of seedfill	Efficiency of yield accumulation	
8. Biomass day^{-1} of plant growth	Efficiency of photosynthesis	
9. Harvest index	Endpoint efficiency of partitioning to yield	

[1]Traits 3, 9 and 2 are the three major (penultimate) components of the process of accumulating the crop yield, and traits 1, 5, 6, 7, and 8 are subcomponents of these major components, i.e. they are the ante-penultimate components of the yield system.

by the genotype. But, the levels of all gene activities that collectively implement this required time (the DTHM) are strongly modulated by the *P* and *T* (Fig. 9.1, compare Figs 2.1, 6.1). The yield, its three major components, and the calculable subcomponent traits measured for a complete YSA are listed in Table 9.1.

9.1.4 Statistical analysis of the traits measured by yield system analysis

A first goal of YSA is quantification of the system-established interconnections among levels of the three major components of the process of accumulating the crop yield. The calendar-time based YSA also quantifies the subcomponents of the three major components, by measuring the level of each of the nine traits listed in Table 9.1. The system-established interconnections among these traits are quantified best as the correlation between the expressed levels of each pair among the nine traits. The correlations across the compared genotypes and across the compared environments quantify, respectively, the system-established interconnections due to the genetic instructions and due to the environmental modulations of them.

The second major goal of a YSA is quantification of the control through the $G \times E$ interaction over the expressed phenotypic level of the nine traits measured by the YSA. Many statistical methods for quantifying the $G \times E$ interaction have been reviewed in Kang (1990) and Kang and Gauch (1996). Zobel (1990) and Gauch (1990) described the Additive Main effects and Multiplicative Interaction effects (AMMI) analysis procedure used within this chapter. The most thorough mathematical description of AMMI is in Gauch (1992). Description of AMMI for persons with minimal mathematical background is in Zobel and Wallace (1995). Cooper and DeLacy (1994), Kang and Gauch (1996) and Cooper and Hammer (1996) present more recent comparisons of methods to quantify $G \times E$ interaction. AMMI analyses of the $G \times E$ interaction control over one or more of the physiological–genetic components of the accumulation of yield have previously been reported (Wallace *et al.*, 1991, 1995a). Annicchiarico and Perenzin (1994), and Gauch and Zobel (1996) reviewed many published applications of AMMI to yield trials. AMMI captures more of the $G \times E$ interaction than the longer recognized (Finlay and Wilkinson, 1963) linear regression analysis (Yau, 1995; Gauch and Zobel, 1996, 1997).

Each of the nine traits measured by YSA will be statistically analysed using the mathematical description of the AMMI analysis, which is:

$$Y_{ger} = \mu + \alpha_g + \beta_e + \sum_{i=1}^{N} \lambda_i \gamma_{ig} \delta_{ie} + \rho_{ge} + \varepsilon_{ger}$$

The terms are defined as follows:

Y_{ger} = the expressed level of the trait for the specified genotype in the specified environment and replication.
μ = the grand mean of all observations. It is the average level of the trait due to all the different genetic directions and environmental modulations of all gene activities and physiological processes that implement the expressed level of the trait.
α_g = the deviation from the grand mean due to the genotype. It is the additive effect by the genotype on the expressed level of the trait. Statistically, it is also called the main effect by the genotype.
β_e = the deviation from the grand mean due to the environment. It is the additive effect by the environment on the expressed level of the trait. It is also called the statistical main effect by the environment.
γ_{ig} = the multiplicative effect due to genotype *g* through the $G \times E$ interaction effect of axis *i*.
δ_{ie} = the multiplicative effect for environment *e* through the $G \times E$ interaction effect of axis *i*.
λ_i = the singular value, a scaling factor that makes the term $\sum_{i=1}^{N} \lambda_i \gamma_{ig} \delta_{ie}$ approximate the total $G \times E$ interaction effect. The square of the singular value (λ_i^2) equals the sum of squares accounted for by the interaction principal component axis $(\text{IPCA})_i$.
N = the number of interaction axes derived by the AMMI model before relegating higher axes to the model's residual. Ordinarily, N is chosen to optimize the AMMI model's predictive accuracy, or to include those axes that are significant at a specified level. The interaction axes are labelled 1 to N, in the order of their progressively smaller sums of squares from within the total sums of squares due to $G \times E$ interaction.
ρ_{ge} = the model's residual $G \times E$ interaction, which equals the sum of all axes beyond N that are not included in the model.
ε_{ger} = the random error for replicate *r* of genotype *g* in environment *e*.

AMMI analysis uses a two-way classification of the data. This assigns all control over the phenotype (expressed level) of the trait to the holistic genotype and to the holistic environment. This whole-system control is illustrated by Fig. 1.1.

The first procedure within AMMI analysis is an analysis of variance (ANOVA). It quantifies the: (i) grand mean (μ), (ii) average of each genotype across all the environments ($\mu + \alpha_g$), which is that genotype's additive control over the phenotype; and (iii) the average of each environment across all the genotypes ($\mu + \alpha_e$), which is the environment's additive control. In addition, the ANOVA quantifies the (iv) total sums of squares across the entire data set, and the sums of squares due to variation across: (v) all the compared genotypes; (vi) all the compared environments; (vii) the total $G \times E$ interaction; and (viii) the error due to the replications (ε_{ger}).

The second procedure within AMMI analysis is a principal component analysis (PCA) applied to just the total $G \times E$ interaction sums of squares quantified by the ANOVA. Using pattern analysis, the PCA differentiates 1 to N statistically significant sources (patterns) of a $G \times E$ interaction effect within the total $G \times E$ interaction. Therefore, N different γ_{ig} and δ_{ie} can be compared. The PCA terms for these largest through progressively smaller statistically significant patterns within the total $G \times E$ interaction are the: Interaction Principal Component Axis 1 (IPCA1), followed by IPCA2, IPCA3, . . . to IPCAN. Within the sums of squares due to each of IPCA1 to IPCAN, the PCA quantifies a multiplicative control due to each genotype (γ_{ig}), and another due to each environment (δ_{ie}).

Rather than using the statistical term *multiplicative effect* for each γ_{ig} and δ_{ie}, through the rest of this chapter, each γ_{ig} and δ_{ie} is called a *multiplicative control*. The word control is used to incorporate physiological–genetic thinking and interpretation. The *multiplicative control* due to a genotype (γ_{ig}) compared with that due to an environment (δ_{ie}) indicates the relative control strength over the expressed phenotypic level through the specified $G \times E$ interaction (IPCA1 or IPCA2, etc.).

A YSA that is followed by an AMMI analysis of all nine traits it has measured will be called a YSA/AMMI. All nine traits are quantitatively variable traits. A quantitative trait may be directly and strongly controlled by as few as one gene, as detailed for control by the photoperiod genes in Chapters 4 to 8 (Figs 1.1, 2.1, 5.1, 9.1, 6.7). Nevertheless, all quantitative traits are indirectly controlled by many genes (Figs 11.1, 11.4, 12.3), with their activity levels being strongly modulated by the environment as best detailed within Chapter 11. That chapter indicates indirect control of quantitative traits arises through many genes, because all quantitative traits must competitively share the small, or medium or large available intraplant metabolites that the environment can provide. This system-established sharing of the metabolites interconnects the genetic control of each quantitative trait with the genetic controls over all the other quantitative traits. The environmental limitations of the gene activities that implement plant growth determine whether the pool of intraplant metabolites available for this inevitable sharing is small, medium or large.

A detail integral to interpreting the AMMI analysis follows. For every combination of a given genotype grown in a given environment (one of the experimental treatments), the estimated level of that trait is indicated by the following:

$$\hat{Y}_{ger} = \mu + \alpha_g + \beta_e + \sum_{i=1}^{N} \lambda_i \gamma_{ig} \delta_{ie}$$

9.1.5 Multiple objectives make YSA/AMMI basic whole-system research

An individual yield trial has an economic objective. It is to determine:

1. Which genotype yields more than the others. This economic objective is achieved by comparing the average yield of each of the genotypes.

Three basic whole-system research-objectives for applying YSA to each individual yield trial are to quantify:

2. The additive genetic control by each genotype over the differences in yield among the genotypes. This basic objective is facilitated by achieving the economic objective 1.

3. The additive control by each genotype over the differences in average level of each of the eight (Table 9.1) physiological-genetic components of the holistic process of accumulating the crop yield. Quantifying this requires comparison for each of the eight component traits of its average for each genotype measured across all of the compared environments.

4. The system-established interconnections (negative or positive correlations) between the genotype-directed changes in level of the yield, and the attendant shifts in level of each of yield's eight physiological-genetic components measured by the YSA, plus the interconnections (correlations) between all possible pairs among the eight components. Quantification of all these interconnections requires the researcher to compare all 36 correlations among pairs of the nine traits measured by the YSA across both all genotypes and across all environments.

YSA of a multilocation yield trial achieves the above four objectives while adding a fifth to eighth objective. They are to quantify:

5. The additive control by each environment over the differences in yield, and the additive control over the difference in the level of each of the eight physiological–genetic components. This requires comparison by the researcher for each of the environments of its average yield across all of the compared genotypes, plus this average for each of the eight components of the process of accumulating the yield.

6. The multiplicative control due to each genotype over the yield through the $G \times E$ interaction effect, and this control over the level of each of the eight physiological components of the accumulation of the yield. This requires comparison for each genotype of its AMMI analysis-quantified γ_{ig} for yield, plus of its γ_{ig} for each of the eight components of the process of accumulating the yield.

7. The multiplicative control due to each environment over the yield level through the $G \times E$ interaction, and this control over the differences in level of each of the eight physiological components of the accumulation of the yield. This requires comparison for each environment of its AMMI analysis-quantified δ_{ie} for yield, plus the δ_{ie} for each of the yield's eight components.

8. The level of the $G \times E$ interaction control over the yield of each genotype when grown in each environment, i.e. of each treatment. For each genotype this requires prior achievement of both of objectives 6 and 7. The total number of these treatments is the number of genotypes multiplied by the number of environments. And, the total number of these $G \times E$ interaction effects is the number of treatments multiplied by the number of statistically significant sources of a $G \times E$ interaction within the total $G \times E$ interaction sums of squares.

In achieving the above eight objectives, a large number of numeric values arise that need to be compared. This is best illustrated using the 30 genotypes × 20 environments (600 treatments) of the complete pigeonpea YSA/AMMI analysis detailed in section 9.3 below. Its 600 treatments require direct comparison of 600 averages. Direct comparison of the $G \times E$ interaction control over each of all 600 average levels would require comparison of another 600 values, bringing the total to comparison of 1200 numeric values. But, a superior comparison and easier physiological-genetic interpretation of the largest $G \times E$ interaction is achievable by presenting the AMMI analysis result as just 50 points on a biplot graph. The 50 points will consist of one for each of the 30 genotypes plus one for each of the 20 environments. The advantages from presentation of the 50 points on a biplot graph will occur in section 9.2 about YSA/AMMI of ten bean cultivars and four or five environments. The advantage will be far more apparent in section 9.3 when the largest $G \times E$ source for all nine traits measured by the YSA is interpreted using AMMI analysis of the 600 treatments for pigeonpea. Direct comparison for each of the successively smaller $G \times E$ interaction effects would require comparison of another 1200 numeric values for each trait.

9.2 Temperature Control over Yield of Bean

The bean yields analysed with YSA/AMMI have ten genotypes × five environments = 50 treatments (Fig. 9.2B). This AMMI biplot reduces the number of values needing comparison from 50 to just 15 points on the biplot graph.

9.2.1 Control by temperature over yield per se

Yields at harvest maturity in 1983 were measured in the field trials at the different elevations, for which U-shaped curves of the DTF across five temperatures (T) between 15.7°C and 28.3°C (Fig. 5.3) were discussed in section 5.4. The focus in 1983 was on control by T over the photoperiod response of the DTF so, regrettably in retrospect, the near-zero yields of all the genotypes were not measured at the 28.3°C site.

Interpretation of the $G \times E$ interaction effect on bean yield begins with recognition that the photoperiod sensitivity versus insensitivity of the genotypes in Fig. 9.2 were determined independent of and before the experiment. The AMMI analysis result on a biplot graph (Fig. 9.2B) for yield indicates large positive multiplicative control due to both the 15.7°C and 18.0°C, and large positive multiplicative control also due to the photoperiod sensitive genotypes. Multiplying the large positive multiplicative control of 15.7 or 18.0° C by the

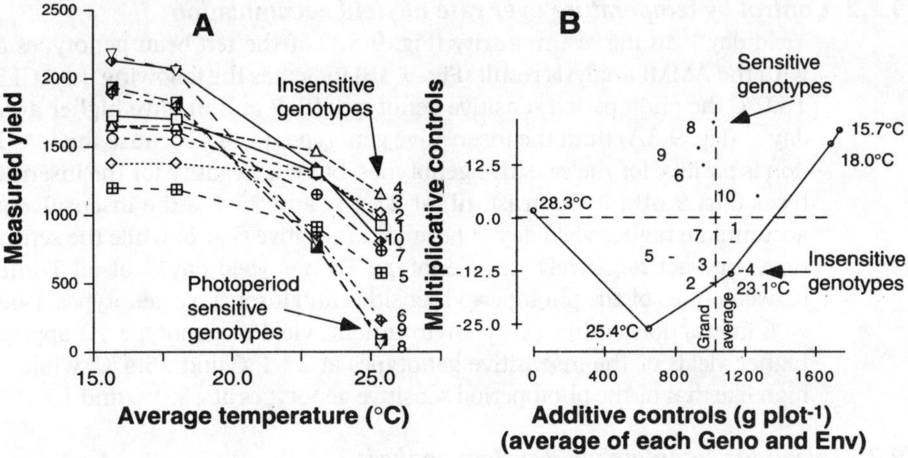

Fig. 9.2. The (A) yields of ten bean genotypes grown at five temperatures (field elevations) under the same 13 h photoperiod, and (B) the AMMI analysis result of these yields. (Modified from Wallace, *et al.*, 1995a.)

also large positive multiplicative controls of the photoperiod-sensitive genotypes (8, 9 and 6) indicates a large positive $G \times E$ interaction effect. The large positive $G \times E$ will strongly enlarge the yield of the sensitive genotypes at 15.7°C and 18.0°C above their average across all the *T*. In contrast, multiplying the large positive multiplicative controls of 15.7°C and 18.0°C by the negative multiplicative controls due to the photoperiod insensitive indicates negative $G \times E$ interaction, which will reduce the yield in 15.7°C and 18.0°C of the sensitive genotypes below their average across all the *T*. The positive $G \times E$ at the two lower *T* accounts for the yields of the photoperiod sensitive genotypes being the highest of all the ten genotypes at 15.7°C and 18.0°C, while their negative $G \times E$ at 23.1°C and at 25.4°C accounts for their yields being the lowest of the ten genotypes at these higher *T*. The measured yields of the photoperiod sensitive genotypes (Fig. 9.2A) agree with these $G \times E$ interaction effects on the yield.

For the insensitive genotypes, multiplying their negative multiplicative control at 23.1°C or their yet larger negative multiplicative control at 25.4°C indicates a large positive $G \times E$ interaction effect on the yield, which will raise the yield above the average for the genotype across all the *T*. Multiplying their negative multiplicative control times the positive multiplicative due to the 15.7°C and 18.0°C indicates negative $G \times E$, which will lower their yield below their average. The measured yields (Fig. 9.2A) of the insensitive genotypes agree with these $G \times E$ interaction effects on the yield.

The AMMI analysis result shows both the lowest of all average yields (additive controls) and the nearest to zero of all multiplicative controls to occur at 28.3°C. The two control levels indicate that at 28.3°C the yield will always be close to zero for all ten of the genotypes. The control levels for 28.3°C were obtained by using yields measured for three of the ten cultivars in 1984, calculating missing plot values for the other seven, and then applying the AMMI analysis to the yield (Fig. 9.2B) and the yield day^{-1} (Fig. 9.3B).

9.2.2 Control by temperature over rate of yield accumulation

Yield day^{-1} to harvest maturity (Fig. 9.3A) of the ten bean genotypes compared with the AMMI analysis result (Fig. 9.3B) indicates the following. (i) At 15.7°C and 18.0°C the photoperiod sensitive genotypes (8, 9 and 6) have higher average yield day^{-1} (Fig. 9.3A) than the insensitive genotypes (1–4), because the $G \times E$ interaction is positive for the sensitive genotypes, but it is negative for the insensitive genotypes (Fig. 9.4B). In contrast, (ii) at 23.1°C and 25.4°C the insensitive genotypes accumulate higher yield day^{-1} because of positive $G \times E$, while the sensitive genotypes interact negatively. (iii) Genotype 10 has yield day^{-1} at all T intermediate between rates of the photoperiod sensitive and insensitive genotypes. Due to low $G \times E$ interaction within every environment, yield of genotype 10 approaches the higher yields of the insensitive genotypes at 23.1°C and 25.4°C, while also being high like that of the photoperiod sensitive genotypes at 18.0°C and 15.7°C.

9.2.3 Need for complete yield system analysis

As the yields and rates of yield day^{-1} were analysed and interpreted, the following was realized. If the biomass had been measured also, the HI (partitioning) and biomass day^{-1} could have been calculated. These would complete the YSA described in Table 9.1.

The first complete YSA applied to bean yield trials and published after peer-review were in Wallace *et al.* (1993a). The correlations among the traits measured within those YSA (Table 9.2) reinforced evidence (Tables 4.3 and 4.4) of the system-established negative or positive correlation among the levels of the three major, genetically directed, and environmentally modulated physiological components of the process of accumulating yield. The average correlation was 0.01 between yield of the genotypes and their DTHM. This $r = 0.01$ was the average of a correlation across all cultivars within each of 51 bean yield trials. The correlations within the 51 trials ranged from 0.70 to −0.74.

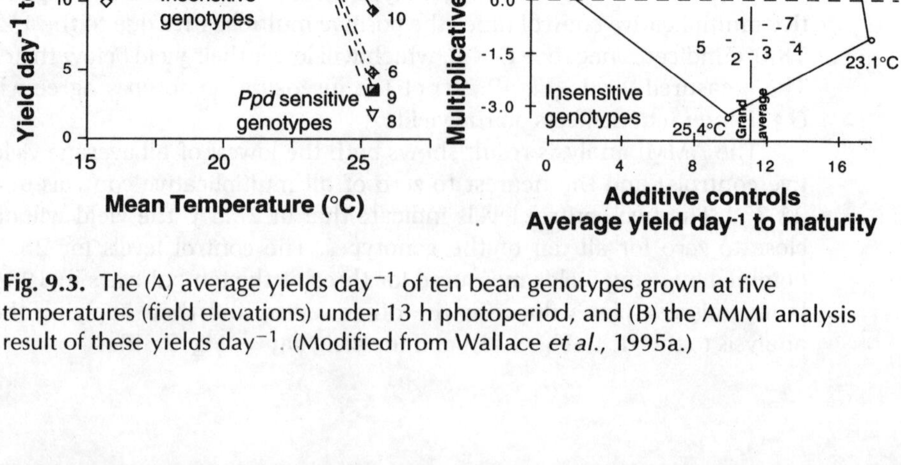

Fig. 9.3. The (A) average yields day^{-1} of ten bean genotypes grown at five temperatures (field elevations) under 13 h photoperiod, and (B) the AMMI analysis result of these yields day^{-1}. (Modified from Wallace *et al.*, 1995a.)

Table 9.2. The average correlation between all possible pairs among nine output traits from yield system analyses measured across 20 to 51 yield trials. The yield trials were conducted in 12 different US states, two Canadian provinces, and Puerto Rico, Peru, Colombia and Belgium. (Modified from Wallace *et al.*, 1993a.)

	Days used for development to indicated stage or by that stage			Rate of accumulation of the indicated biomass			Biomass and its partitioning to vegetative versus reproductive growth	
	Days to flower	Days to maturity	Days of seedfill	Yield day^{-1}	Seedfill day^{-1}	Biomass day^{-1}	Biomass	Harvest index
Yield	0.07	0.01	−0.01	0.93	0.84	0.72	0.700	0.54
Probability that the number of positive and negative correlations is equal across yield system analysis of many yield trials	0.37 IND	0.48 IND	0.578	Zero	Zero	Zero	0.000	0.0001
Days to flowering		0.66	−0.11	−0.01	0.070	0.11	0.300	−0.17
Probability of equal number of positive and negative correlations as determined by χ^2 analysis		0.00 IND	0.095	1.00	0.194	0.179	0.002 IND	0.016 IND*
Days to maturity			0.57	0.27	−0.16	−0.07	0.340	−0.45
Probability of equal number of positive and negative correlations			0.0002	0.0001 *	0.131	0.157	0.002 IND	0.0001 IND**
Days of seedfill				−0.14	−0.3	−0.19	0.040	−0 23
Probability of equal number of positive and negative correlations				0.336	0.001	0.398	0.655 IND	0.025 IND**
Yield day^{-1} of plant growth					0.92	0.76	0.540	0 67
Probability of equal number of positive and negative correlations					0.0001	0.0001	0.0001 IND	0.0001 IND**

(*Continued*)

Table 9.2. *Continued*

	Days used for development to indicated stage or by that stage			Rate of accumulation of the indicated biomass			Biomass and its partitioning to vegetative versus reproductive growth	
	Days to flower	Days to maturity	Days of seedfill	Yield day^{-1}	Seedfill day^{-1}	Biomass day^{-1}	Biomass	Harvest index
Yield day^{-1} of seedfill						0.730	0.59	0.69
Probability of equal number of positive and negative correlations						0.0001	0.0001 IND	0.0001 **
Biomass day^{-1}							0.86	0.21
Probability of equal number of positive and negative correlations							0.0001	0.003
Total biomass								−0.05
Probability of equal number of positive and negative correlations								0.602
Harvest index								

IND = Measurements of the correlated traits are done independently so calculation does not arithmetically introduce auto-correlation; all other correlations include, some auto-correlation. For interpreting the biological cause of the correlations, cells with * or ** are the most informative. See Brand, 1966 and Kacser *et al.* 1995 for a similar whole-system approach to quantification of the system-established interconnected control over complex biochemical pathways.

9.2.4 *The three component continuum of the accumulation of yield*

Chapters 4 and 5, Tables 4.3, 4.4 and 9.2, and the 0.70 to −0.74 range of correlations of the previous paragraph show the following for bean. The levels of all three major components of accumulating yield – and the rate of accumulation of the biomass and yield – are all interconnected. The system-established interconnections are due to the competitive partitioning of differential proportions of the photosynthate toward growth and development of the organs that become the yield – versus – partitioning to continuation of growth and development of organs that do not become part of the yield. High yield can arise through either extreme of the continuum of this partitioning, i.e. through a high HI in association with high yield day^{-1} and a short DTHM (Figs 9.1, 5.1 and 2.1), or through a low HI in association with small yield day^{-1} and longer DTHM.

9.2.5 *AMMI analyses of days to flowering of bean*

9.2.5.1 Relation of days to flowering to yield and rate of yield accumulation

For the field trials of bean for which yield and rate of yield accumulation are analysed in sections 9.2.1 and 9.2.2 above, AMMI analysis of their already discussed DTF (Fig. 5.3) is presented next. For comparison of the DTF of bean directly with its AMMI analysis result (Fig. 9.4B), the U-shaped DTF curves from 1983 are repeated in Fig. 9.4A. Of the ten genotypes, the most photoperiod sensitive ones 8, 6 and 9 flowered latest at 28.3°C, where every genotype had near zero yields (Fig. 9.2A). The DTF of genotypes 8, 6 and 9 are among the earliest at only 14.7°C and 14.9°C, at which they also gave the highest of all yields (Fig. 9.2) and the highest of all rates of yield accumulation (Fig. 9.3).

In agreement with genotypes 5 and 7 having earlier DTF than the other genotypes at all but the lowest 15.7°C, the AMMI analysis result shows their

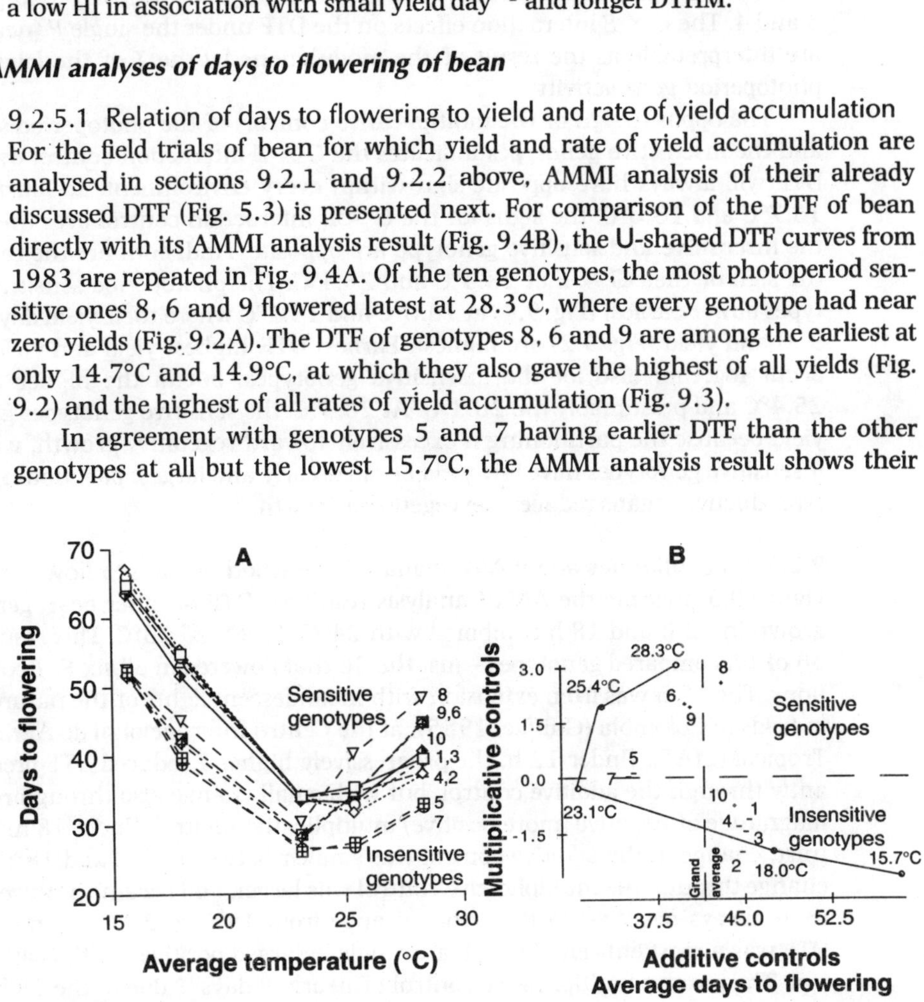

Fig. 9.4. The (A) days to flower of ten bean genotypes grown at five temperatures (field elevations) under 13 h photoperiod, and (B) the AMMI analysis result of these days to flowering. (Modified from Wallace *et al.*, 1995a.)

additive controls to be lower than those (the average DTF) of all eight other genotypes (compare Figs 9.4A and 9.4B). Their multiplicative controls are also closest to zero, indicating that across 15.7–28.3° they have the smallest *G* × *E* interaction control of all the genotypes over their expressed DTF. Genotype 10 has the next smallest multiplicative control and therefore, a relatively small change of its DTF through the *G* × *E* interaction. The small multiplicative control and therefore small *G* × *E* reflects genotype 10 having intermediate DTF at 28.3°C and 25.4°C and also at the lower *T* of 18.0°C and 15.7°C (Fig. 9.4A). The photoperiod sensitive genotypes (8, 6 and 9) cause large positive multiplicative control at 28.3°C and 25.4°C, which accounts for their latest DTF at 25.4°C and 28.3°C, but early DTF at both 18.0°C and 15.7°C. The 18.0°C and 15.7°C temperatures cause large negative multiplicative control as do the less sensitive genotypes 1, 2, 3 and 4. The *G* × *E* interaction effects on the DTF under the single *P* (near 13 h) are interpretable as the result of the modulations by the *T* of the level of the photoperiod-gene activity.

The opposite sign of the multiplicative controls of the photoperiod sensitive and the insensitive genotypes indicates the *G* × *E* interaction control over their DTF will always have opposite sign within every environment. To illustrate, at 15.7°C and 18.0°C, the signs for the *G* × *E* interaction control over the DTF of the insensitive and sensitive genotype are opposite. And, both are the reverse of the sign of their *G* × *E* at 28.3°C and 25.4°C. The photoperiod sensitive genotypes flower earliest (Fig. 9.4) at 18.0°C and 15.7°C where their yield day^{-1} (Fig. 9.3) and yield (Fig. 9.2) are highest. Earliest DTF, highest yield and yield day^{-1} occur together also for the insensitive genotypes, but at the higher 23.1°C, 25.4°C and presumably the 28.3°C. At 28.3°C, the sensitive genotypes gave low yield because the partitioning is excessively toward vegetative growth, while the insensitive genotypes have low yield because early and largest partitioning to the reproductive organs reduces the vegetative growth.

9.2.5.2 Peculiarities about AMMI analysis revealed by days to flowering

Figure 9.5 presents the AMMI analysis result for DTF of other bean genotypes grown in 12 h and 18 h combined with 24°C, 18°C and 13°C. This result is for 36 of 68 compared genotypes – just the 36 that flowered in all six *P*–*T* combinations. The 18 h was from extension with incandescent lights of the natural 12 h in fields in Colombia (Gniffke, 1985), at the Centro Internacional de Agricultura Tropical (CIAT). Under 12 h, the progressively higher *T* reduced DTF predominantly through the additive control, but with small change also through reduced magnitude of negative (more positive) multiplicative control. Under 18 h, in contrast, change in the additive control was smaller between 13°C and 18°C, while change through the multiplicative control was larger, and became positive rather than always negative. With further change from 18°C to 24°C, change of the DTF was almost entirely through a strongly enlarged positive multiplicative control. The positive multiplicative control of nearly 8 days$^{0.5}$ due to the 18 h/24°C multiplied by the also positive multiplicative control of 2 to near 4 days$^{0.5}$ of the photoperiod sensitive genotypes indicates a *G* × *E* interaction caused delay of their DTF in the range of 14–28 days. Contrarily, the positive multiplicative control by 18 h/24°C multiplied by the negative multiplicative control of about

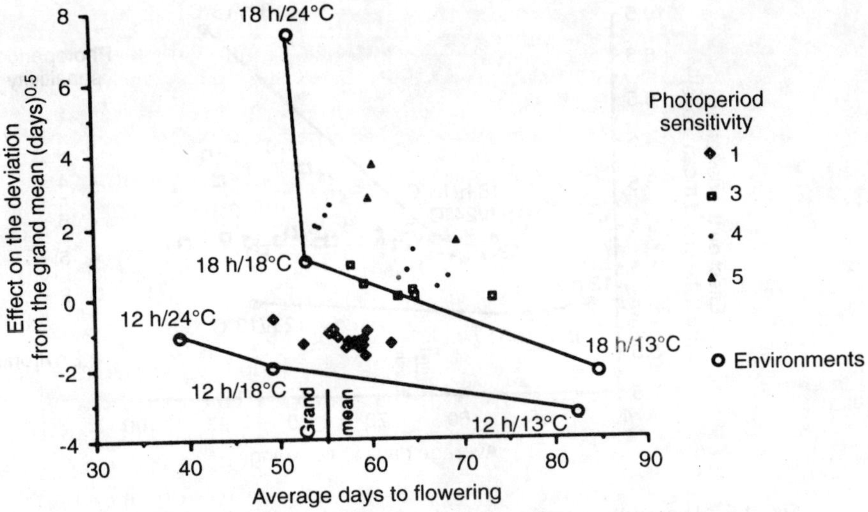

Fig. 9.5. The multiplicative controls plotted against the additive controls for the largest genotype × environment interaction effect on the days to flowering of the 36 of 68 genotypes that flowered in all of the six specified photoperiod–temperature combinations. (Reprinted from Wallace *et al.*, 1991.)

-1 days$^{0.5}$ for the insensitive genotypes indicates $G \times E$ interaction that reduces their DTF by about half of the 14–28 days. The photoperiod sensitivity scores presented for the 36 genotypes in Fig. 9.5, and those for the ten genotypes in Fig. 5.3 – the latter being those of the genotypes in Figs 9.2, 9.3 and 9.4 – were all assigned by CIAT before the AMMI analysis.

Another AMMI result is for the largest $G \times E$ interaction control over the DTF determined across all 68 genotypes of bean, rather than just the 36 of 68 that flowered in all six *P–T* combinations. For the 68 (Fig. 9.6) compared with 36 genotypes (Fig. 9.5), the slopes of lines that connect the *P–T* combinations are reversed on the biplot graph. Reversed slopes occur because the AMMI analysis of all 68 (Fig. 9.6) includes the larger average DTF (additive controls larger than 70 days) and larger positive multiplicative controls (larger than 4 days$^{0.5}$) due to some but not all of the genotypes classified by CIAT as being highly photoperiod sensitive with a score of 8. A much larger multiplicative control of 10.5 days$^{0.5}$ is quantified for the 18 h/13°C environment. The reversed slopes and different additive and multiplicative controls provide further insight about AMMI analysis results. Each AMMI-quantified multiplicative and additive control is quantitatively correct for just the specific subset of compared genotypes grown in the specific subset of the compared *P–T* combinations. The quantified multiplicative control levels due to the same genotypes (or due to the same environment) will likely be positioned similarly relative to their negative to positive level in another YSA/AMMI. But, the exact value of both the multiplicative control and the additive control will change – if either any of the genotype(s) is dropped or previously non-compared ones are added, or if any *P–T* combination differs. The values will

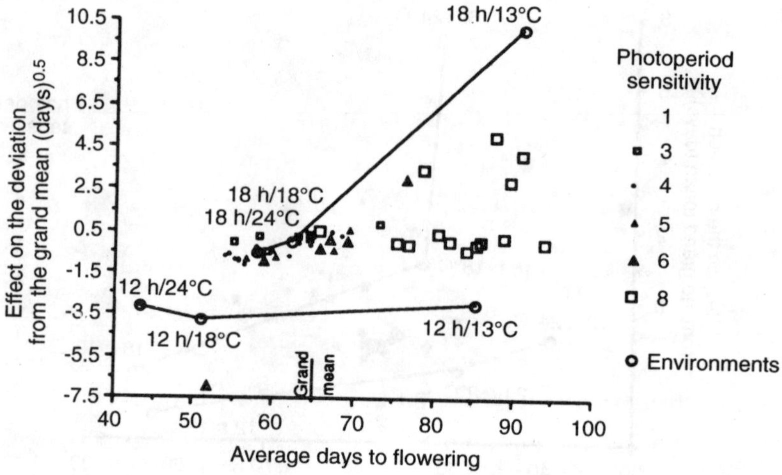

Fig. 9.6. The multiplicative controls plotted against the additive controls for the largest genotype × environment interaction effect on the days to flowering of 68 genotypes grown within six specified photoperiod–temperature combinations. (Reprinted from Wallace *et al.*, 1991.)

almost always change for field trials, because it is highly unlikely for any environment to be exactly repeated.

A second insight arises through recognition of the extensive and more or less continuous range of insensitivity to sensitivity of the 68 compared genotypes. The AMMI result of Fig. 9.5 is for genotypes spanning sensitivity scores of 1 to 5, while that of Fig. 9.6 includes additional genotypes with higher sensitivity scores of 6 to 8. That some of these most photoperiod sensitive genotypes were much more delayed by 18 h at the lowest *T* of 13°C than others indicates much remains to be learned about the holistic range of the genotype × photoperiod × temperature (GPT) interaction controls over the DTF. This is a message also of Fig. 4.2. A conclusion in Chapters 7 and 8 is that the lowest *T* that allows any photoperiod-gene activity needs to be quantified. As shown by AMMI analysis of bean (Mmopi *et al.*, 1995, see Fig. 9.8), there are: (i) the photoperiod gene(s), (ii) at least one enhancer gene that enlarges the photoperiod-gene activity and consequent delay of the DTF; and (iii) at least one gene that causes early versus late DTF with no modulation of its activity by the *P*, but which does cause definite early versus late DTF and interacts with the photoperiod-gene activity to mask its control over the DTF. Finally (iv), there is control by the background genes. The level of all quantitative traits is interconnected by their competitive sharing of the available intraplant metabolites. Through this, the controlling background genes become the collective majority of all the genes of the plant.

9.3 Complete Yield System Analysis of a Multilocation Pigeonpea Yield Trial

The complete YSA for pigeonpea was independently planned, planted and its data collected by Y.S. Chauhan, C. Johansen and Laxman Singh at the ICRISAT at Patancheru, India. Wallace was invited to apply AMMI analysis to these YSA data. This is the only complete calendar-day-based YSA/AMMI analysis we know of that was not initiated by Wallace and summarized previously (Table 9.2), except a second for pigeonpea at ICRISAT (Prasad, 1996).

9.3.1 Materials and methods of a yield system analysis of pigeonpea

Multilocation yield trials were conducted on four soil-locations (Table 9.3). Two were at the ICRISAT, Patancheru, India (17°N and 72°E); one was at ICRISAT's Cooperative Research Station at Hisar (29°N and 75°E). The fourth was at Gwalior (26°N and 78°E). At Hisar, only super phosphate was applied at 125 kg ha^{-1}. On the other three, a basal dose of 100 kg ha^{-1} of diammonium phosphate was incorporated before sowing.

Table 9.3. Mean grain yield (t ha^{-1}), biomass (t ha^{-1}), and mean temperatures (°C) from sowing to flowering (MTFL) and sowing to maturity (DTHM), and rainfall (mm; RF) of pigeonpea grown in 20 environments.

Environment	Sowing	Soil	Yield	Biomass	MTFL	DTHM	RF
B1I	Jun	Vertisol	1.08	7.36	26.5	26.0	886
B1N	Jun	Vertisol	1.02	7.26	26.4	25.9	886
B2I	Oct	Vertisol	0.76	2.77	21.7	21.7	7
B2N	Oct	Vertisol	0.49	1.73	21.7	21.6	7
B3I	Jan	Vertisol	0.71	3.06	22.8	24.6	86
B3N	Jan	Vertisol	0.40	1.91	22.9	24.5	86
G1I	Jul	Inceptisol	1.90	8.45	28.5	27.2	735
G1N	Jul	Inceptisol	1.47	6.90	28.6	27.5	735
G2I	Aug	Inceptisol	1.64	5.95	28.4	25.3	447
G2N	Aug	Inceptisol	1.40	5.14	28.4	25.7	447
H1I	Jul	Entisol	1.39	17.01	29.6	28.4	613
H1N	Jul	Entisol	1.38	15.83	29.6	28.6	613
H2I	Aug	Entisol	1.60	12.26	29.0	25.5	378
H2N	Aug	Entisol	1.58	12.98	28.8	25.3	378
R1I	Jun	Alfisol	2.09	6.95	26.3	25.9	909
R1N	Jun	Alfisol	1.72	5.90	26.3	25.9	909
R2I	Oct	Alfisol	0.72	2.82	21.7	21.7	7
R2N	Oct	Alfisol	0.55	1.73	21.7	21.6	7
R3I	Jan	Alfisol	0.72	2.81	22.8	24.4	81
R3N	Jan	Alfisol	0.36	1.60	22.8	24.4	81
LSD (5%)			0.285	2.331			
CV %			30.8	28.9			

Environment: B = black soil (Vertisol), R = red soil (Alfisol), G = Gwalior (Inceptisol), H = Hisar (Entisol); I = irrigated, N = unirrigated; 1, 2, and 3 indicate the different sowing dates.

The sowing dates at Patancheru were 16 June 1988, 12 October 1988 and 4 January 1989 on its black soil of volcanic origin, and on 14 October 1988, 4 January 1989 and 22 June 1989 on its red soil of granite origin. At Hisar the sowings were 1 July 1988, and 11 August 1988. At Gwalior they were 1 July 1988 and 6 August 1988. For all ten sowing dates, a split plot design included irrigation as main plots, and genotypes as sub plots, with three replications. The ten planting dates × irrigation versus non-irrigation provided 20 environments.

Hereafter, the soils at Patancheru will be referenced as B (for black) and R (for red). The most highly fertile soil, at Hisar, will be referenced as H, and the soil at Gwalior as G (compare Table 9.3). Successive planting dates on each soil are numbered 1, 2, and 3. The soil/planting-dates are symbolized as B1, B2, B3, R1, R2, R3, H1, H2, G1 and G2. Irrigation versus non-irrigation is stated as, for example, irrigated B1 or non-irrigated soil/planting-dates R3 and H2. Disease and insect damage were minimized, to maximize focus on the abiotic factors of the environment.

Twenty determinate and ten indeterminate pigeonpea genotypes (Table 9.4) were sown at 30 cm row and 10 cm plant spacing. Two seeds were planted and thinned to one. The plots consisted of four 4-m rows. The determinate and indeterminate genotypes were separated by a buffer plot of a determinate genotype ICPL 87.

Days to 50% flowering (DTF) and to 80% grain maturity were recorded. The latter, the DTHM was recorded as when approximately 80% of the pods had turned completely brown and those remaining were fast approaching browning. The central two rows (2.4 m^2) were harvested to determine the oven-dry biomass and grain yield. The HI and days of seedfill, and the biomass and yield day^{-1} to harvest maturity, and yield day^{-1} of seedfill were calculated.

9.3.2 *The holistic genetic, environmental and* G × E *interaction controls*

The ANOVA procedure within the AMMI analysis quantified the total sums of squares across the 600 treatments and partitioned it among the sums of squares across: (i) all 30 genotypes; (ii) all 20 environments; and (iii) due to the total $G \times E$ interaction. Respectively, (i) to (iii) quantify the relative *control strengths* due to differences among the genotypes, among the environments and due to the total $G \times E$ interaction. On each biplot graph presenting the AMMI analysis result for a trait measured by the YSA (Fig. 9.7), these three control strengths are presented as the percent each accounts for of the total sums of squares (variation) across all 600 treatments. The AMMI result for each of the nine YSA traits, and for the GDDs to flowering, are all presented adjacently within Fig. 9.7. The adjacent presentations facilitate visualization of the system-established interconnections among the expressed levels of all the traits quantified by the YSA.

The ANOVA followed by PCA within the AMMI analysis quantified: (i) *the additive control* over the phenotypic level of the trait that is caused by each of the 30 compared genotypes; and (ii) *the additive control* that is caused by each of the 20 compared environments. The PCA quantified also: (i) *the multiplicative control* over the trait due to each of the genotypes; and (ii) *the multiplicative control* due to

Table 9.4. Mean accumulated °C days to flowering (TFL) and °C days to maturity (Tmat), days to 50% flowering (DTF) and days to grain maturity (Mat), grain yield (t ha^{-1}), biomass (t ha^{-1}), and mean temperatures (°C) from sowing to flowering (MTFL) and sowing to grain maturity (Mtmat) of 20 determinate (DT) and 10 indeterminate (NDT) pigeonpea genotypes grown in 20 environments.

Genotype	Code	Habit	TFL	Tmat	DTF	Mat	Yield	Biomass	MTFL	Mtmat
ICPL 4	1	DT	1032	1651	66	108	1.09	5.21	25.6	25.3
ICPL 87	2	DT	1128	1810	72	122	1.23	7.44	25.7	24.8
ICPL 151	3	DT	1060	1722	68	114	1.18	6.16	25.6	25.1
ICPL 83006	4	DT	1018	1694	65	111	1.17	6.26	25.6	25.2
ICPL 83015	5	DT	996	1630	64	107	1.07	5.24	25.5	25.3
ICPL 83019	6	DT	950	1624	61	107	0.96	5.13	25.6	25.2
ICPL 84023	7	DT	945	1570	61	103	1.08	4.97	25.5	25.3
ICPL 85010	8	DT	1004	1661	65	109	1.02	5.13	25.6	25.2
ICPL 85014	9	DT	1024	1681	66	110	1.15	6.24	25.6	25.2
ICPL 85024	10	DT	934	1566	60	102	0.75	3.68	25.5	25.3
ICPL 86005	11	DT	1139	1829	73	124	1.19	8.44	25.7	24.7
ICPL 86009	12	DT	1048	1730	67	115	1.21	5.89	25.6	25.1
ICPL 86010	13	DT	1118	1799	71	120	1.02	5.78	25.7	25.0
ICPL 87095	14	DT	942	1609	60	105	0.93	4.77	25.6	25.3
ICPL 87097	15	DT	1000	1654	64	109	1.08	5.53	25.6	25.1
ICPL 87098	16	DT	1066	1727	69	114	1.13	6.26	25.5	25.2
ICPL 87101	17	DT	1120	1820	71	123	1.24	7.37	25.7	24.8
ICPL 87104	18	DT	1090	1800	70	120	1.35	7.32	25.7	25.0
ICPL 87105	19	DT	1159	1858	74	126	1.18	8.24	25.7	24.7
ICPH 9	20	DT	1068	1736	68	116	1.37	7.23	25.7	25.0
UPAS 120	21	IDT	1101	1763	70	117	1.30	7.25	25.7	25.1
MANAK	22	IDT	1122	1796	72	120	1.31	7.63	25.7	25.0
ICPL 86019	23	IDT	1065	1713	68	113	1.20	6.26	25.7	25.2
ICPL 86020	24	IDT	1050	1740	67	115	1.03	6.93	25.6	25.2
ICPL 87110	25	IDT	1091	1717	70	114	0.98	5.71	25.7	25.1
ICPL 87111	26	IDT	1037	1693	67	111	0.93	5.01	25.5	25.3
ICPL 87113	27	IDT	1116	1795	71	120	1.27	7.74	25.7	24.9
ICPL 87117	28	IDT	1134	1773	73	117	1.15	6.87	25.6	25.1
ICPH 8	29	IDT	1233	1844	78	125	1.53	9.49	25.7	24.8
ICPH 149	30	IDT	1299	1892	82	130	1.39	10.44	25.8	24.5
LSD (5%)			15.29	16.85	0.8	1.4	0.127	0.668		
Mean			1070	1730	68	115	1.15	6.48		
CV%			4.0	2.7	3.7	3.5	30.8	28.9		

each of the environments. The multiplicative controls account for the genotype's or environment's control over the expressed level of the trait through the AMMI-differentiated $G \times E$ source (the IPCA1 or IPCA2 or . . . IPCAN) within the total $G \times E$ interaction effect (sums of squares).

In addition to quantifying and interpreting the $G \times E$ interaction control over the trait levels, another objective of YSA/AMMI is to quantify the system-established interconnections between the levels of the traits. The interconnectivity is measured as the correlation between the levels of each pair among the nine

YSA-measured traits. These correlations will quantify (as examples of the interconnections) the system-established reduction in level of the HI: (i) as the biomass is enlarged, or the (ii) DTF and (iii) DTHM are enlarged; and (iv) as the biomass day^{-1} is enlarged but the yield day^{-1} decreases (Table 9.2). Quantifying the system-established interconnected controls over (correlations between) the levels of the major components of the process of accumulating crop yield is the objective of the YSA *per se*. The AMMI analysis enlarges understanding of these system-established interconnections by facilitating interpretation of the control through the $G \times E$ interaction.

9.3.3 Interpretation of the YSA/AMMI analysis result of pigeonpea

9.3.3.1 Relative main and $G \times E$ interaction controls

The percentages of the total control over the phenotypic level represented by the sums of squares: (i) across all 30 genotypes, (ii) across all 20 environments, and (iii) across the total $G \times E$ interaction are presented on the biplot graph for yield, and the graph for each of the physiological components of accumulating the yield (Fig. 9.7). The control strengths of 1, 2 and 3 are all highly significant ($P < 0.001$) for all traits.

9.3.3.2 Procedural steps for interpretation of the $G \times E$ interaction

The PCA within the AMMI analysis differentiated as few as three but as many as seven statistically significant ($P < 0.001$) patterns (sources) of $G \times E$ interaction control over the yield, and over the $G \times E$ control of each physiological component of accumulating the yield. The two-way classification of the AMMI analysis designated total control over each trait as being by only (i) the holistic genotype and (ii) the holistic environment (as illustrated in Fig. 1.1). Therefore, each successively smaller $G \times E$ source is interpretable as representing: (i) a specifiable gene action (or a subset of gene actions) that implements a physiological process, and (ii) the modulations of that gene action and/or physiological process by differential levels of specifiable factor(s) within the holistic environment.

Neither any gene action, physiological process, nor modulating environmental factor is specified within the two-way classification of control for the AMMI analysis. To interpret the gene action(s) and environmental modulations causal of a $G \times E$ interaction pattern, the researcher must compare the possibilities as described next.

The AMMI-quantified multiplicative controls – one due to each genotype and one due to each environment – are plotted (Fig. 9.7) along the vertical axis of the biplot graph against, along the horizontal axis, the additive control due to the same genotype, or the same environment. To better differentiate among the points of the 30 genotypes and the 20 environments, by minimizing overlapping among their 50 points, the AMMI result for each of the 30 genotypes is plotted above, and the result for each of the 20 environments is plotted below against duplicate copies of the biplot axes and their scale (Fig. 9.7). Duplicate axes were not needed for the AMMI analyses presented in section 9.2 and Figs 9.2–9.4 above for bean, because (i) their but 15 points were spaced far enough apart that

(ii) their relevant genetic or environmental factors were easily labelled without overlapping.

This chapter will interpret only the largest statistically significant physiological–genetic source (IPCA1) of the $G \times E$ interaction control over each YSA-measured trait. On the biplot graph for each trait, the percentage of the total $G \times E$ interaction sums of squares represented by this largest source is listed, below the percent of the total treatment sums of squares accounted for by the total $G \times E$ interaction.

The control over each trait through the largest $G \times E$ interaction source will be interpreted as illustrated next for the phenological traits. For all traits, a $G \times E$ interaction source can be interpreted using physiological–genetic terms, but only after the researcher has labelled: (i) all points for the genotypes with their identity and relevant known traits, and (ii) all environmental points with their relevant variable factors – their *P* and *T*, irrigation versus non-irrigation, and their soil.

This is the first pigeonpea YSA/AMMI analysis of pigeonpea ever conducted. *For this reason, the starting point for interpretation of the largest* G × E *interaction source of the pigeonpea data will be the prior interpretation of YSA/AMMI results presented for the DTF of bean in section 9.2 above.* The photoperiod sensitivity versus insensitivity of the pigeonpea genotypes will be derived, based on responses of their DTF that essentially duplicate those for the DTF of bean with insensitivity versus sensitivity to *P–T* combinations determined before and independently of the YSA/AMMI analysis.

9.3.3.3 Phenological traits

The expressed DTF of bean was interpreted, in section 9.2, as resulting from control through a statistically quantified GPT interaction. The GPT interaction was interpreted as control due to only any photoperiod-gene activity and its delay of development. Control through the photoperiod response is a correct but incomplete interpretation. It did not incorporate the full insight available from the two-way classification for the AMMI analysis of all control being by (i) the holistic genotype interacting with (ii) the holistic environment. The interpretation was that the largest source of the $G \times E$ interaction involves modulation of any photoperiod-gene activity by both the *P* and the *T*. But, the holistic control by *T* has to incorporate also that the DTF is controlled simultaneously through the thermal response. Control by *T* over both the photoperiod and the thermal response was detailed in Chapters 6 and 7, and see Squire (1990), Hodges (1991) and Jordan (1993).

About half of the interpreting of the $G \times E$ interaction control over yield of pigeonpea and over its major physiological components will deal with the phenological traits, the DTF and the DTHM, i.e. with the usage of time. A forced reason for major emphasis on the DTF is that this is the first YSA/AMMI analysis of pigeonpea ever conducted. The photoperiod sensitivity versus insensitivity of the pigeonpea genotypes was not known in advance; it will be derived based on responses of DTF of the genotypes to the *P–T* combinations, and interpreted based on responses reported for bean above in section 9.2. Biological necessity for the YSA/AMMI giving great attention to the DTF and DTHM follows. Richards (1991) stated: 'Crop phenology is the most important single factor influencing

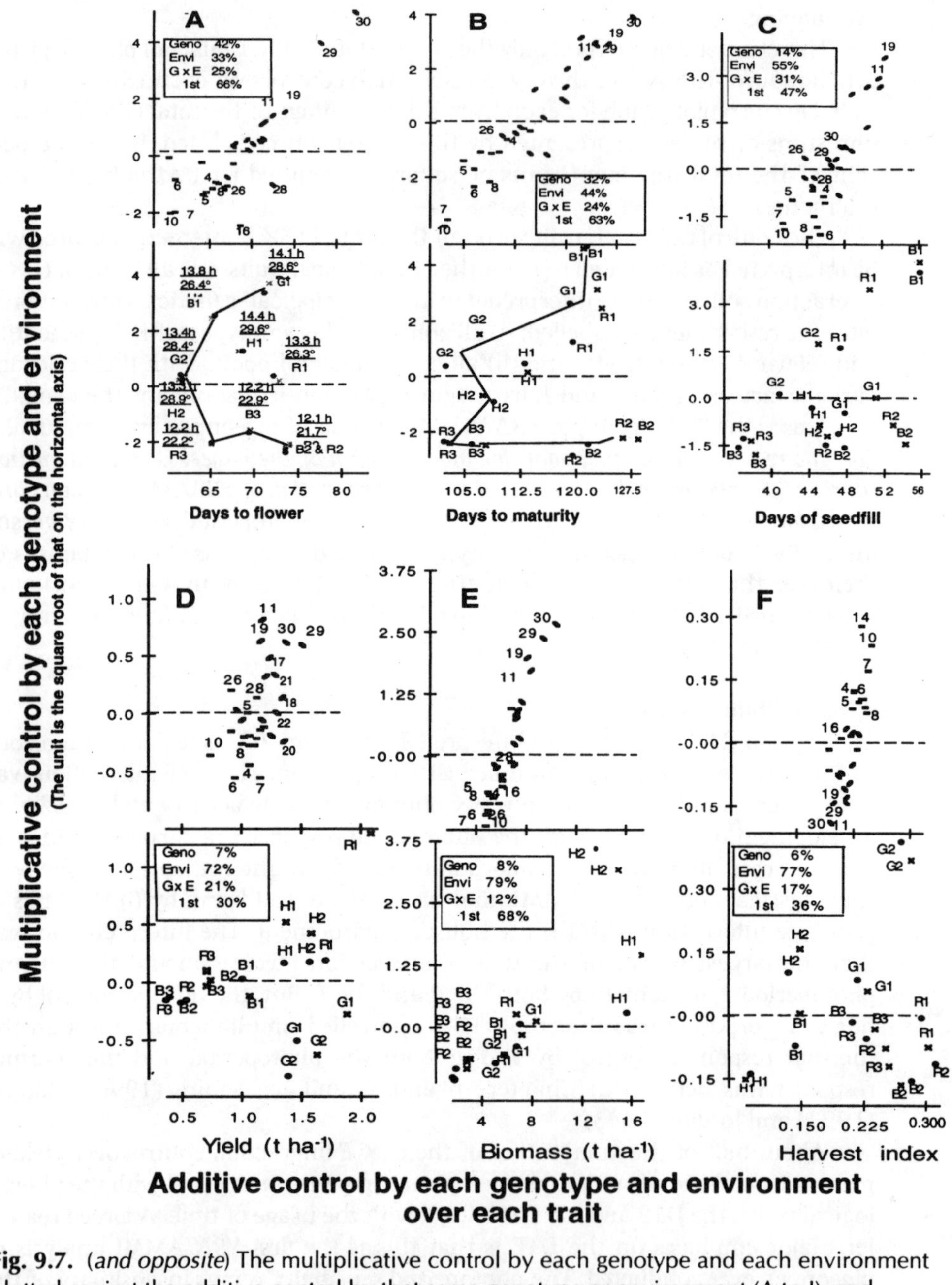

Fig. 9.7. (*and opposite*) The multiplicative control by each genotype and each environment plotted against the additive control by that same genotype or environment (A to J), and relationships between biomass and harvest index (K) and days to harvest maturity and harvest index (L). In each biplot graph, the insensitive determinate genotypes are represented by horizontal dashes (–); sensitive determinate genotypes by spindles slanted to the right (⟍); indeterminate genotypes by spindles slanted to the left (⟋); non-irrigated environments by large solid dots; and irrigated environments by crosses (x).

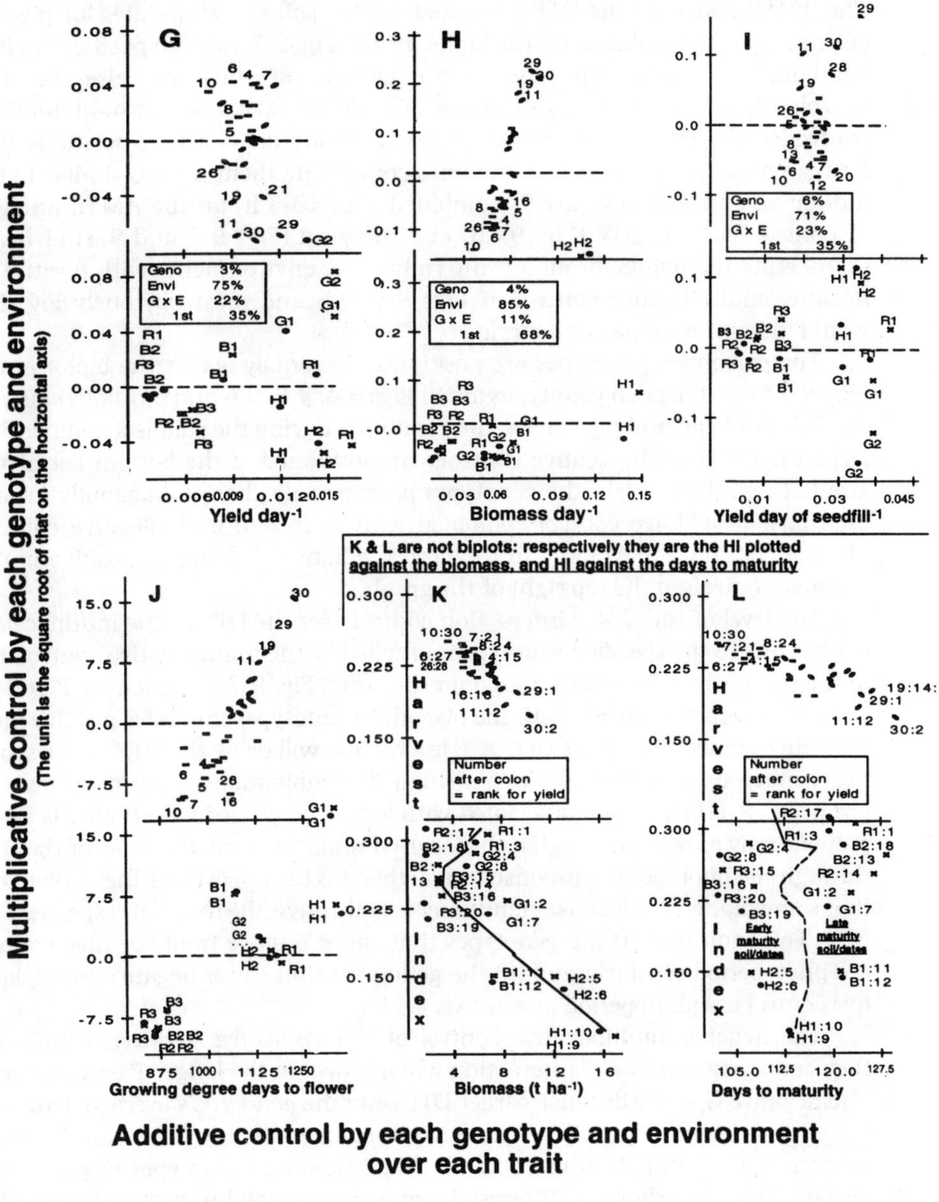
Multiplicative control by each genotype and environment
(The unit is the square root of that on the horizontal axis)
G
H
I
Geno 3%
Envl 75%
G x E 22%
1st 35%
Geno 4%
Envl 85%
G x E 11%
1st 68%
Geno 6%
Envl 71%
G x E 23%
1st 35%
Yield day-1
Biomass day-1
Yield day of seedfill-1
K & L are not biplots: respectively they are the HI plotted against the biomass, and HI against the days to maturity
J
K
L
Harvest Index
Number after colon = rank for yield
Early maturity soil/dates
Late maturity soil/dates
Growing degree days to flower
Biomass (t ha-1)
Days to maturity
Additive control by each genotype and environment over each trait

yield and adaptation.' This pigeonpea trial will show the controls over the DTF to establish system-interconnected controls over the levels of every trait measured by the YSA.

AMMI QUANTIFIED PHYSIOLOGICAL GENETIC CONTROLS OVER DAYS TO FLOWERING. The AMMI result for the DTF separated the ten soil/planting-dates for pigeonpea among eight positions on the biplot graph (Fig. 9.7A). Each position included both the irrigation and non-irrigation of one or else of two soil/planting-dates. The irrigation was after the flowering, so it would not have controlled the DTF. Eight of the ten soil/planting-dates are at positions that form a half-circle, as shown by the lines connecting them. This is similar to the half circle positioning of five *T* (combined with 13.3 h) for the AMMI analysis of control over the DTF (Fig. 9.4B) and the yield (Figs 9.2 and 9.3) of bean. Connecting the points for all but the right-most environments with positive or negative multiplicative control of Figs 9.5, 9.6 and most obviously Fig. 9.8 would indicate a similar half circle.

The pigeonpea genotypes are positioned diagonally across the biplot graph (Fig. 9.7A), as the bean genotypes are in Figs 9.4, 9.5, 9.6 and most obviously in Fig. 9.8. For bean and pigeonpea, the genotypes having the smallest additive and largest negative multiplicative controls are positioned at the bottom left of the graph. From these levels, the genotypes progressively change diagonally toward intermediate additive control combined with near zero multiplicative control. Change continues diagonally to the largest additive and largest positive multiplicative controls at the top right of the graph.

The level of the $G \times E$ interaction control over the DTF is: the mutiplicative control due to the specified genotype multiplied by the multiplicative control due to the specified environment. Based on this, from Fig. 9.7A, the longer *P* and/or higher *T* interact positively with the pigeonpea genotypes that have positive multiplicative control. The positive $G \times E$ interaction will delay the DTF of the sensitive genotypes within these long *P*/high *T* combinations. Any delay by the positive $G \times E$ interaction associated with long *P* combined with high *T* is interpretable only as being through a photoperiod response. And, the delay of the DTF through the photoperiod response by a higher *T* is the opposite of the earlier DTF that same higher *T* will cause simultaneously through the thermal response (Fig. 6.7). This indicates: (i) the genotypes that cause positive multiplicative control are photoperiod sensitive; and (ii) the genotypes that cause negative multiplicative control are photoperiod insensitive.

The negative multiplicative control of the insensitive genotypes indicates they cause negative $G \times E$ interaction when grown in the longer *P* and higher *T*. The negative $G \times E$ will cause earlier DTF than the genotype's average across all the environments (Fig. 9.7A). The environments with shortest *P* and lowest *T* cause negative multiplicative control as the insensitive genotypes of pigeonpea also do. Thus, the shortest-*P*/lowest-*T* combinations will interact positively with the photoperiod insensitive genotypes to delay their DTF, but will interact negatively with the photoperiod sensitive genotypes to decrease their DTF. Any delay of the DTF of the insensitive genotypes by the shortest (non-delaying *P*) and lowest of all *T* can be interpreted only as delay by the low *T*, i.e. as a thermal

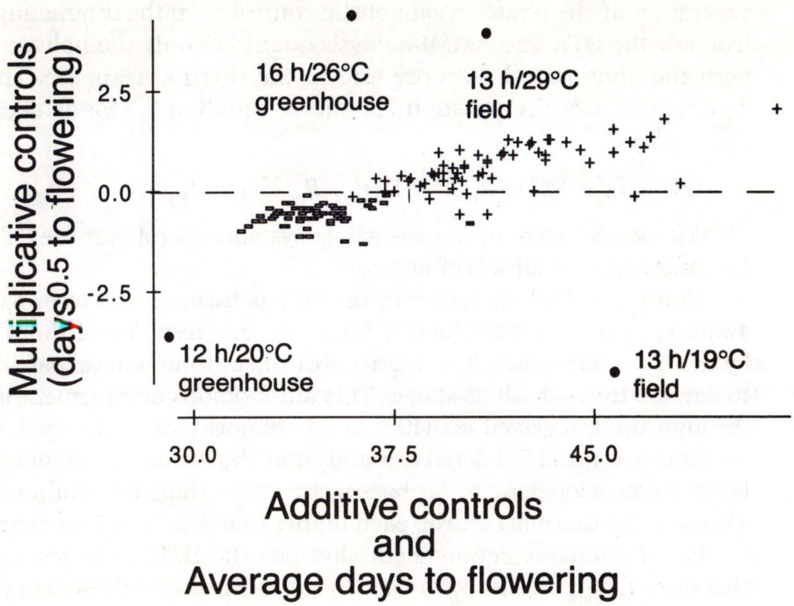

Fig. 9.8. The multiplicative controls through the genotype × environment interaction effect on the days to flowering (DTF) plotted against the additive controls for some F_8 progenies derived by single seed descent from a cross between a photoperiod sensitive and an insensitive bean cultivar. A bimodal distribution of the DTF like that given in Fig. 4.3 indicated control by a single photoperiod gene. Designation of each progeny as being photoperiod sensitive or else insensitive was based on its being within the early or the delayed mode of the bimodal distribution. Thus designated insensitive F_8 progenies are identified by the symbol (–) and the sensitive progenies by the symbol (+). (Reprinted from Mmopi *et al.*, 1995.)

response. All $G \times E$ effects on the DTF of pigeonpea (Fig. 9.7A) are just as they are for bean (Figs 9.4B and 9.8)

The smallest average DTF of the five T for bean combined with 13 h occurs at the intermediate 23.1°C (Fig. 9.4). Similarly, for pigeonpea the smallest average DTF of the ten P–T combinations occurs at the intermediate P of 13.3–13.4 h (Fig. 9.7A).

CONTROL OVER THE DAYS TO FLOWERING BY JUST TWO SUBSETS OF GENES. From sections 5.5, 6.3, and illustrated succinctly in Fig. 6.7, any observable optimum T has to be considered as only the apparent optimum T (T_{aopt}). This is because the T causal of the smallest observed DTF will be lowered by a more delaying P. This is shown for bean (Figs 4.1, 4.2, 5.3, 5.4, and 5.5), for pea (Fig. 6.5), and less clearly for sorghum (Fig. 6.3). The more delaying P is longer for SDP, but shorter for LDP. The photoperiod response and the thermal response are both quantified by the GPTmodel's simulation of the DTF, which uses the mathematical equation 6.3 based on the physiological–genetic control over the DTF derived in Chapters 4 and 5. This GPTmodel's equation is presented again below, because it facilitates separation of the photoperiod holistic control from the interacting thermal control over the DTF. The AMMI analysis quantifies only the holistic control due to both the photoperiod response and to the thermal response. The GPTmodel's equation follows. See equation 7.4 and/or equation 6.3 for detailed definitions of the symbols.

$$\mathrm{DTF} = D_b + S_t(T - T_{opt})^2 + S_p(P - P_{opt})^2(T - T_{bp})^2$$

This GPTmodel accounts for the whole-system control over the DTF being by just the following two subsets of genes.

Control-1. This control over the DTF is through the term $S_t(T - T_{opt})^2$. Its quantity of control over the DTF is the thermal response of the DTF. It is caused by the Q_{10}-acceleration by a higher T of the autonomous capability of the plant to develop through all its stages. This autonomous development is implemented through the integrated activities of the majority of all the genes of the plant. Sections 6.4.1 and 7.3 detail this and show the concept of an autonomous capability for development to be backed by other than the authors of this book. Through the thermal control, each higher than the base T allowing any advance of the autonomous development shortens the DTF. Each genotype establishes this base T, T_{opt} and its S_t to thereby establish the highest rate of its potential development (shortest).

Control-2. This control over the DTF is through the term $S_p(P - P_{opt})^2(T - T_{bp})^2$. Its DTF quantity arises through both (i) any deviation of the applied P from the true genotype-established optimum P $(P - P_{opt})$, multiplied by (ii) any deviation of the applied T above the lowest T (the T_{bp}) that allows any photoperiod gene activity. This is detailed in Chapters 6 and 7. Each genotype establishes its level of both its P_{opt} and T_{bp}. Their levels jointly establish the magnitude of the delays of the DTF of photoperiod sensitive genotypes of pigeonpea (Fig. 9.7A), bean (Figs 4.1, 4.2, 5.3, 5.5), pea (Fig. 6.5), sorghum (Fig. 6.3). All delays are all enlarged by both: (i) a more delaying P; and (ii) the Q_{10} enhancement of any photoperiod-gene activity due to each progressively higher than the T_{bp} that first allows this activity.

Interpretation of the control of the DTF of pigeonpea based on the AMMI analysis result (Fig. 9.7A) uses the always opposite and therefore additive changes of the DTF by the T through the thermal response (control-1) – versus – through the photoperiod response (control-2). Interpretation requires re-emphasizing the following additional consequences. First, all observable change of the DTF is due to only that part of the larger of the two opposite and additive changes of the DTF in response to the T that is not cancelled by the smaller. Second, a more delaying P enlarges delay of the DTF through only the photoperiod response. It does not alter the thermal response. Third, interpretation is aided by understanding of the *biological interaction* between the thermal response and the photoperiod response (Fig. 6.7) – as contrasted with – the statistically quantified $G \times E$ interaction.

Biologically, as shown in Fig. 6.7, the thermal and photoperiod responses interact as follows. Beginning from the lowest T, the reduction of the DTF by a higher T through the thermal response (i.e. through the autonomous development) is larger per degree centigrade than any delay of the DTF through the photoperiod gene activity and photoperiod response (compare Fig. 5.5). With each T progressively higher than the lowest T, however, the delay of DTF per degree centigrade through the photoperiod response becomes progressively larger, while the earlier DTF per degree centigrade through the thermal response becomes progressively smaller. This progressively increasing effect due to the photoperiod and thermal responses continues until these always opposite (and therefore additive) responses become equal. At this T, the temperature-modulated opposite changes of the DTF each cancel the other to give the smallest observable DTF. This is the apparent optimum T (the T_{aopt} in Fig. 6.7). All progressively higher T than the T_{aopt} continue to cause progressively larger delays of DTF per degree centigrade through the photoperiod-gene activity, while reduction of the DTF through acceleration of the autonomous development by the thermal response becomes progressively smaller per degree centigrade. Every observed optimum T for development to flowering must be interpreted as only a T_{aopt}, because it will be made smaller by either a more delaying P or a more photoperiod sensitive genotype (Fig. 4.2).

Being only a T_{aopt} is also shown statistically by the AMMI analysis of the DTF for both bean and pigeonpea. It is indicated by the smallest average DTF of all the compared P–T combinations. The average DTF of an environment is simply the average DTF of all the genotypes grown within it. From the AMMI results, this smallest average DTF (the average T_{aopt} for the environment measured across all the genotypes) occurs at an intermediate P–T combination for pigeonpea (Fig. 9.7A), and at an intermediate T under the single P (near 13 h) for bean (Fig. 9.4B). The biological GPT interaction (Fig. 1.1) totally determines the DTF of each genotype within each P–T combination (Chapter 1, paradigm shift 27). The average expressed DTF across all the genotypes within each P–T combination becomes the average DTF of that environment.

The following applies whether an optimum T is indicated by the *biologically expressed* DTF of one genotype grown across multiple T (Figs 4.1, 4.2, 5.3 and 5.5 for bean, Fig. 6.5 for pea, and Fig. 6.3 for sorghum), or is indicated by the AMMI analysis result across multiple P–T combinations and genotypes for bean (Fig. 9.6) and pigeonpea (9.7A). Any observed optimal T will only be the true genotype-established T_{opt} for development of the genotype – if and only if simultaneously – the applied P is also the true genotype-established optimal photoperiod (P_{opt}). This is because only the P_{opt} will prevent all photoperiod gene activity. Experimental comparisons must be conducted of factorial combinations of a range of P that spans the P_{opt} and a range of T that spans the T_{opt} to prove that both optima did occur simultaneously.

Above, the photoperiod response has been shown to interact in the *biological sense* with the thermal response to jointly strongly control the expressed DTF. In the following sections, comparison of the GDDs with the DTF and the AMMI-quantified $G \times E$ interaction controls over both will show that: the ever-present control by the thermal response and any photoperiod response strongly if not fully controls the DTF of all genotypes – provided the genotypes (i) possess no vernalization genes or (ii) do not have any vernalization requirement that has not been completed in full.

STATISTICALLY QUANTIFIED COMPARED WITH BIOLOGICALLY INTERPRETED $G \times E$ INTERACTION. From the *biological viewpoint* of the $G \times E$ interaction, every phenotype arises through the genetic direction for one or more gene action(s) and the environmental modulations that determine the actually expressed levels of these gene actions (Fig. 1.1). From this perspective, the holistic process of expressing a phenotype is always 100% controlled by a $G \times E$ interaction. Our inability to observe directly and quantify the multiple component biological processes ($G \times E$ interactions as illustrated most simply for plant development in Fig. 6.7, but compare Fig. 2.1) forces us to quantify the holistic result using statistical analysis. The procedure used for statistical quantification of the $G \times E$ interaction partitions the holistic control among: (i) a statistical main effect, which is the average of the genotype and its additive effect on the trait level; (ii) a statistical main effect, which is the average of the environment and its additive effect; and (iii) the statistically quantified $G \times E$ interaction effect(s) on the trait level. From a purely biological perspective, the additive main effect due to the genotype and the additive effect due to the environment are artefacts of the statistical procedure used to quantify the $G \times E$ interaction. Biologically, there is only $G \times E$ interaction because the gene action causal of a phenotype cannot occur if the environment has not provided every resource needed for the expressed level of that gene action (Chapter 1, paradigm shift 27). These biological versus statistical viewpoints require recognition with respect to all the interpretations that follow.

PHYSIOLOGICAL GENETICS OF GROWING DEGREE DAYS TO FLOWERING. Rather than the eight soil/planting-dates positioned along a half-circle, and two (H1 and R1) positioned central within this half circle as occurred for the DTF (Fig. 9.7A), the GDDs to flowering of this trial's ten soil/dates are aligned diagonally (Fig. 9.7J). The GDDs of the 30 pigeonpea genotypes are aligned diagonally also, like those of the bean genotypes (as shown best in Fig. 9.8 but also in Fig. 9.5 and Fig. 9.6). The diagonal positioning of the pigeonpea environments consists of two compact groups, and a third widely dispersed group (G1, H1 and B1). The most compact group (soil/dates R1, R3, B2, and B3) is due to the shortest P and lowest T of all the environments (all near 12 h and 22°C, see Fig. 9.7A). These shortest P and lowest T cause the largest negative multiplicative control and smallest additive control over the GDDs to flowering of any of the P–T combinations (Fig. 9.7J).

When grown in the non-photoperiod delaying P–T, the photoperiod sensitive genotypes cause negative $G \times E$ interaction. This will: (i) lower the DTF (Fig. 9.7A) and also the GDDs to flowering (Fig. 9.7J) of these sensitive genotypes, and thereby; (ii) lower the average DTF and GDDs of these environments. Magnitudes of the positive multiplicative control due to both the sensitive genotypes and the photoperiod delaying P–T extend to about twice those of the negative multiplicative control due to the insensitive genotypes and the non-photoperiod delaying P–T (Figs 9.7A, 9.7J). Because of these larger magnitudes, the average DTF and also GDDs of the non-delaying P–T will be reduced more by their negative $G \times E$ interaction with the photoperiod sensitive genotypes, than their averages will be enlarged by their positive $G \times E$ interaction with the insensitive genotypes. All delay of the DTF of the insensitive genotypes in the non-photoperiod-delaying P–T, and the consequent enlargement of their average GDDs, can be interpreted only as due to the thermal response. The photoperiod sensitive genotypes cause greater enlargement of the average DTF and GDDs for the photoperiod delaying *P–T*, while also causing the larger change toward earlier average DTF and toward smaller average GDDs for the non-photoperiod delaying *P–T* combinations.

The photoperiod and thermal responses, respectively, are the biological processes designated as control-2 and control-1 over the DTF above in section 9.3.2.1. These two responses are additive within the GPTmodel for simulating the DTF, but within the AMMI-analysis quantified holistic GPT interaction (the joint control over the DTF through the photoperiod response and the thermal response) the $G \times E$ interaction is a multiplicative effect. Based on the entirely biological interpretation, the photoperiod response and the thermal response are simply, totally and entirely a GPT interaction.

THE HOLISTIC GENOTYPE × PHOTOPERIOD × TEMPERATURE INTERACTION CONTROLS BOTH DAYS AND THE GROWING DEGREE DAYS TO FLOWERING. Both the DTF and GDDs to flowering of the genotypes are controlled through the holistic GPT interaction consisting of the photoperiod response interacting with the thermal response. Control through this holistic GPT interaction is reinforced by three high correlations. Across all 20 environments, correlation between the multiplicative controls over the DTF (Fig. 9.7A) and the multiplicative controls over the GDDs (Fig. 9.7J) is 0.986 (Table 9.5). Across all 30 genotypes this correlation is 0.987. Across the genotypes, correlation between their DTF and their GDDs to flowering is also high (0.999). The three correlations between 0.986 and 0.999 indicate near-perfect system-established interconnection between the $G \times E$ interaction controls over the DTF and over the GDDs. On the biplot graphs (Figs 9.7A and 9.7J) the high correlations reflect: (i) an identical pattern of the diagonal positioning on the AMMI-result biplot-graph of the genotypes for both their DTF (Fig. 9.7A) and their GDDs to flowering (Fig. 9.7J); and (ii) a *genotype-established* diagonal positioning also for the average GDDs (Fig. 9.J) of the environments as described in the next section. In contrast to the correlation of 0.999 between the DTF and the GDDs across all 30 genotypes, across all 20 environments the correlation between their average DTF and their average GDDs is a low and negative −0.182 (Table 9.5).

GROWING DEGREE DAYS GIVE NEAR PERFECT INDICATION OF THE CURVILINEAR RESPONSES OF THE DAYS TO FLOWERING OF ALL GENOTYPES TO BOTH PHOTOPERIOD AND TEMPERATURE. Each of the 30 pigeonpea genotypes had a U-shaped response of its DTF across the tested range of *P–T* combinations. This is illustrated for six in Fig. 9.9. The U-shaped curve of the DTF is contrary to the theoretical requirement for the GDDs to indicate the DTF only if the DTF responds to the *T* linearly.

In Fig. 9.9, three photoperiod sensitive genotypes of pigeonpea each have a different level of positive multiplicative control, and the smallest DTF of all three is at either 13.3 h /28.9°C or 13.4 h/28.4°C. The DTF is delayed slightly by 12.1 h/22.7°C, and strongly by 14.1 h/28.6°C. The largest delay of the DTF of the sensitive genotypes being due to the longest *P* indicates this delay is through the photoperiod response.

Three insensitive genotypes have progressively smaller levels of negative multiplicative control. They also express their earliest DTF at the intermediate

Table 9.5. Correlations among: (1) growing degree days to flowering, (2) multiplicative control over the growing degree days to flowering, (3) days to flowering and (4) multiplicative controls over the days to flowering across all environments and across all 30 pigeonpea genotypes, and also across only those with negative or positive multiplicative control over the genotype × environment interaction.

The four traits	1 GDDs to Fl.	2 Mul. Con. GDDs	3 DTF	1 GDDs to Fl.	2 Mul. Con. GDDs	3 DTF
1. Growing degree days to flowering						
	Across all environments			Across all genotypes		
2. Multiplicative control over GDDs	0.858			0.898		
3. Days to flowering	−0.182	−0.182		0.999	0.882	
4. Multiplicative control over days to flowering	0.852	0.986	−0.221	0.842	0.987	0.823
	Across environments with positive multiplicative control			Across sensitive genotypes with positive multiplicative control		
2. Multiplicative control over GDDs	0.480			0.983		
3. Days to flowering	0.559	0.400		1.000	0.981	
4. Multiplicative control for days to flowering	0.314	0.965	0.395	0.969	0.989	0.966
	Across environments with negative multiplicative control			Across insensitive genotypes with negative multiplicative control		
2. Multiplicative control over GDDs	0.583			0.419		
3. Days to flowering	0.428	0.473		0.998	0.361	
4. Multiplicative control over days to flowering	0.640	0.989	1.000	0.362	0.907	−0.052

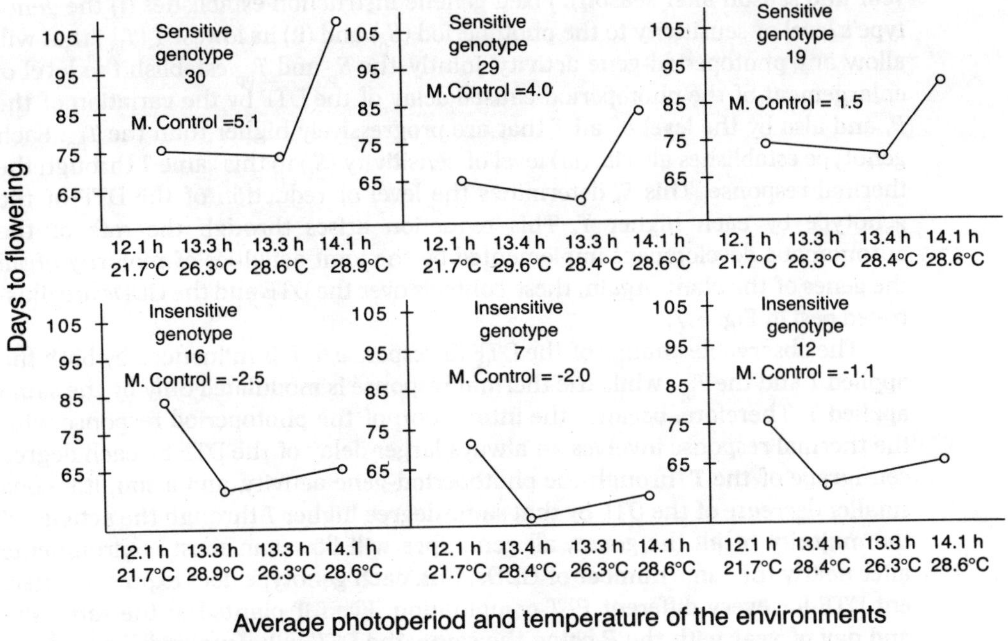

Fig. 9.9. Days to flowering of three photoperiod sensitive to less sensitive and three insensitive to less insensitive genotypes grown in four environments with short to longer photoperiods and low to higher temperatures. (From unpublished data of the International Crops Research Institute for the Semi Arid Tropics.)

13.3 h/28.9°C or 13.4 h/28.4°C, but their larger delays are by 12.1 h/22.7°C, with smaller delay by 14.1 h/28.6°C. The larger delay being due to the non-photoperiod-delaying 12.1 h/22.7°C indicates this delay is through the thermal response, as expected for photoperiod insensitive genotypes.

Table 9.5 presents the six correlations between all possible pairs among the following four traits, the GDDs, DTF, multiplicative controls over the GDDs and the multiplicative control over the DTF. All six correlations are shown for each of three subsets of the genotypes: (i) across all 30 genotypes, (ii) across just photoperiod sensitive genotypes, and (iii) across only the photoperiod insensitive genotypes. All six correlations are shown also for three subsets of the environments: (iv) across all 20 environments, (v) across just the photoperiod-delaying *P–T* combinations, and (vi) across only the non-photoperiod-delaying *P–T*. Why is it that across (i) the three subsets of the 30 genotypes, 14 of the 18 correlations are high and positive (1.000 to 0.823) and only four are low (0.419 to −0.052), while (ii) across the three subsets of the 20 environments, 12 correlations are low (0.640 to −0.221) and only six are large and positive (1.000 to 0.852)?

The 14 of 18 high correlations between 0.823 and 1.000 across the three subsets of the genotypes reflect that each genotype has a fixed genetic instruction. This same genetic instruction functions in every environment (year after year and season after season). Fixed genetic instruction establishes (i) the genotype's level of sensitivity to the photoperiod (S_p) and (ii) its lowest T (T_{bp}) that will allow any photoperiod-gene activity. Jointly, the S_p and T_{bp} establish the level of enlargement of the photoperiod-caused delay of the DTF by the variation of the P, and also by the level of all T that are progressively higher than the T_{bp}. Each genotype establishes also its (iii) level of sensitivity (S_t) to this same T through the thermal response. This S_t determines the level of reduction of the DTF of the genotype by each higher T. This reduction arises through the rate of the autonomous development implemented by the joint activities of majority of all the genes of the plant. Again, these controls over the DTF and the GDDs are illustrated best in Fig. 6.7.

The observable change of the DTF in response to P is influenced by both the applied T and the T_{bp}, while the thermal response is modulated only by the same applied T. Therefore, because the interaction of the photoperiod response with the thermal response involves an always larger delay of the DTF by each degree centigrade of the T through the photoperiod-gene activity, and a simultaneous smaller decrease of the DTF by that same degree higher T through the activity of the majority of all the genes, all genotypes will flower in most environments after nearly the same number of GDDs. But, each genotype does express a different DTF for every different P–T combination. Even if planted at the same site and day of year with the P being the same, the DTF will usually differ because the T will differ. For plantings at different dates, both difference of P and of T will differentially modulate the DTF. Whatever the genotype-established S_p, S_t and base-T_{bp}, the interacting control by the P and T shown in Fig. 6.7 will cause a different DTF of the genotype, while causing near-identical GDDs to flowering of each genotype, regardless of whether it is an insensitive or a photoperiod sensitive genotype.

The single correlation that is always large and positive for both all three subsets across the environments and for all three subsets across the genotypes is between the multiplicative control over the DTF and the multiplicative control over the GDDs. The range of these correlations across all six subsets is 0.907 to 0.989. This indicates virtual identicalness of the statistically quantified multiplicative control over the DTF with the multiplicative control over the GDDs, measured across both all environments and all genotypes.

Due to the interaction between the photoperiod and the thermal responses, each different P–T combination alters the DTF of every genotype, while causing minimal if any alteration of the GDDs to flowering of the genotype. The different average DTF of each environment (Fig. 9.7A) that the given subset of genotypes establishes causes the low correlation between the average DTF and average GDDs to flowering of the environments. The AMMI analysis has not and cannot quantify the S_t, the P_{opt} or the T_{bp}. These parameters are derivable only by the GPTmodel. Within an AMMI analysis, they remain confounded because the YSA has and can quantify only the holistic control due to all controls over the DTF. These whole-system caused changes of the DTF and GDDs are the only control over the DTF that can be observed directly. Therefore, applications of AMMI analysis and of the GPTmodel to the same DTF data provide complementary insights. Their combined interpretations are more informative than either alone.

Wallace *et al.* (1995a) compared the DTF of ten bean genotypes (Fig. 9.4A) with their GDDs (Fig. 9.10A). They concluded: 'When GDDs are used *in lieu of* calendar days additional interpretations emerge. Nearly 100% more days are needed to develop to flowering) . . . at 15.7 and 18.0° than at 23.1°, but only about 50% more GDDs are needed' (Fig. 9.10A compared with Fig. 9.4A). For both DTF and GDDs, development to flowering is most rapid at an optimum *T* near 23.1°C. About 100% more GDDs were needed at 28.3 than at 23.1°C, but only about 60% more days. Thus, for both bean (Fig. 9.4B) and pigeonpea (Fig. 9.7J), AMMI analysis shows photoperiod sensitive genotypes grown at high *T* to enlarge the GDDs more than the DTF, while low *T* reduces the GDDs more than the DTF is reduced. For both crops, differential change of GDDs versus DTF occurs because magnitudes of positive multiplicative controls and consequent *P–T* caused *G* × *E* interaction are larger than magnitudes of the negative multiplicative controls and the consequent *G* × *E* interaction through the thermal response.

THE ACCUMULATED GROWING DEGREE DAYS INDICATES THE DAYS TO FLOWERING OF ALL GENOTYPES SO ITS THEORETICAL BASIS NEEDS UPDATING. After reviewing responses of the DTF of corn, wheat and barley Shaykewich (1995) concluded: 'The development rate of most species is sigmoidal rather than a linear function of temperature. Consequently, phenological models assuming a linear relationship (e.g. degree-days) are inappropriate.'

The theoretical basis for the thermal-time response requires updating to recognize its ever-functioning interaction with the photoperiod response. The theory (Squire, 1990; Hodges, 1991; Jordan, 1993) for using the GDDs to indicate the DTF totally avoids rather than incorporates the (i) delay of DTF by P, and (ii) enlargement of this photoperiod-caused delay by each higher *T*. The GDD theory

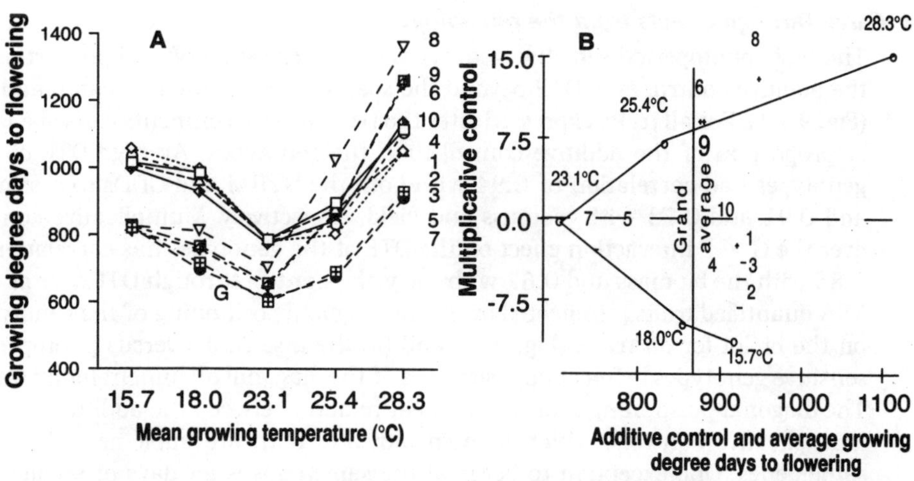

Fig. 9.10. The (A) growing degree days to flower of the bean genotypes 1–10 grown at the different temperatures of five field elevations all with a 13 h photoperiod, and (B) the AMMI analysis result from these growing degree days to flowering. (Modified from Wallace *et al.*, 1995a.) Compare with the corresponding graphs for the days to flowering in Fig. 9.4.

assumes that the rate of development to flowering must be linearly related to the T. But, illustrated in Fig. 6.7 for any given P and genotype is that neither the (i) change of the DTF across a range of T through the photoperiod response or (ii) the change of the DTF through the thermal response is ever linear. The opposite effects by T through the photoperiod and temperature responses additively but also interactively control the expressed DTF. The result is the U-shaped response of the DTF to a sufficient range of T.

The theory for the GDDs has called for ignoring all high T that delay rather than reduce the DTF. This theory has not recognized that (ii) a delay of the DTF begins (Fig. 6.7) at the lower apparent optimum T ($T_{aopt)}$) – if there is any photoperiod-gene implemented delay of the DTF in response to a progressively yet higher T. This delay begins as soon as the higher T has enlarged the photoperiod response to cause larger delay than the earlier DTF that the same higher T causes through the thermal response.

DAYS TO MATURITY. A more distorted half circle, but reflecting that connecting the P–T combinations on the biplot for DTF of pigeonpea (Fig. 9.7A), connects the midpoint positions of the DTHM of the same soil/dates (Fig. 9.7B). The average DTHM of soil/date H1 is intermediate as is its DTF. All P–T combinations with smaller average DTF than H1 have early average DTHM. All with later average DTF have later DTHM. Effects by the early versus late phenology (DTF) are apparent upon the GDDs to flowering (Fig. 9.7J), yield day^{-1} (Fig. 9.7G), yield day^{-1} of seedfill (Fig. 9.7I), and yield (Fig. 9.7D). But, as compared with the environmental positions for the DTF (Fig. 9.7A), all show subsequent shifts in position of the multiplicative plotted against the additive controls. Control due to the early versus late DTF is apparent also for both the biomass (Fig. 9.7K), and HI (Fig. 9.7L) plotted against the DTHM.

9.3.4 *Carry-through effects from the phenology*

The most photoperiod sensitive genotypes extend the span of and proportion of the additive control over DTF beyond the span caused by the P–T combinations (Fig. 9.7A). For all traits expressed after the DTF, the environments caused a larger proportion of the additive control than the genotypes. Average DTF of the genotypes had correlation of 0.94 with both the DTHM and GDDs to maturity, and 0.91 and 0.73 with biomass and yield, respectively. Multiplicative control over the $G \times E$ interaction effect on the DTF of the genotypes has correlation of 0.85 with the biomass and 0.67 with the yield. Control through DTF over all the YSA quantified traits is indicated by: (i) the diagonal positioning of the genotypes on the biplot for all traits (Fig. 9.7); and (ii) the insensitive versus photoperiod sensitive genotypes being at opposite ends of the diagonal alignment for all traits. The diagonal positioning indicates a linear relation between the additive and the multiplicative controls, which in turn indicates both are implemented by the same genes. One exception to being at the same ends is for days of seedfill, for which all genotypes with indeterminate habit (21–30) are shifted to a central position, while for genotypes with determinate habit (1–20) the sensitive versus insensitive genotypes remain at opposite ends (Fig. 9.7I). There is also: (i) slight reversal of the diagonal slope of the genotypes for yield day^{-1} (Fig. 9.7G); and (ii) reversal of ends of the slope occupied by sensitive versus insensitive genotypes for yield day^{-1} and also for the HI (Fig. 9.7F).

9.3.4.1 Control by phenology over biomass and harvest index

For all traits the most photoperiod insensitive and sensitive genotypes are at opposite ends of their diagonal positioning on the biplots – with the one exception of the days of seedfill for only the indeterminate habit (Fig. 9.7C) described in the previous paragraph. Usual positioning at opposite ends indicates that when photoperiod-gene activity delays the DTF (Fig. 9.7A), this also enlarges the DTHM (Fig. 9.7B) and the biomass (Fig. 9.7E) and the yield (Fig. 9.7D). Greater leaf area will occur as the growth duration and biomass are enlarged (Tables 4.1, 4.2, 9.1), while the HI is reduced (Fig. 9.7F). The greater leaf area that results from photoperiod-gene directed partitioning predominantly to continued vegetative growth accounts for higher rates of biomass accumulation of the sensitive genotypes (Fig. 9.7H), and this partitioning competitively reduces the HI. Genetically established indeterminate growth habit in contrast to determinate habit, and irrigation in contrast to non-irrigation, enlarged the biomass to a lesser extent (Fig. 9.7E). Across the 50 averages for 30 genotypes and 20 environments, the biomass has correlation of 0.62 with the range of negative to positive multiplicative controls over the $G \times E$ effect on the DTF. By causing larger DTHM the longer P and higher T interact to raise yield of the most photoperiod-sensitive genotypes, but not nearly as much as the larger positive $G \times E$ interaction due to post-flowering irrigation at soil/date R1/13.3 h (Fig. 9.7D). Negative interaction of these long-P/high-T combinations with the insensitive genotypes lowered yields of these genotypes below their already smallest of all averages (Fig. 9.7D). Two soil/dates (G1/14.1h and G2/13.3 h) interacted with the insensitive genotypes to raise yield above their averages (Fig. 9.7D).

Comparison of aerial biomass and the HI (Fig. 9.7E, F) reveals large negative correlation (Fig. 9.7K) between these two of the three major physiological-genetic components of accumulating yield. The days used to accumulate the measured yield and biomass, i.e. the DTHM is the third major component. Correlation between biomass and HI is −0.80 across averages of both all genotypes and all environments, −0.79 across just the 20 environments, and −0.84 across the 30 genotypes. The most photoperiod sensitive genotypes 11, 19, 29 and 30 have the largest of all average yields and the largest positive multiplicative controls over the $G \times E$ interaction effect on yield (Fig. 9.7D), and have these same highest ranks for aerial biomass (Fig. 9.7E). These sensitive genotypes also have the smallest average HI, and largest negative multiplicative controls over the $G \times E$ interaction effect on the HI (Fig. 9.7F). In contrast, the photoperiod insensitive genotypes 10, 4, 6 and 7 have among the lowest average yields (Fig. 9.7D) in association with the largest negative multiplicative control over the $G \times E$ interaction effect on the yield, while having among the highest average HI and third, fourth and fifth largest negative multiplicative control by a genotype over the $G \times E$ interaction effect on the HI. The positive $G \times E$ interaction between the insensitive genotypes and the non-delaying P–T combinations enlarge the HI (Fig. 9.7F) while reducing the average aerial biomass (Fig. 9.7E, K).

As it did for yield (Fig. 9.7D), irrigation caused change toward more positive multiplicative control through the $G \times E$ interaction effect on the aerial biomass (Fig. 9.7E), the largest irrigation-modulated changes being for soil/dates H1 and R1 (Fig. 9.7E). But, the largest of all $G \times E$ effects over biomass occurred between soil/date H2, which had the largest positive multiplicative control strength, and soil/dates R3, B3, R2 and B2, which had the largest of all negative multiplicative control strengths.

CONTROL OVER BIOMASS AND YIELD BY PHENOLOGY. The difference between the largest positive multiplicative control by sensitive genotype 30 and largest negative multiplicative control by insensitive genotype 10 over the $G \times E$ interaction effect on the biomass is comparable to the extreme difference for these controls of soil/date R3/12.2 h compared with H2/13.3 h (Fig. 9.7E). This indicates the genotype-caused $G \times E$ interaction control over biomass is similar to the magnitude of the $G \times E$ due to the environmental variations. Across genotypes only, biomass has correlation of 0.91 with DTF and 0.92 with DTHM. This indicates gene action with strong control of the phenology and consequent control also over the biomass.

DURATION AND RATE OF SEEDFILL. Duration and rate of seedfill constitute the actual process of accumulating yield. Both begin after flowering. Indeterminate genotypes were buffered such that they caused near zero interaction control over the days of seedfill (Fig. 9.7C). Post-flowering irrigation increased average days of seedfill for all soil/dates except H1/13.3 h and B1/13.8 h. The greater $G \times E$ caused enlargements are for R2/12.1 h, B2/12.1 h, R1/12.1 h and G1/14.1 h. The $G \times E$ due to irrigation increased the days of seedfill most for genotypes having determinate habit combined with sensitivity to photoperiod, and caused smallest decreases for the determinate but insensitive genotypes. With or without irrigation, soil/date B1/13.8 h caused nearly the same and largest of all positive $G \times E$ interaction controls over the days of seedfill of the sensitive genotypes, and nearly the same also but about 50% smaller negative $G \times E$ interaction for the insensitive genotypes. Soil/dates R3/12.2 h, B3/12.2 h, R2/12.1 h and B2/12.1 h, all with 22°C, cause smaller and always opposite (negative rather than positive, or vice versa) $G \times E$ interaction control over the days of seedfill of all genotypes in comparison to that caused by soil/date B1/13.8 h.

Highly photoperiod sensitive genotypes interacted with soil/dates R1/13.3 h and H2/13.3 h to give greatest increases of yield day^{-1} of seedfill (Fig. 9.7I). Insensitive genotypes interacted at these soil/dates with about half these magnitudes of decreases. G2/13.3 h caused the largest positive interaction effect on the yield day^{-1} of seedfill of the insensitive genotypes, and about twice as large a decrease for the photoperiod sensitive lines. H2/13.3 h and G2/13.3 h are the two environments with the smallest average DTF (Fig. 9.7A) across all of the 30 genotypes. Soil/dates B2/12.1 h, B3/12.1 h, R2/12.1 h and R3/12.2 h, all with near 22°C, caused near zero $G \times E$ interaction effect on the yield day^{-1} of seedfill.

9.3.4.2 Rates of biomass and yield accumulation

When aerial biomass was discussed above, emphasis was on any photoperiod-gene activity that enlarged the DTF and DTHM, thereby extending the duration of the accumulation of biomass and yield, and enlarging negative correlation between the biomass and the HI. Rates of accumulation of biomass and yield were not emphasized.

There was an 11-fold difference between the smallest and largest average aerial biomass across the 20 environments (Fig. 9.7E, Table 9.3), and tenfold difference of average biomass day^{-1} (Fig. 9.7H). Smallest average rates were for soil/dates R2/12.1 h, B2/12.1 h, R3/12.2 h and B3/13.8 h (all with 22°C). Largest rates were for H1/13.3 h (Fig. 9.7H) in association with small positive multiplicative control. Soil/date H2/13.3 h had rank 2 for biomass day^{-1} and rank 1 as cause of the $G \times E$ interaction control over it. Its multiplicative control (0.35 t ha^{-1} day^{-1})$^{-0.5}$ had magnitude about ninefold larger than the next ranked (−0.04 t ha^{-1} day^{-1})$^{-0.5}$ for non-irrigated soil/date R2/12.1 h and sixfold larger than the (0.07 t ha^{-1} day^{-1})$^{-0.5}$ for irrigated H1/14.4 h (Fig. 9.7H). Highly photoperiod sensitive genotypes interacted with soil/date H2/13.3 h with large positive $G \times E$ over biomass ha^{-1} day^{-1}, while insensitive genotypes interacted with smaller negative $G \times E$. Soil/date H1/13.3 h also had rank 1 for average biomass ha^{-1} day^{-1} (Fig. 9.7H), H2/13.3 h had rank 2, B1/13.8 h, R1/12.3 h, G1/14.1 h and G2/13.4 h had intermediate averages, and R2/12.1 h, B2/12.1 h, R3/12.2 h and B3/12.2 h had the lowest of all the averages for biomass ha^{-1} day^{-1}. Irrigation caused only small changes of the average biomass ha^{-1} day^{-1} across the full growing season and the interaction effect on the biomass.

Soil/dates R2/12.1 h, B3/12.2 h, B2/12.1 h and R3/12.2 h with both the shortest P and lowest (22°C) T (Fig. 9.7A) had the lowest average yield ha^{-1} day^{-1} to maturity (Fig. 9.7H), yield ha^{-1} day^{-1} of seedfill (Fig. 9.7I), and biomass (Fig. 9.7E). With and without irrigation they caused near zero $G \times E$ interaction effect on the yield ha^{-1} day^{-1} of seedfill (Fig. 9.7I), but caused a small negative interaction effect on the biomass ha^{-1} day^{-1} of the sensitive genotypes (Fig. 9.7H). Their $G \times E$ control over yield ha^{-1} day^{-1} of the sensitive genotypes was near zero to slightly negative without irrigation, but slightly positive with irrigation (Fig. 9.7G). Their $G \times E$ controls over yield ha^{-1} day^{-1} of the insensitive genotypes were of smaller magnitude and negative. Irrigation changed their interaction control toward enlargement of the yield ha^{-1} day^{-1} of seedfill of the photoperiod sensitive genotypes, and smaller reduction for insensitive genotypes. Soil/dates R1/13.3 h, G2/13.4 h, G1/13.3 h, H2/13.8 h and H1/14.4 h had ranks 1, 3, 4, 2 and 5 for yield ha^{-1} day^{-1} of seedfill, respectively (Fig. 9.7I). Soil dates H1/14.4 h, H2/13.3 h, and R1/13.3 h interacted positively with the most sensitive genotypes to enlarge both yield ha^{-1} day^{-1} to maturity and yield ha^{-1} day^{-1} of seedfill, while G1/14.3 h and G2/13.4 h interacted negatively with these sensitive genotypes, but positively with the insensitive genotypes

9.3.4.3 Control of partitioning by genotype

Partitioning of the net accumulated aerial biomass between yield and vegetative organs, as measured by the HI across the genotypes, had correlation of −0.84 with the biomass, −0.83 (Fig. 9.7K) with DTF and −0.81 with DTHM (Fig. 9.7L). These negative correlations reinforce the following prior interpretation for bean (Tables 4.1, 4.2 and sections 4.2 and 4.6, plus Table 9.2 and Figs 4.1 and

9.1; see also section 9.2). Photoperiod-gene activity controls whether the available photosynthate is partitioned more toward initiation and continued growth and development of reproductive organs, which become the yield, or is competitively partitioned more to continued growth and development of additional vegetative organs (Chapter 4). Witzenberger *et al.* (1988) have shown also for groundnut (*Arachis hypogaea*, section 4.2.5) that photoperiod controls partitioning as measured by the HI, as have Linnemann *et al.* (1995) for bambara groundnut (*Vigna subterranea*). Section 4.2.5 describes that the control over partitioning in peanut occurs only after, and not before flowering. That section cites yet other evidence that the *P* may modify any to most if not all stages of development of the reproductive organs. The stage(s) controlled vary with the plant species.

9.3.4.4 Control of partitioning by environmentally enhanced growth

Soil/date H2/13.3 h, most strongly soil/date G2/13.4 h, and to a lesser extent G1/14.1 h interacted with photoperiod sensitive genotypes to lower their HI (Fig. 9.7F). Soil/dates R2/12.1 h, R3/12.2 h, and B2/12.1 h, all at 22°C, interacted with these sensitive genotypes to raise their HI. Soil/date H1/14.4 h (and 29.6°C) also interacted to raise the HI of the photoperiod sensitive genotypes. All environments caused the opposite (negative versus positive) interaction effect on the HI of insensitive genotypes. Soil/date H1 with 14.4 h and 29.6°C interacted positively as expected to delay DTF (Fig. 9.7A), and to raise yield (Fig. 9.7D). But, in opposition to overall negative correlation between HI and biomass, irrigation at H1 interacted with sensitive genotypes to raise their biomass (Fig. 9.7E), and increase their DTHM (Fig. 9.7C), but lower their yield day^{-1} to harvest maturity (Fig. 9.7G). Biomass of non-irrigated H1 was 22% higher than biomass of rank-3 for non-irrigated H2, and 29% higher than rank-4 biomass of irrigated H2. Thus, all four environments with highest biomass were on soil H. The $G \times E$ interaction of soil/date H2/13.3 h raised biomass of the sensitive genotypes about fourfold more than it was raised by H1/14.4 h (Fig. 9.7E), and lowered their HI (Fig. 9.7F) about the same quantity as it was raised by H1.

From the previous paragraph, all four environments involving the highly fertile soil H produced far larger biomass than any other environment. From the materials and methods, only soil H received no application of nitrogen. Paradoxically, in spite of accumulating far the largest biomass (Fig. 9.7E) but smallest average HI of all soil/dates, soil/date H1 interacted to enlarge the HI of the photoperiod sensitive genotypes, even though this soil date had the lowest HI of all 20 soil/planting dates, and sensitive genotypes had the lowest HI among the 30 cultivars (Fig. 9.7F). The rank-2 biomass of H2/13.3 h was attended by the second smallest HI of a soil/date (Fig. 9.7F), and its average biomass was 66% larger than that of rank-3 for soil/date G1/14.1 h. Soil/date G1/14.1 h shared the third smallest HI (Fig. 9.7F) with B3/12.2 h, which had rank-4 biomass (Fig. 9.7E, J). The four soil/dates with smallest biomass (B2/12.1 h, B3/12.2 h, R2/12.1 h and R3/12.2 h) all had smaller average biomass but larger average HI than the largest of any genotype (Fig. 9.7K, L).

Clearly, as shown for bean (Tables 4.1, 4.2, 9.2) and pigeonpea (Fig. 9.7E and F), negative correlation between HI and aerial biomass is controlled by the genotype for sensitivity to photoperiod. But, in addition, HI was lowered also by other factor(s) of the environment that strongly promoted growth (biomass accumulation). Lowest HI of all and largest biomass of all for soil/date H1, and adjacent ranks of both for soil/date H2, suggest control over the HI occurs through the capability of the environment to modulate small versus intermediate to large plant growth (accumulation of biomass, compare Fig. 9.7K and L).

Many crop scientists recognize that environments that lead to vigorous vegetative growth often do not give higher yields, because of lower HI. Grashoff (1990) studied the effect of time of irrigation on yield of faba bean (*Vicia faba* L.). He concluded: 'Plenty of water during and after flowering resulted in stimulation of vegetative growth, but reduced initial reproductive growth as well as final seed yield, compared with mild water shortage during flowering but plenty of water after flowering.' Irrigation during flowering not only lowered seed yield as compared with limited water at flowering and excess water afterward; it also lowered HI.

Average biomass of the soil/dates had correlation of 0.55 with the total rainfall. This correlation was smaller than with the average *T* to flowering (0.85) and to maturity (0.76), and with the average *P* to flowering (0.80) and to maturity (0.71). These larger correlations, and the range of photoperiod-gene established sensitivities to *P*–*T* combinations, justify emphasis on increase by both longer *P* and higher *T* of the photoperiod-gene activity. They emphasize this activity's negative control (Chapter 5) over partitioning toward reproductive organs, with consequent enlarged biomass (Figs 5.1, 9.1).

Only positive correlations occur among the rainfall, *P* and *T*, all three having negative correlation with HI. The incident radiant energy would also be positively correlated. Thus, all these environmental factors are linked within the climates spanned. However, the results indicate the *P*–*T* combinations and the varied genotypic sensitivities to them most strongly control the phenology, and thereby also partially control the DTHM, biomass and yield. These controls arise through photoperiod response and its interaction with the thermal response. Water availability, soil fertility, and probably the incident radiant energy would most strongly control the total biomass accumulation and its effect on the yield and the HI.

Negative correlation between yield and the HI in this pigeonpea trial is unlike virtually always positive correlation for bean grown in productive environments. Average correlations for bean yield of 0.54 with HI and 0.70 with biomass (Table 9.2) were each across 33 yield trials. Of the 33 correlations, yield of only one trial had negative correlation with HI, and yield of only another trial had negative correlation with the biomass.

A second pigeonpea yield trial at ICRISAT has now been subjected to complete YSA/AMMI analysis of its four environments representing factorial combinations of the red and black soils and irrigation versus non-irrigation (Prasad, 1996). Correlation between HI and other traits (Table 9.2) were −0.54 with biomass, −0.72 with DTF, −0.75 with DTHM, −0.74 with seedfill duration and −0.43 with biomass day^{-1}. And, across the four soil-moisture environments, HI had correlations of −0.84 and −0.79, respectively, with DTF and DTHM, but 0.83 with seed yield. Across these four environments biomass had correlation of 0.99 with seed yield, and across genotypes this correlation was 0.81. Figure 4.5 shows the highest HI to also be associated with the smallest DTHM of sorghum.

Previously reviewed evidence suggests this association of the highest HI combined with smallest DTHM applies across at least 26 crops (Wallace, 1985; Wallace *et al.*, 1993c).

9.4 Yield System Analysis and AMMI Analysis Illustrate a Procedure and Philosophy for Whole-system Research

Whole-system research of crop yield requires quantifying the correlations among levels of the smallest number of near-fully integrated components the yield system can be divided among, while these component traits remain inclusive of all the processes essential to accumulation of the yield (or of other system outputs). Also required is the correlation of the level of each component with the level of the yield *per se*. Ideally, these correlations should be obtained from yield trials of multiple genotypes grown across multiple environments. Ultimately, the results should be compared across multiple plant species.

The requirement that whole-system research deal with the smallest number of components the system can be divided into while still accounting in full for the system output is fulfilled by every procedure of a complete YSA/AMMI-analysis. How all the genetic and environmental controls are accounted for will be detailed for the 600 treatments of the pigeonpea trial.

1. The complete YSA did measure the yield, did compare the yield level with the level of each of the smallest number of near-fully integrated components the process of accumulating the yield can be divided among, and did remain inclusive of all the processes that account for accumulation of the yield. In addition, the YSA measured these relationships for subcomponents of the major components such as the average biomass day^{-1} and yield day^{-1}.

2. For the ANOVA followed by a PCA that constitute the AMMI analysis, the 30 genotypes were labelled simply 1, 2, 3, . . . 30. And, the 20 environments were labelled B1I, B1N, B2, B3, R1 . . . R3I. R3N. The successive three characters indicate the soil, the planting date and the 'Irrigation' versus 'Non-irrigation'. Their designation facilitates interpretation of the AMMI analysis result. The labelling of only different genotypes and different environments for the AMMI analysis classified all control as due to the holistic genetic direction of the specified one of the 30 genotypes, or due to the holistic modulation of the genetic direction of the genotype by one of the 20 environments. The specified genotype and specified environment constitute the smallest possible number of the all-inclusive controls over the yield system as indicated in Fig. 1.1. For the pigeonpea experiment, the measured levels of each trait have arisen through 600 different genotype–environment combinations (treatments).

3. The AMMI analysis considered the smallest number of components of the holistic genetic direction of each different genotype to consist of: (i) an additive control (that genotype's average across all of the compared environments); and (ii) a multiplicative control over any $G \times E$ interaction involving that genotype. And, the AMMI analysis considered the smallest number of components of the holistic modulation of the level of the expressed genetic direction by each environment to consist of: (iii) an additive control (that environment's average across all of the compared genotypes); and (iv) a multiplicative control over the $G \times E$ interaction due to that environment.

4. An additive and a multiplicative control due to each of 30 genotypes constitutes 60 genetic controls. An additive and a multiplicative control for each of the 20 environments constitutes 40 controls by the environments. The AMMI analysis quantifies this total of 100 controls due to the 30 genotypes and 20 environments.

5. All 100 controls are all quantified and presented. But, they are reduced back to 50 points when the additive control and the multiplicative control for each genotype and of each environment are plotted one against the other on a biplot graph that has two axes. One holistic control due to a genotype and to an environment agrees with the biological perspective that the holistic genetic direction and the holistic modulation of the direction are causal of the phenotype (Fig. 1.1). The 50 points from plotting the 50 multiplicative against the 50 additive controls on a biplot graph represent the smallest number of a holistic genetic control – combined with a holistic environmental control – that can account in full for all the biological control over the phenotypic level of the 600 treatments.

6. By far the greatest contribution to whole-system research provided by the AMMI analysis is an easy interpretation of the always-complex and multiple $G \times E$ interaction controls over the expressed level of all quantitatively variable traits. Achieving easy interpretation has as a first requirement that the patterns of all the statistically quantified additive and multiplicative controls of the AMMI analysis be plotted one against the other on the biplot graph. Next, the researcher must designate the possible gene action(s) or physiological process(es) and the possible negative versus positive modulation(s) thereof by the levels of likely environmental factors, as described in section 9.3.3.2. Using these steps to interpret the $G \times E$ has been demonstrated in this chapter. Although used, an aspect of the method that most reduces the effort needed to interpret the $G \times E$ interaction has not been sufficiently emphasized. It is that it is the several extreme positive and extreme negative multiplicative controls – due to both the genotypes and the environments – that facilitate easy biological interpretation of the $G \times E$ interaction controls. All multiplicative controls that are near zero can be largely ignored, because zero or near zero $\times$ any other number is zero or near zero. Ignoring the near zero multiplicative controls reduces the 50 points on the biplot graph to be used to interpret all control over the phenotypic level of all 600 treatments by about half, i.e. to about 25 points. These largest of the 50 multiplicative controls provide the easy interpretation of the $G \times E$ interaction.

7. That the AMMI analysis categorized all control as being due a holistic genotype and a holistic environment is another way it contributes to whole-system research. The environments have involved the variables of four soils, two or three planting dates (*P–T* combinations) on each soil, and irrigation versus non-irrigation of each planting date. The largest and most statistically significant pattern of the $G \times E$ interaction effect on all the YSA measured traits was interpretable as due to differential modulations by the *P–T* combinations of the DTF and DTHM of the photoperiod insensitive versus the photoperiod-sensitive genotypes. And, delay of the DTF by positive $G \times E$ interaction due to the positive multiplicative control of a genotype multiplied by the positive multiplicative control of an environment was interpretable as delay through the photoperiod response, while positive $G \times E$ interaction due to the negative multiplicative control of a genotype multiplied by the negative multiplicative control of an environment was interpretable as delay through the thermal response. The relative control due to the soil and to the irrigation could be compared with that due to this holistic GPT interaction control over the trait level.

8. The complete YSA facilitated quantification of the system-established interconnections (correlations) between the expressed levels among all pairs of the smallest number of components that accounted in full for the system output. The system outputs focused on most in this chapter were the yield and/or the DTF. Focus on the DTF has shown it and the DTHM (the phenological traits) to be a major control over the yield and over all of its major physiological components. The interconnections among all the component traits show the control to be by the system as a whole.

9. The final concept that renders the YSA/AMMI whole-system research follows. It is that all control over every trait was for the experimental study assumed to be due to a set of (i) genotype-established genetic directions and (ii) the modulated level of implementation of the genetic directions allowed by one or more environmental factors. The researcher hypothesized these controls and evaluated their biological validity and their statistical reliability.

Chapter 9 Summary

The calendar-day based YSA measured directly the levels of traits 1 to 4 below, and from them calculated the levels of traits 5 to 9.

1. Days to flowering	Time used for development to flowering	
2. Days to harvest maturity	Time used to develop to harvest maturity	
3. Aerial biomass	The overall photosynthetic efficiency	
4. Yield	The economically important output	
5. Days of seedfill	Time used for actual yield accumulation	
6. Yield/day to maturity	Efficiency of yield accumulation	*The rate of partitioning*
7. Yield/day of seedfill	Efficiency of yield accumulation	
8. Biomass/day of plant growth	Efficiency of photosynthesis	
9. Harvest index	Endpoint efficiency of partitioning to yield	

An AMMI analysis was applied to each of the nine traits. The statistical term 'effect' was replaced by the term 'control', because control refers to the genetic directions and the environmental modulations of the level of the gene actions that establish each trait level. The additive control due to each genotype is the average level of the trait across all the compared environments. The additive control due to each environment is its average level of the trait measured across all the compared genotypes. The multiplicative effect due to each given genotype times the multiplicative control of the given environment quantifies the $G \times E$ interaction control over the trait level for the given treatment (one of the genotypes grown in one of the environments). The conclusions follow.

1. Differential modulation by P and T of the DTF of photoperiod insensitive versus photoperiod sensitive genotypes caused the largest source within the total $G \times E$ interaction control over the DTF. The AMMI analysis quantified this

largest $G \times E$, and it was shown to arise through the photoperiod response – interacting with – the simultaneous thermal response.

2. For bean and pigeonpea, this photoperiod response interacting with the simultaneous thermal response (i.e. the holistic control over the DTF) carried through to become a control also over the DTHM, yield per DTHM, net accumulated biomass per DTHM, the total aerial biomass, the HI and the yield. This holistic control over the DTF is implemented by three subsets of genes: (i) the majority of all the genes of the plant implement growth and an autonomous capability of the plant to develop to all its stages, while at high temperatures (ii) any photoperiod gene activity and/or (iii) any vernalization gene activity delays the autonomous development. With all these genes involved, the control is by the system as a whole.

3. The AMMI analysis result was presented as one point on a biplot graph for each genotype and one for each environment. Each point indicated the control through the $G \times E$ due to the genotype or to the environment plotted against the additive control due to the same genotype or environment. These controls jointly determined the expressed level of the trait. The benefit gained from the AMMI analysis is an easier interpretation of (i) the variation in the genetic instruction and (ii) the variation in the level of factor(s) of the environment causal of the larger source(s) of the $G \times E$ interaction. Such informative physiological-genetic interpretation of the $G \times E$ interaction controls has not been as readily achievable in the past.

4. The accumulated GDDs gave near-perfect indication of the DTF of both the insensitive and the photoperiod sensitive genotypes. This was shown as follows. The AMMI-analysis-quantified control by the genotypes through the $G \times E$ interaction had correlation of 0.97 to 0.999 with both the DTF and the GDDs and control by the environments through the $G \times E$ interaction for both the pigeonpea and bean genotypes. These correlations show near-perfect indication of the DTF of the photoperiod sensitive genotypes for which – across some of the range of normal growing T – each higher T increased the DTF. This increase of the DTF by a higher T is contrary to the linear reduction of the DTF in response to a higher T that the theory of the GDDs expects for successful indication of the DTF by the accumulated GDDs. According to that theory, the GDDs can predict the DTF of the photoperiod insensitive but not the sensitive genotypes. Successful indication by the GDDs of the DTF of both photoperiod sensitive and insensitive genotypes requires a shift of the GDD theory to recognize as illustrated best in Fig. 6.7: (i) the interaction between the photoperiod and the thermal response; and that (ii) the two responses always cause opposite changes of the DTF through the applied T.

5. There can be no cheaper way to acquire progressively more complete understanding of the biological functioning of the whole plant system than measuring in yield trials already being conducted the (i) level of the yield (the system output); and (ii) the level of each of the smallest number of near-fully integrated processes that account in full for the biological expression of this yield. Measures of (i) and (ii) fulfil the definition for whole-system research described as paradigm shift 25. This is amply illustrated for yield as one system output, and for the DTF as another system output.

6. It was shown that within the yield system the level of every quantitatively variable component trait is interconnected with the levels of the other quantitatively variable traits.

7. From the biological viewpoint, every phenotype results from two holistic controls; (i) a specific subset of genetic direcuons, and (ii) the modulations of the implemented level of these genetic directions by the levels of specific subset of environmental factors. Thus, biologically, every phenotype is 100% due to a $G \times E$ interaction control. Due to man's inability to differentiate directly, follow and quantify the internal biological activities, statistical analysis is used to quantify the $G \times E$ interaction, and it divides the 100% $G \times E$ interaction control among (i) a statistical main (average) effect due to each genotype, (ii) a statistical main (average) effect due to each environment, and (iii) a statistically quantified $G \times E$ interaction effect. Both the statistical and biological viewpoints need to be applied for maximal interpretation of the $G \times E$ interaction control.

10 Interplant Competition and Breeding for Higher Crop Yield

Paradigm shift 29 is derived in this chapter. It is that, from the viewpoint of interplant competition, the maximum potential yield per plant of a genotype is one component of the potential crop yield of a genotype. The second and only other component from this viewpoint is the relative reduction of the yield per plant as the plant density is increased to that which maximizes the crop yield per land area. Selection for these two components can assist breeding for higher yield.

Darwin (1869) recognized two major forces that control the evolution of domesticated plants and animals: natural selection and selection by man. Both function and interact to determine what can be selected in plant breeder's fields. Natural selection results from the effect on each plant of the physical environment, and the effect by the neighbouring plants. Thus, natural selection consists of (i) selection for fitness of the genotype to the physical environment; and (ii) selection through competition with its neighbouring plants (Donald and Hamblin, 1983). The latter is the subject of this chapter.

A broad definition of competition includes all the interactions of a plant with its neighbouring plants. Direct interaction is when a plant directly interferes with another plant by shading or crowding. Indirect interaction occurs when two or more plants of a community each seek the measure it can use of any environmental resource factor(s), but the supply available is below the combined demand of the plants (Donald, 1963). This narrow-sense (Connell, 1990) competition is the focus of this chapter, because it is highly relevant to plant breeding.

Progress toward selection and release of a new cultivar will always involve selections during the early segregating generations, followed by further selection during the advancing generations with progressively greater homozygosity. Limited understanding about the relation between the yield plant^{-1} within segregating progenies and the crop yield derivable from the genotype has – in the past – allowed direct selection for yield only within the advanced generations, because the progenies can be evaluated as a homogeneous population that will breed true. On the contrary, during the early generations, the different genotypes are represented by only single individual(s) most of which will continue to segregate. New insights presented within this chapter, based on the system thinking by Yan and Wang (1992), Fasoulas (1973), Fasoulas and Fasoula (1995) and Fasoula and Fasoula (1996) suggest enlarged future opportunity to select for the crop yield potential of the early generation segregates.

Selection for yield during the advanced generations is comparatively simple and reliable. It requires only that the environment in which the selection is done is representative of the environments where the genotype will be grown as a crop. Selection for yield *per se* of homogeneous populations is effective because all the interplant competition is between individuals of the same genotype; the selection is directly for the ultimate output of the system which is the yield, and; the expressed yield results from the integration of all the functioning components of the system. During the earlier segregating generations, both differential competitive abilities of the multiple genotypes and the continued segregation of most of the individuals for most of all the component traits make the selection effectiveness problematic. For these reasons, the segregating genotypes of the F_2 generation can be evaluated only by the performance of each individual of the plant population. Every early-generation plant has high probability of being a different genotype and of not breeding true. Effective selection toward higher crop yield potential based on the yield of the individual plants of the early generation segregates requires understanding the following questions:

1. What is the relationship between the yield of the individual plant of a genotype grown in heterozygous populations and that genotype's crop yield when grown in a monoculture?
2. What is the influence on yield of the individual plant due to increases of the plant density, i.e. by the interplant competition?
3. During the segregating generations, what plant traits can be selected for that will result in higher crop yield of the monoculture crop?

10.1 Concepts about Competition and Crop Yield

There have been three major approaches to studying plant competition. The first is to grow populations of mixed identifiable genotypes, and to determine the change in proportions of the genotypes after several generations. Genotype(s) that increase in percentage are regarded as strong competitors. Those that decrease are weak competitors. The second approach is to study the effect of plant density on the mean yield plant^{-1}. The third is intercropping experiments, a kind of density experiment that involves two or more genotypes or plant species.

Australian scientist C.M. Donald was the first to draw serious attention to the effects of interplant competition on plant breeding and selection. Donald (1963, 1968) and Donald and Hamblin (1983) extensively reviewed the research about interplant competition of several crops. They drew three conclusions.

1. When seeds of several genotypes are mixed in equal proportion and grown successively for several years, one genotype will come to dominate the mixture, suggesting this genotype is a strong competitor. This genotype is selected for its superior interplant competition, and/or for its smaller seed size but greater seed number per unit of the yield.
2. Truly strong competitors are poor yielders when grown in a monoculture. Therefore high yielding cultivars should be weak competitors.
3. Strong competitors are in general tall, leafy, and freely tillering (or branching) genotypes. They have competitive advantage because, being taller and above the other genotypes, they capture the radiant energy and limit access of the shorter genotypes to the light needed for high levels of photosynthesis.

Based on these conclusions, Donald (1968, 1981) proposed a now famous (Sedgley, 1991; Hamblin, 1993) weak-competitor ideotype for wheat and other cereals. Donald's ideal plant should have a single, short, and strong stem with few, small, narrow and upright leaves. It is known as the uniculm ideotype. Later this ideotype was extended to a general ideotype for all annual seed crops (Donald and Hamblin, 1983). The following characteristics were listed for such an ideotype: (i) strictly annual habit; (ii) erect growth form; (iii) dwarf stature; (iv) strong stems; (v) no branches, i.e. non-tillered habit; (vi) reduced foliage (smaller, shorter, narrower, or fewer leaves); (vii) erect leaf disposition; (viii) determinate habit; (ix) high harvest index; (x) photoperiod insensitive (for most but not all environments); and (xi) early flowering (for most but not all environments).

Donald's uniculm ideotype has not been successful (Belford and Sedgley, 1991; Balkema-Boomstra and Masterbroek, 1993). It is necessary, therefore, to re-examine Donald's reasoning. First, a weak competitor may not be a good yielder, because competitive ability and crop yield are largely independent, in contradiction to the second sentence of conclusion 2 above. Conclusion 2 came from an extensive review about competition among crops and pasture plants (Donald, 1963), 5 years before Donald published his uniculm model (Donald, 1968). Second, some doubt the assumption that the uniculm ideotype is a weak competitor (Marshall, 1991).

Nevertheless, Donald's concepts have important implications for breeding. First, in a segregating population, the individuals that have the larger yield often do not give high crop yield when grown in pure stands. They are more likely to give low yield. Therefore, as stated by Donald and Hamblin (1983), breeders need to avoid selection based on individual plant performance during the early segregating generations grown at plant densities approaching those of crop stands. During these early generations, the interplant competition occurs mostly among plants of different genotypes. In the presence of strong competition, the poor competitors will have depressed yields, whatever their yield potential in pure stands. Donald and Hamblin suggested plants with poor competitive ability must be allowed to express their yield potential in pure culture. Because high plant density plantings will be used on the farm, the segregating generations should also be grown and the selection done at relatively high plant density. This belief was negated by Fasoulas (1988, as cited by Reynolds *et al.*, 1994).

Two recent reviews of competitive ability in plant breeding and agronomy suggest the following management. Yan and Wallace (1995a) indicated that if segregating plants are spaced far enough apart there is no competition between the individuals, this will allow selection of the genotype having the greatest potential for high yield plant^{-1}. This selection will raise the gain in crop yield potential, because the potential crop yield of a genotype can be fully accounted for by, (i) its yield plant^{-1} in an environment with no interplant competition, and (ii) the genotype's tolerance to the increase of the plant density required for high crop yield per land area. Fasoula and Fasoula (1996) used terms suggested by Donald and Hamblin (1976) to classify spacing between plants. Closer spacing is a competitive environment – if unlike genotypes are competing. The unlike genotypes may be either the segregates after a cross, or intentional (or unintended) mixtures of two or more genotypes. However, the closer spacing is a crop environment if all the plants are of the same genotype. According to Fasoula and Fasoula (1996), all the plants of modern crop cultivars usually have the same or nearly the same genotype, because experience has shown this genetic uniformity contributes to high crop yield. Fasoula and Fasoula (1996) derived their conclusions from comparing the coefficient of variation (CV) of the yield plant^{-1} of genotypes grown at different population densities. They found the CV of a genotype to be lowest in isolation environments, i.e. with spacing distant enough to largely reduce interplant competition, intermediate in the crop environments with highest yields, but largest in competitive environments that give low crop yield. They concluded that the CV level of the genotype indicates the magnitude of inequality of the competition among its plants. They found the crop yield to be highest when the competition due to each plant is equal, so that all have equal access to the environmental resources required for plant growth and development.

Fasoula and Fasoula (1996) concluded that the low magnitude of competition in the isolation environment, which is associated with a low CV, enlarges the heritability of and consequent yield gain due to selection for yield on an individual plant basis. Their review of the literature includes examples for many crops that support gain in the effectiveness of selection for the yield of individual plants. Yan and Wallace (1995a) used a different approach to arrive also at the conclusion that selection for yield plant^{-1} will enhance the efficiency of genetic gain in crop yield if, but only if, the plants are spaced sufficiently apart to remove all interplant competition. Their conclusion was mathematically deduced. The crop yield potential of a genotype is positively related to the square of the potential yield of its individual plants. Reynolds *et al.* (1994) arrived at the same conclusion from research on wheat, and they cited other supportive research results on bread wheat, durum wheat, rye and cassava. All show selection for crop yield potential during the early generations to be made more effective by reducing the competition between the plants of the many different segregating genotypes from crosses. The reduction of interplant competition requires increased interplant spacing, the isolation environment being an extreme example.

10.2 Yield of Individual Plants Relative to Interplant Competition

One argument against selecting for weak competitive ability has been lack of a reliable measure for this trait (Marshall, 1991). Actually, the statement of Donald and Hamblin (1983) suggested the following simple and comprehensive measure of competitive ability.

If any plant within a crop acquires more water, nutrients, or light at the expense of another plant, it will have the potential to be advantageously represented by a larger number of progeny in the subsequent generations. This plant therefore has a higher probability of being selected, i.e. being a successful competitor (Donald and Hamblin, 1983). Thus, the number of seeds of a plant and the consequent number of progeny in a competing population are a measure of one aspect of the competitive ability. Seed number and seed size tend to be negatively correlated. But, large seeds confer some competitive advantages over small seeds, because they produce larger seedlings (Donald and Hamblin, 1983). The *grain yield plant*$^{-1}$, which is the product of the seed number plant^{-1} times the weight seed^{-1}, therefore, can be a more comprehensive measure of the competitive ability. Donald and Hamblin (1983) concluded that selection based on yield of the individual plants of a segregating population grown at a competitive plant density is actually selection for high competitive ability. The above statements by Donald and Hamblin will be re-examined.

10.2.1 The hyperbolic yield-density models

Donald (1963) wrote: 'Plant physiologists have studied single plants, and agronomists have looked at the whole crop, but individual plants within the community have scarcely been investigated. This area promises both scientific depth and great potential reward in terms of crop production.' Perhaps the best way to study 'the plant within the community' is density experiments to acquire information on interplant competition. The effects on crop yield by plant density have been studied mostly by agronomists rather than breeders. Their objective has been to determine the optimum sowing rate for maximizing the crop yield of the tested cultivars (Donald 1963), rather than to study interplant competition for the benefit of plant breeding. Nevertheless, density-yield relationships provide information about plant-to-plant competition within the population. Comparison of several densities facilitates a plant-density yield model. Donald (1963) considered the hyperbolic model to be the best among models that describe the yield-density relation, because it almost always gives high goodness of fit. The hyperbolic model is:

$$1/Y = \mathrm{a} + \mathrm{b}D \qquad [10.1]$$

where Y is the average yield plant^{-1}; D is the plant density, and; a and b are fitted parameters. Parameter b is a measure of the sensitivity of the genotype to increasing plant density. Since plant density is the primary element determining interplant competition (Goldberg, 1990), it is equally accurate to say that b is a measure of the sensitivity of the genotype to increasing magnitude of competition. Since all the plants in a crop are of the same genotype, b is also a measure of the competitive capability of the genotype.

Mathematical deduction reveals that 1/b estimates the asymptotic crop yield, i.e. the yield when the density is extremely high. However, this estimates the potential (maximum) crop yield of the genotype – if and only if – the data are compatible with the 'law of constant final yield' (Helenius and Jokinen, 1994). This is usually true for the biomass yield, but generally is not true for the grain yield (Donald, 1963; Donald and Hamblin, 1983). The expression 1/a estimates the asymptotic (maximum) yield plant^{-1} when the density is extremely low. However, in most cases this estimation is statistically insignificant or is negative and, therefore, is biologically meaningless. Consequently, the most important information to be learned from the hyperbolic model (equation 10.1) is its estimation of b, the sensitivity of the genotype to increasing density of its plants, which is also a measure of the competitive ability of this genotype.

For a mixture of two genotypes, the relation between the yield plant^{-1} of the genotype of interest (genotype 1) and the densities of the two genotypes is described by equation 10.2 (Roush *et al.*, 1989; Jokinen, 1991; Pantone and Baker, 1991; Helenius and Jokinen, 1994; Rezende and Ramalho, 1994):

$$1/Y_1 = a_1 + b_1 D_1 + b_2 D_2 \quad [10.2]$$

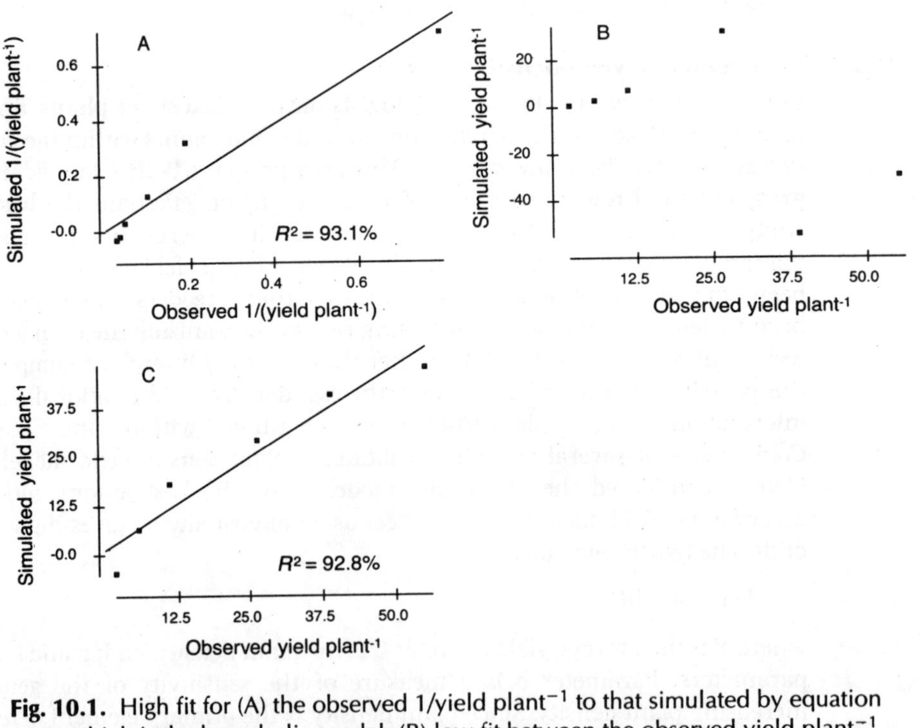

Fig. 10.1. High fit for (A) the observed 1/yield plant^{-1} to that simulated by equation 10.1, which is the hyperbolic model; (B) low fit between the observed yield plant^{-1} and that simulated by the hyperbolic model after conversion from 1/yield plant^{-1} back to yield plant^{-1}; and (C) high fit of the observed and simulated yield^{-1} by equation 10.8, which is the linear logarithmic model. (The cultivar is 'Local', and the data are from a density experiment of pigeonpea reported in Akinola and Oyejola, 1994.)

where Y_1 is the average yield plant^{-1} of genotype 1; D_1 and D_2 are the number of plants land area^{-1} of genotypes 1 and 2, respectively, and a_1, b_1 and b_2 are fitted constants, b_1 being the sensitivity of genotype 1 to the density of its kind and b_2 being the sensitivity of genotype 1 to the density of genotype 2. Mathematically, $1/a_1$ estimates the asymptotic yield plant^{-1} of genotype 1 at extremely low densities; $1/b_1$ estimates the asymptotic crop yield of genotype 1 in monoculture (when $D_2 = 0$), and: $1/(b_1 + b_2)$ estimates the asymptotic crop yield of genotype 1 in the mixture of the two genotypes which give the highest total yield. The probability of a_1 being statistically insignificant is very high. Therefore, again, the sensitivities of yield plant^{-1} of genotype 1 to the competition from both genotypes are the only reliable parameters that can be obtained from the hyperbolic equation 10.2.

There is an additional problem with hyperbolic models. Their goodness of fit for $1/Y$ is usually high (above 90%), but this does not mean the model will give high prediction of the yield plant^{-1}. The goodness of descriptive fit of hyperbolic models to $1/Y$ and its goodness of prediction of the yield are not the same. The goodness of prediction may be very poor, even when the goodness of fit is high. For example, the fitness to $1/Y$ for a set of pigeonpea density data of Akinola and Oyejola (1994) is higher than 90%, yet the goodness of prediction of yield plant^{-1} of the individual genotypes is as low as 2.8%. The high goodness of fit of the hyperbolic models results from linearization of the yield data by changing it to $1/Y$. It does not necessarily give as much biologically meaningful information as the fit to $1/Y$ suggests, however, because the fit can be much lower after $1/Y$ is converted back to Y (Fig. 10.1A and B). Further, it is not intuitively easy to understand the relation between the yield plant^{-1} and the density when $1/Y$ is used in place of Y *per se*.

10.2.2 *The yield of an individual plant and its response to plant density*

We applied the following model toward describing the yield–density relation when two or more genotypes are grown in mixtures.

$$Y_1 = a_1 - b_1 (D_1 - 1) - b_2 D_2 \qquad [10.\ 3]$$

where Y_1 is the average yield plant^{-1} of the genotype of concern (genotype 1). D_1 is the number of plants of genotype 1 land area^{-1} in the mixture, D_2 is the number of plants land area^{-1} of the other genotype(s) in the mixture, and a_1, b_1 and b_2 are fitted constants. The parameter a_1 estimates the potential yield plant^{-1} of genotype 1; b_1 estimates the sensitivity of genotype 1 to the density of the plants of genotype 1, while; b_2 estimates the sensitivity of genotype 1 to the density of the plants of all other genotype(s).

Equation 10.3 typically gives goodness of fit no lower than 80%, but not as high as the typical fit of equation 10.2. One advantage of equation 10.3 is that in addition to being easier to understand, all the parameters of this model are usually statistically valid, and therefore biologically plausible. Equation 10.3 implies that the yield plant^{-1} of genotype 1, symbolized as Y_1, will reach its maximum a_1 when there is no competition from any other plant, and that this maximum will be reduced by increased competition from progressively more plants of either its own genotype (b_1D_1) or of other genotype(s) (b_2D_2).

Assume a monoculture where there are no other genotypes. Now, $D_2 = 0$, and equation 10.3 becomes:

$$Y_1 = a_1 - b_1(D_1 - 1) = Y_{pure} \quad [10.4]$$

In contrast, if only one plant of genotype 1 is in the population and all other plants are of the other genotype(s), equation 10.3 becomes:

$$Y_1 = a_1 - b_2 D_2 = Y_{mix} \quad [10.5]$$

If D_2 is made up of many different genotypes, the situation is like that of a segregating population.

If we want to compare Y_{mix} and Y_{pure} at comparable plant densities, i.e. assuming $D_1 = D_2 = D$, then we have the following relation:

$$Y_{mix} = Y_{pure} + b_1 + (b_1 - b_2)D$$

or

$$Y_{mix} = a_1 - b_1 D + (b_1 - b_2)D \quad [10.6]$$

Equation 10.6 implies that, under comparable densities, the yield plant^{-1} of a given genotype in a mixture (Y_{mix}) is the yield plant^{-1} of that genotype in a monoculture (Y_{pure}), modified by its sensitivity to competition from plants of its kind minus its sensitivity to competition from the plants of other kind(s), i.e. modified by the term $[b_1 - b_2]$. If the yield plant^{-1} of the genotype of concern is more sensitive to competition from its kind than to competition from other kind(s), i.e. $b_1 > b_2$, then Y_{mix} will be greater than expected from Y_{pure}, and vice versa. Therefore, $[b_1 - b_2]$ is a measure of the relative competitive ability of a genotype over other genotype(s).

According to Donald and Hamblin (1983), Y_{mix} is a measure of competitive ability, and selection for Y_{mix} within a segregating population is selection for competitive ability, as discussed at the beginning of section 10.2. However, equation 10.6 reveals that the Y_{mix} is determined by four factors: (i) the potential yield plant^{-1} of the genotype (a_1); (ii) the sensitivity to competition from plants of its kind (b_1), which is the measure of that genotype's competitive ability in monoculture; (iii) the relative sensitivity to competition from plants of other kind(s), which is that genotype's relative competitive ability in mixture, and will disappear when the genotype is grown in pure stands; and (iv) the plant density. Under a given density, the greater yield of a plant in a segregating population (a greater Y_{mix}) may be due to either (i) a larger potential yield plant^{-1} of the genotype, and/or (ii) a smaller sensitivity to competition from plants of its kind, and/or (iii) a greater relative competitive ability over the other genotype(s). Thus, an observed yield plant^{-1} in a segregating population under competitive densities represents the effects by all three factors; it does not reflect just the competitive ability of the genotype.

Two points are important according to equation 10.6. First, selection based on individual plant performance within the segregating generations grown at competitive densities is likely, but not necessarily, to be selection for competitive ability. In agreement with Donald and Hamblin (1983), the individual performance under competitive density can reflect the competitive ability of a genotype, but only when the density (D) is so high that the term $[b_1 - b_2]D$ becomes the predominant part of Y_{mix}. The possibility that the individual performance reflects the competitive ability of a genotype decreases as D becomes lower and lower. Second, if the selection is based on the yield performance of individual plants at non-competitive densities (or isolation populations), it is selection for the potential yield plant^{-1} (a_1). Therefore, it is the plant density (D) that confounds the selection for yield among the individuals within segregating populations.

The relation between the potential yield plant^{-1} of a genotype and that genotype's potential crop yield when grown in monoculture is discussed next.

10.3 Relation Between the Potential Yield of Individual Plants and the Potential Crop Yield of a Genotype

10.3.1 Deduction from the linear yield-density model

The crop yield in a pure culture (Z) is, unquestionably, the product of the average yield plant^{-1} (Y_1) times the number of plants (D_1) in a land area. Considering the relation given in equation 10.4, we have:

$$Z = Y_1 D_1 = [a_1 - b_1(D_1 - 1)]\, D_1$$

Mathematically it can be proven, given the relationship of equation 10.4, that the maximum crop yield (Z_{max}) of a given genotype, which can be achieved when the density is optimum [$D_{opt} = (a_1 + b_1)/(2b_1)$] under a given environment, is:

$$Z_{max} = (a_1 + b_1)^2/(4b_1) \qquad [10.7]$$

where a_1 is the yield potential of the individual plants, and b_1 is the sensitivity to increasing density of the genotype. Simulation by equation 10.7 (Fig. 10.2) indicates the following. (i) When a_1 is held constant, Z_{max} responds to b_1 in a hyperbolic shape. That is, the potential crop yield of a genotype in monoculture (the Z_{max}) is negatively related to the sensitivity to the density, and therefore to the competition from the plants of its own genotype, b_1, but this relation becomes weaker and weaker as b_1 becomes larger (Fig. 10.2A). (ii) When b_1 holds constant, Z_{max} responds to a_1 in a parabolic shape. The potential crop yield of a genotype in monoculture (the Z_{max}) is positively related to the potential yield of its individuals (a_1) (Fig. 10.2B). A similar statement was made by Yan and Wang (1992) and Yan and Wallace (1995a).

Equation 10.7 conveys the following clear messages. Selection of individual plants during the segregating generations should be: (i) for high potential yield plant^{-1} under a non-competitive density (a_1), and (ii) against sensitivity to crop density, i.e. against sensitivity to increasing density of plants of its own genotype (b_1). Selection for yield plant^{-1} (a_1) is possible during the segregating generations, but direct selection against sensitivity to its own genotype (b_1) is not possible within a segregating population. Nevertheless, indirect selection for this sensitivity can be done by selecting against any morphological or phenological traits known to contribute toward greater sensitivity to increasing plant density.

Knowledge about the relation between the potential yield plant^{-1} (a_1) and

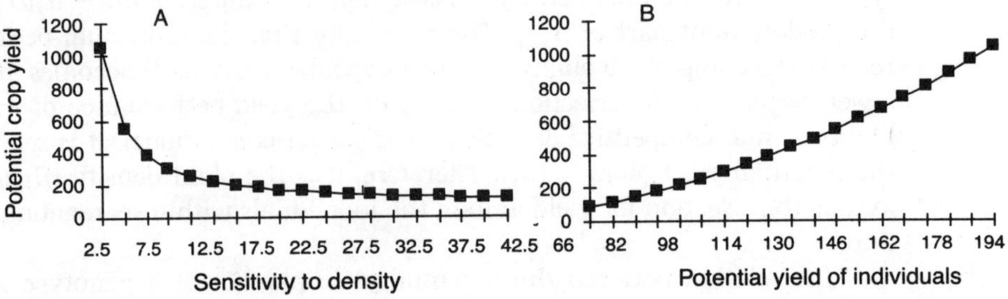

Fig. 10.2. Simulations by equation 10.7 of: (A) influence of the sensitivity to plant density of a genotype on its crop yield potential when the maximum potential yield plant^{-1} is fixed; and (B) influence of the potential yield plant^{-1} of a genotype on its crop yield potential when the sensitivity to plant density is fixed. The simulations assume the sensitivity to density and the potential yield plant^{-1} are independent, but usually they are negatively correlated. The assumed fixed values are arbitrarily chosen.

the sensitivity to density (b_1) is rare. Preliminary investigation suggested they are not independent; rather, they are strongly and positively correlated (Yan and Wang, 1992). But, the yield plant^{-1} and its sensitivity to increasing density alter the crop yield oppositely, thereby demanding selection for an optimal compromise of their levels. Further research is needed of these optimal levels.

Admirably, Donald (1968) pointed out: 'The capacity of a genotype to yield well in a community can be analyzed in terms of two parameters, namely, i) the yield plant^{-1} in the absence of competition from neighbors, and ii) the change of yield plant^{-1} in response to crowding by other plants of like genotypes.' Respectively, these are exactly a_1 and b_1 in equation 10.7.

The following citation presents the reasoning of Donald (1968) that led to his ideotype of weak competitors.

> Clearly the individual plant within the community will express its potential for yield most fully if it suffers minimum interference from its neighbors. Thus its neighbors should be weak competitors. And since, for the purpose of discussion, all plants in the crop are of like genotype, then the ideotype itself must be of low competitive ability.

Here, Donald equated the competitive ability of a genotype with its sensitivity to competition from plants of like genotype, the latter being more easily measured. Thus, a weak competitor is a genotype whose individual yield is relatively less sensitive to competition from its neighbours of like genotype; it is more tolerant to increasing crop density; that is, it has a small b_1. However, while emphasizing the importance of this low competitive ability – low sensitivity to increasing plant density being our preferred terminology – Donald neglected the at least equally important trait – the potential yield of the individual plants of the genotype (a_1).

Equation 10.7 implies that to improve the crop yield potential, a large maximum yield plant^{-1} (a_1) may be even more important than a small competitive ability (b_1); both should be considered in designing an ideotype for high yield, and for the selection of individual plants. A tall genotype is not likely to have a high crop yield potential, since it tends to be sensitive to increasing plant density. Also, a uniculm genotype is unlikely to have a high crop yield potential, because of low potential yield plant^{-1} compared with multiculm plants. This explains why selection for Donald's uniculm ideotype has not advanced yield. It explains also why Rasmusson, a strong advocate of ideotype breeding, prefers an ideotype for barley that has many tillers rather than a single main culm (Rasmusson, 1987, 1991; Loomis, 1993).

10.3.2 Deduction from a linear logarithmic yield-density model

As briefly discussed in section 10.2.1, the hyperbolic model, by using $1/Y$ (equations 10.1 and 10.2), has the advantage of almost always giving high goodness of fit. But, it has the following disadvantages. (i) Predicting the yield requires transformation of the yield plant^{-1} to its crop yield *per se*; consequently, the high goodness of fit does not always give a high goodness of prediction (compare Fig. 10.1A and B). (ii) The estimation of 'a' in equation 10.1 or of 'a_1' in equation 10.2, the inverse of which predicts the potential yield plant^{-1}, is usually statistically insignificant and thus biologically meaningless. (iii) The hyperbolic model cannot give reliable information about the crop yield potential of a genotype, if the density–crop–yield relation does not fit the 'law of final constant yield'. And, this failure occurs whenever the yield is a portion of rather than the total biomass (Donald, 1963). Thus, deduction from the hyperbolic models of the potential crop yield of a genotype, which is inversely related to its sensitivity to increasing density, is invalid for most cases. (iv) Therefore, the hyperbolic models give no information about either the potential yield of the individual plants of a genotype or that genotype's potential crop yield. Such information is required, since individuals are the basic unit of selection (and evolution), while the potential crop yield is a final objective of the breeding.

The linear model (equations 10.3 to 10.5) is more informative than the hyperbolic model. However, its disadvantage is that it usually does not give as high a goodness of fit. This is a constraint, because the level of the goodness of fit determines the validity of equation 10.7, and the conclusions derived from it. Therefore, a third model is needed, one that combines the advantages of both the hyperbolic and linear models. We tried the following linear logarithmic model:

$$Y = \mathrm{a} - \mathrm{b}\ln(D) \qquad [10.8]$$

where Y is the yield plant^{-1}, D is the crop density, and a and b are fitted constants. Respectively, a and b estimate the potential yield plant^{-1} and the sensitivity to increasing crop density. Although not shown, extensive tests based on published data in the literature indicated this model usually gives goodness of fit as high as, or only slightly lower, than the hyperbolic models. Although the hyperbolic model (equation 10.1) and the linear logarithmic model (equation 10.8) give comparable goodness of fit, the latter describes much better the *true* relation between the yield plant^{-1} and the crop density. This is shown in Fig. 10.1.

The yield land area^{-1} (Z) is the product of the yield plant^{-1} times the crop density. Therefore:

$$Z = Y \times D = [a - b\ln(D)]\,D$$

Mathematical deduction gives the following relation among the crop yield potential (Z_{max}), the potential yield plant^{-1} (a), and the sensitivity to the density (b):

$$Z_{max} = b\exp(a/b - 1) \qquad [10.9]$$

Verbally, under a given environment, the potential crop yield of a genotype is exponentially and positively related to its potential yield plant^{-1}, and exponentially but negatively related to its sensitivity to crop density. This conclusion and simulation by equation 10.9 (Fig. 10.3) is qualitatively identical to that for equation 10.7 (compare Fig. 10.3 with Fig. 10.2). That is, selection for greater crop yield potential should include: (i) selection for higher potential yield plant^{-1} and (ii) simultaneous selection against sensitivity to increasing crop density.

To illustrate the need for these simultaneous selections, the data of a maize density experiment of Tollenaar (1989) are re-analysed (Table 10.1). Eight cultivars released between 1959 and 1988 were compared under four crop densities. Equation 10.8 fits the data with R^2 ranging from 91.7% to 99.9%. The maximum crop yield potential estimated by equation 10.9 was very close to the observed maximum for all eight cultivars. The new cultivars had consistently higher potential yield plant^{-1} than the old ones, while the sensitivity to density tended to increase or to remain at the level of the old cultivars.

That lower sensitivity to high plant density can contribute toward higher yield is indicated by comparing cvs. Pioneer 3978 and Pioneer 3902. Both had high yield potential. Pioneer 3902 had a much higher potential yield plant^{-1}, while Pioneer 3978 was much less sensitive to density. Thus, both a higher potential yield plant^{-1} and greater tolerance to crowding can increase the potential crop yield. Multiple and complementary pathways to higher yield exist, but their comparative levels must be optimized.

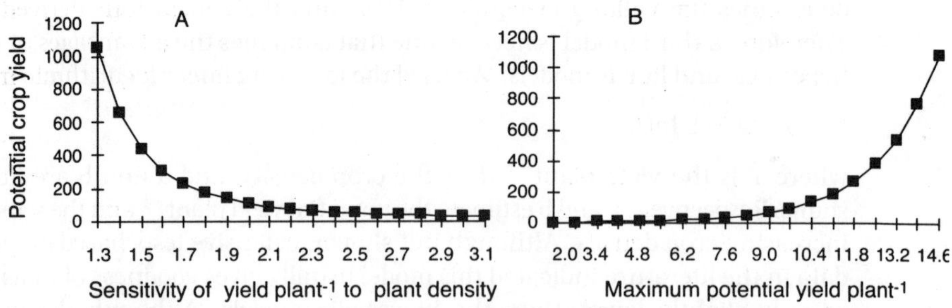

Fig. 10.3. Simulations by equation 10.9 of: (A) influence of the sensitivity to plant density of a genotype on its crop yield potential when the maximum potential yield plant^{-1} is fixed; and (B) influence of the potential yield plant^{-1} of a genotype on its crop yield potential when the sensitivity to plant density is fixed. The simulations assume the sensitivity to density and the potential yield plant^{-1} are independent, but usually they are negatively correlated. The assumed fixed values are chosen arbitrarily.

Table 10.1. Potential yield plant^{-1} (a) and sensitivity to density (b) of eight maize cultivars released from 1959 to 1988. (The parameters are by equation 10.8 and the data are from Tollenaar, 1989.)

Cultivar	Year of release	R^2 fit (%)	a (g plant^{-1})	b (sensitivity)	Estimated Y_{max} (t ha^{-1})	Observed Y_{max} (t ha^{-1})
Pride 5	1959	93.40	269.38	88.29	6.87	6.97
Warwick 263	1961	98.00	273.53	86.36	7.54	7.34
Dekalb 29	1962	97.30	249.80	78.66	6.93	6.80
United 106	1970	91.70	216.91	65.42	6.63	7.01
7PAG SX111	1973	98.30	383.79	132.03	8.89	8.55
Pioneer 3978	1978	97.40	304.68	88.54	10.17	10.09
Pioneer 3851	1983	99.90	321.68	99.59	9.26	9.24
Pioneer 3902	1988	99.80	407.04	130.83	10.80	10.87

Russell (1986) presented similar evidence for maize. Four original-population varieties and four single-cross hybrids for each 10-year era of 1930 to 1980 were compared for yield at low, intermediate, and high plant densities across 8 location-years. The parameters for the yield–density relations for each of the seven cohorts of varieties estimated using equation 10.8, are presented (Table 10.2). Again, the goodness of fit is high, from 98.8% to 100%, and the predicted maximum yields are very close to the observed highest yields. It can be seen that: (i) all the improved cultivars have higher yield potential than the original population because of higher potential yield plant^{-1}, rather than because of less sensitivity to density; (ii) among just the improved cultivars, however, their highest crop yield potentials are associated with less sensitivity to crop density. Thus, it is increased potential yield plant^{-1} that accounts for the superiority of the modern over the older varieties, but less sensitivity to crop density accounts for the highest yields among the modern cultivars. Consequently, a higher a : b *ratio* (from equation 10.9) best describes the improved yield potential of the newer cultivars.

Table 10.2. Potential yield plant^{-1} (a) and sensitivity to density (b) of seven successive subsets of maize varieties released during the 1930s to 1980s. (The parameters are estimated by equation 10.8 and the data are from Russell, 1986.)

Cultivars released during	R^2 fit (%)	a (g plant^{-1})	b (sensitivity)	Estimated Y_{max} (t ha^{-1})	Observed Y_{max} (t ha^{-1})
Non-hybrid	98.80	381.07	158.83	6.44	6.34
1930s	99.70	474.83	190.20	8.49	8.27
1940s	99.70	465.24	190.62	8.05	7.80
1950s	99.40	448.37	181.95	7.87	7.60
1960s	99.90	462.81	173.27	9.21	9.11
1970s	100.00	447.04	156.21	10.05	10.10
1980s	99.70	478.79	170.43	10.41	10.60

Unpublished data for three Chinese wheat cultivars, which were important cultivars in the 1960s, 1970s, and 1980s, respectively, indicate that the relatively high crop potential yields of the newer cultivars were associated with a higher potential yield plant^{-1}.

10.4 Strategies of Selection

10.4.1 Selection during the segregating generations

The above discussion of interplant competition explicitly indicates that the crop yield potential of a genotype is determined by: (i) its potential yield plant^{-1} and (ii) its tolerance to increasing plant density. This relation leads to the following suggestions for strategies of selection during the segregating generations. First, effective selection for yield during the early segregating generations is possible. Second, this selection should be: (i) for higher potential yield plant^{-1}, and (ii) for less sensitivity to increasing plant density, as quantified by a smaller reduction of the yield plant^{-1} as the plant density of the given genotype is increased. During the segregating generations, the latter can be achieved only by selecting for traits known to influence this tolerance to density, i.e. to decrease the sensitivity to interplant competition. Finally, negative association is expected between these two traits. Therefore, a compromised optimal level of each is required, to achieve the greatest gain in crop yield potential.

10.4.1.1 Selection for higher potential yield plant^{-1}

By definition, the potential yield plant^{-1} of a genotype is its highest possible yield for an individual plant. This potential can be expressed only in environments with no limitation by water, nutrients, pests or diseases, and with no competition from other plants. However, all breeders have a target environment with limited resources for growth. Therefore, it is realistic to evaluate and select for high potential yield plant^{-1} for each different target environment. This requires growing the segregating population at a density low enough that there is no or only minor competition among the individuals, and selecting the individual plants that give the higher yield. Experienced breeders need not measure the plant yield directly. The yield can be judged visually. For cereal crops, for example, many large heads and normal maturation will certainly give high yield plant^{-1}. For pulses, many large pods combined with good seed size will give high individual yield.

10.4.1.2 Selection for less sensitivity to increasing plant density

During the segregating generations, to the extent possible, it is important to select against sensitivity to density, simultaneously with the selection for greater yield potential plant^{-1}. It is not possible during these generations, however, to select specifically against sensitivity to the crop density of the same genotype. It is possible, nevertheless, to select against traits known to contribute to high sensitivity to density. All the traits Donald and Hamblin (1983) consider to confer weak competitive ability (section 10.1) are likely to alter favourably the sensitivity to density. Reduced plant height, and leaves that allow light penetrance into the canopy are among the more favourable traits. Generally, plants that occupy the larger space, either vertically (being taller) and/or horizontally (being wider), tend to increase the sensitivity to crowding. Such plants can be discarded, therefore, if they do not produce a yield proportional to the space they occupy. Further identification is needed of the traits that can be used for indirect selection against sensitivity to higher plant density.

10.4.1.3 Eliminating plants with bad trait(s)

It is safest to select against traits known to be definitely 'bad' traits for the target environment (Yan *et al.*, 1991a). In fact, it is always easier to recognize and eliminate bad plants than to recognize and select the 'good' plants in a segregating population. The most obvious bad plants are those that show susceptibility to serious disease(s) or pest(s). Plants having these often 'independent' (section 11.1.1) non-quantitatively variable traits should be discarded as early as possible. Plants can also be judged from their levels of quantitative traits (Chapter 11). For wheat, the bad quantitative plants include those that are too late, too tall, too droopy, too vulnerable to cold, and plants that bear heads that are too small. With this 'safe discard' policy, Yan *et al.* (1991a) estimated that at least 75% of the plants of a segregating population can be discarded with confidence in each generation. The remaining plants should be allowed to express their crop yield potential in the advanced generations.

10.4.2 Selection during advanced generations

The advanced generations have become almost homozygous, making it possible to evaluate each selected genotype directly for its crop yield capability. Multiple locations and/or years are essential for reliable information. For understanding the variability of crop yield as related to the interplant competition, multiple densities could be incorporated into the advanced-generation yield testing to derive the following information about each tested genotype: (i) its potential yield plant^{-1}, and (ii) its sensitivity of the yield plant^{-1} to crop density. This will provide additional estimations about each genotype's (iii) potential crop yield; and (iv) optimum crop density which determines the seeding rate that will maximize the crop yield.

Carefully planned multiple-density yield trials will give the needed information about the relation between sensitivity to density and the various morphological and phenological traits. Such information can feed back to assist selection during the segregating generations. In section 10.3.2, it was shown that the relationship between yield plant^{-1} and the logarithmically transformed densities is nearly linear (equation 10.8). Therefore, during the advanced generations, as few as three densities are sufficient to quantify the needed information about the potential yield plant^{-1} of the genotypes and the sensitivity of their yield plant^{-1} to increasing plant density. The increased cost can be balanced by reducing the experimental replications.

10.4.3 Coefficient of variation as an indicator of crop productivity

The concept of Fasoula and Fasoula (1996) is highly relevant to breeding for higher yield. Their concept, described briefly in the third paragraph of this chapter, is that crop yield is negatively associated with the CV of the yield plant^{-1}

across the plants of a crop stand. They consider the CV to be enlarged by unequal competition among the plants. This reasoning led us to the following considerations.

1. Crop yield has an optimum level of the interplant variability among the plant population, as quantified by the CV.
2. Large CV results from unequal competition among the plants.
3. Unequal competition arises from any post-emergence variation between the plants for size and/or stage of development. This post-emergence inequality of competition is enlarged by increases of plant density, i.e. by increased intensity of the interplant competition, because any initially disadvantaged and therefore less competitive plants progressively obtain even smaller proportions of the limited available resources. The most disadvantaged plants can be eliminated, or can have their yield reduced to zero by severe competition.
4. The initial interplant variation can be genetically established. Even with no genetic variation, however, variation can be acquired during the original growth and development of the seed at different positions on the inflorescence, and/or acquired due to environmental differences during germination of the seed, or early growth of the seedlings. All non-genetically directed variabilities are called acquired diversity.
5. To improve the genetic potential for crop yield breeders should select within their target environment (i) for high genetic uniformity and (ii) select against any

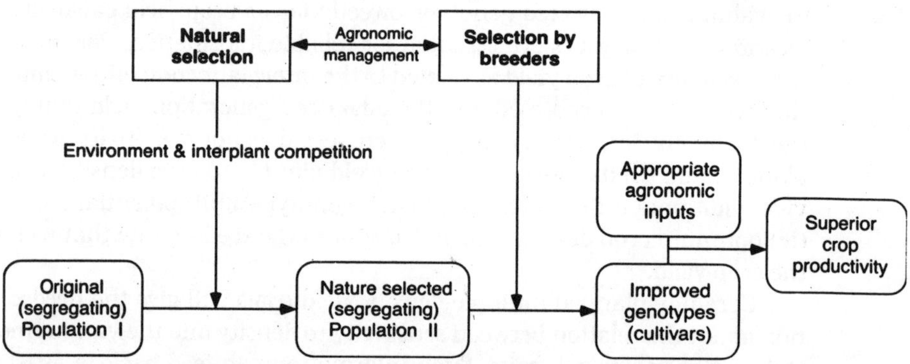

Fig. 10.4. Relationships between natural selection by the environment and selection by breeders. In pre-agricultural times, selection was exclusively by natural selection. This included selection by physical factors such as temperature, photoperiod and water availability, and by biological factors including diseases, pests and competition among the plants. Selection by the physical factors leads to ecological adaptation. Selection through interplant competition leads to stronger competitors but low yield potential. In a breeding programme man incorporates additional and purposeful selection. This selection is for the traits thought to be desirable, and through natural selection due to the agronomic management. If the breeder does not appropriately choose and/or control the environment, natural selection is likely to eliminate the desired genotypes before he/she has the opportunity to identify and select them. Agronomic inputs and management are integral to selection of genotypes that will maximize the gain in potential crop productivity of improved cultivars.

genetically determined tendency for acquired diversity. Selection for genetic uniformity is the common practice, but selection for reduced genetic tendency for acquired diversity has received minimal attention.

6. Traits and environmental factors that confer the tendency for acquired interplant diversity should be identified and studied. Lack of uniform size of the seeds from an individual plant is an example of a genetically established tendency for acquired diversity by the post-emerged plants.

The CV is calculated as follows:

$$CV = S/M$$

where S is the standard deviation among the plants within a population, and M is the average yield across the plants. Therefore, selection for a low CV means selection for high average yield plant^{-1}, and/or for low variation among the plants. Selection for a large M in widely spaced populations corresponds with selection for a greater yield potential of the individual plants.

Chapter 10 Summary

Background

Successful cultivars arise from the combined results of natural and human-imposed selection (Fig. 10.4). A breeder's intentional selection is virtually always preceded and accompanied by natural selection. The natural selection is toward: (i) adaptation to the environment in which the plants are growing and (ii) stronger competitiveness than the neighbouring plants. The gain in crop yield of a breeding programme depends on the natural selection by the environment, in addition to the selection by the breeder. Therefore, the direction and magnitude of the natural selection should be managed by the breeder. He/she should choose or establish testing environments characterized by the temperature and photoperiod and soil conditions of the target environment, and by modifying the water, fertilizer, weeding, plant density, etc.

Breeders focus on the consequences from selection using their own hands. They are often less aware of the selection by the hands of nature. The probability must be minimized that the hands of nature have discarded the desired high yielding genotypes before the breeder can select them with his/her hands. Undesired discarding by nature is highly likely during the early segregating generations, especially when the plants are grown under high competitive pressure, i.e. under high plant densities like those of the crop. At such high densities, the nature-selected plants are the strong competitors against the neighbouring plants, and they are usually poor yielders when homozygosity results in all the plants having the same genotype.

Conclusions

The important applicable conclusion is that in the future the natural selection for stronger interplant competition during the early segregating generations can and should be reduced, by growing the segregating populations at lower plant densities. In these low plant-density environments, the breeder should select the early-generation segregates (genotypes) having the highest yield potential plant^{-1}. From a whole-system viewpoint of interplant competition, this potential yield plant^{-1} of the genotype is one of but two 'major components' that account in full for the potential crop yield of the genotype (equations 10. 7 and 10.9). The second component of the crop yield potential is the tolerance of the yield plant^{-1} of the genotype to increasing plant density, i.e. a low reduction of the yield plant^{-1} as the plant density is increased to the high density needed for high crop yield per land area.

Reduction in yield plant^{-1} is precisely equivalent to the conclusion of Donald (1968) that the breeder should select for low competitive ability. However, rather than selecting for weak competitive ability in a broad sense, as Donald suggested, it is more focused and establishable as a goal to select for greater tolerance to increasing plant density, i.e. for smaller reduction of the yield plant^{-1} increase^{-1} of the plant density. During the early segregating generations a problem is inability to directly and simultaneously select for both maximum yield plant^{-1} and smaller reduction of the yield plant^{-1} per plant density. It is possible to select for traits known to contribute toward lower competitive ability. Such traits include reduced plant height, leaves that allow light penetrance into the canopy, plants that occupy small vertical and/or horizontal space and a high harvest index.

A further conclusion follows. The crop yield is maximized when every plant expresses the same level of interplant competitiveness. Maximization occurs because the reduction of yield of the poorer competitors is greater than the gain in yield of the superior competitors. This is one of multiple reasons that cultivars are selected to be genetically uniform.

Further discussion of selection based on the effects on yield through interplant competition is in section 11.3.3.1. There, interplant competition is considered to be one of the major components that occupy the constant capacity of the plant system. Section 12.4.5.2 details how to apply selection for high harvest index being indirect selection for high tolerance to increasing plant density.

11 System-established Interconnections among Plant Traits and Implications for Plant Breeding Strategies

Paradigm shift 28 is stated and/or implied throughout all chapters of this book. Its physiological basis is most thoroughly derived and detailed within this chapter. Shift 28 is that selection for a physiological–genetic component trait that is intended to give indirect selection toward higher level of a system output must be toward the optimal level of any trait that is a near-fully integrated component of the process of achieving that system output. If the indirect selection is toward either a higher or lower level than is the optimum level of the component trait, the selection will reduce the system functioning, which will lower rather than raise the targeted system output.

When Yan began his career as a wheat breeder, he was asked to breed early-maturing, large-headed wheat cultivars. He experienced no difficulty selecting genotypes which had either very early or very large heads. These were crossed. Several generations later, some segregates had both early maturity and large heads. However, all had spring habit, in spite of the winter habit of much of the parental germplasm. Modest winter habit was essential to withstanding the low *T* of winters in his region in central China. Therefore, another breeding cycle was begun, the objective being cultivars with all three traits: earliness, large heads and winter habit. This objective was not realized; genotypes with any two of the three desired traits were easily found. If not impossible, however, it was extremely difficult to find plants that had all three traits.

The segregates indicated no difficulty in combining any two of the three traits and, therefore, no genetic linkage among the three traits. Nevertheless, the relation between the levels of any pair of the traits depended on the level of the third. This led to interpretation that the three traits can be represented as a circle with a constant radius and therefore area, with the three traits competing for this constant area (Yan, 1992, 1993a; Yan *et al.*, 1993). If one trait was selected to have a higher level, one or both others must occupy a smaller area within the circle. This is a highly simplified statement, since the complete plant system involves many more traits.

Yan (1993a) extended the above mentioned three-trait circle to involve nine traits including: maturity, morphological yield components, product quality, and resistance to environmental stresses. Yan called the interpretation that these multiple traits share the area of a circle and compete for its area 'the hypothesis of constancy'. Yan and Wallace (1995a) called this interrelationship among plant traits 'the hypothesis of a constant-capacity plant system'. This hypothesis is used in this chapter to interpret the interrelationships among all the quantitative traits considered desirable for breeding a superior cultivar, to interpret cultivar improvements of the past, and to suggest strategies for continuing future crop improvement. As discussion of the constant-capacity hypothesis proceeds, readers will find supporting evidence based on their own experiences.

11.1 The Hypothesis of a Constant-capacity Plant System

11.1.1 Independent versus interdependent traits

For each plant species, numerous morphological and physiological 'traits' are desired for a cultivar. These include independent and interdependent traits (Yan, 1993a; Yan and Wallace, 1995a). Independent traits are qualitatively expressed with no dependency on the level of other traits (sections 11.5 and 12.4.4). Resistances to diseases are often, but far from always, independent traits. On the contrary, the level of an interdependent trait is inevitably positively or negatively correlated with the level of one or more other quantitatively variable traits (Figs 2.1, 5.1, 9.1; section 12.1). Breeding progress is constrained by negative associations between desired traits. Biological bases for negative associations among levels of the quantitatively variable traits may include: genetic linkage; pleiotropic genetic controls (Rasmusson, 1987; Marshall, 1991); environmental controls through interconnections among the physiological processes of the plants (Yan *et al.*, 1993; Yan, 1993a; Wallace *et al.*, 1993a, b), or; be due to two or more of these causes. The hypothesis of a constant-capacity plant system and the necessity of its major components to share the available intraplant metabolites constitute a generalized explanation for the system-established interconnections and interdependencies among all the quantitatively variable traits of plants. The pervasiveness of such system-established interconnections is illustrated in every chapter of this book (Figs 2.1, 5.1 and 9.1).

Cultivars are generally evaluated based on their yield, product quality, stability within an ecological region, and range of adaptation across ecological regions. We suggest (Fig. 11.1) five categories of the interdependent and quantitatively variable traits: (i) early versus late maturity (compare Fig. 11.2); (ii) resistance to environmental stresses such as drought, cold, low soil fertility, etc.; (iii) competitive ability of plants of the genotype to access the limited growth resources as compared with other genotypes (weeds and also plants of the same crop species); (iv) morphological (compare Fig. 11.3) and physiological components (compare Fig. 11.4) of yield; (v) product quality. The morphological components of yield vary from species to species. For cereals or pulses, they are the heads or pods land area^{-1}; seeds head^{-1} or pod^{-1}, and the weight seed^{-1}.

Fig. 11.1. Illustration of the hypothesis of a constant-capacity plant system. This hypothesis has the following properties. (i) There are five categories of traits that are essential to the success of crop cultivar. Each may include multiple sub-component traits. Collectively, the categories and their sub-components occupy the near constant capacity plant system represented by the area (and diameter) of the circle. (ii) This near constant capacity of the holistic system is established by the limited resources for growth available from each given environment, and the stresses of that environment. Each environmentally established near constant capacity applies to all genotypes of each crop species grown in the given environment. Superior environments enlarge this near constant capacity across all the genotypes. (iii) All five categories, including their subcomponent traits, compete for their share of the holistic system's constant capacity. Their share becomes the part they receive of the available intraplant metabolites. The competition for the available metabolites renders it impossible to maximize genetically all the categories and their component traits at the same time. Increase of any one trait will necessarily shrink one or more of the other component traits. For the same reason, reduction of the level of any one trait can automatically increase the level of one or more of the others. (iv) Crop yield can be increased by: (a) enlarging the capacity of the system by supplying agronomic inputs; (b) reducing the level of the categories or traits that are at genetically established levels that are excessive for the given environment; and (c) selecting toward that lower or higher level of all the traits that will compensatorily optimize the system functioning. (Modified from Yan and Wallace, 1995a.)

The major physiological components of yield are the biomass accumulated day^{-1}, the harvest index (HI), and the days to harvest maturity (DTHM). These components are universal for nearly all crop species (Fig. 11. 4, compare Figs 2.1, 5.1 and 9.1). The diverse factors of product quality include flavour, texture, organ size and colour, protein content, etc.

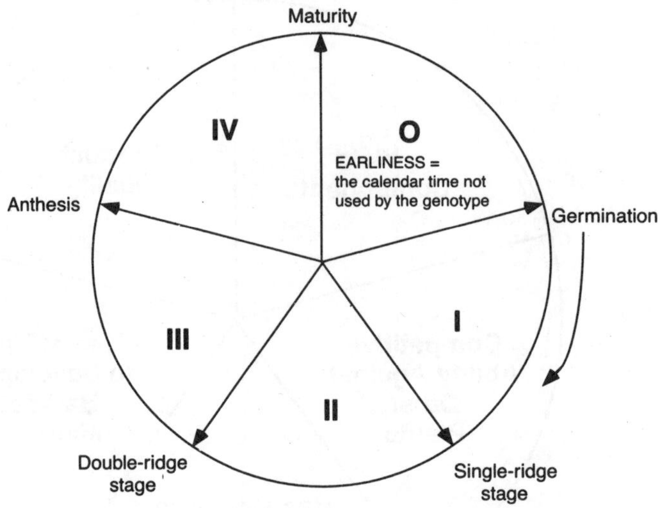

Fig. 11.2. Four morphological phases of wheat development and their usages of calendar time. The circle represents the constant 365 days of a year. The morphological phases I to IV, respectively, occupy the time spans used to complete development to the stages of single-ridge, double-ridge, anthesis and maturation. The time from maturation to sowing and germination of the next crop is a quantitative measure of the earliness required by the cultivar, which is designated as phase O. Increase (or decrease) of the duration of any phase automatically and simultaneously decreases (or else increases) the duration of one or more of the other phases. The time span of phase I is directed by vernalization genes, but the duration required for cultivar adaptation depends on the environmentally established need for winter hardiness or for resistance to other environmental stresses. The duration of phase II is directed by both vernalization genes and the photoperiod genes, and it determines the potential size of the heads of wheat. Most of the crop's photosynthate is accumulated during phases III and IV. (Modified from Yan and Wallace, 1995a.)

11.1.2 The hypothesis of constant-capacity system

The hypothesized constant capacity of a system is illustrated by the fixed diameter and area of a circle (Fig. 11.1). The hypothesis for the system of a successful crop cultivar can be stated as follows.

1. The cultivar's five categories of traits constitute the complete biological system of the plant.
2. The capacity of the system is represented by the total area of the circle.
3. For each given environment, the diameter of the circle is a near-constant for all the genotypes of the specified plant species. The expressed level of each of the five categories of traits that make up the system – including all the component and subcomponent traits within each category – involves a competitive requirement by every category for a large proportion of the constant capacity of the plant system.

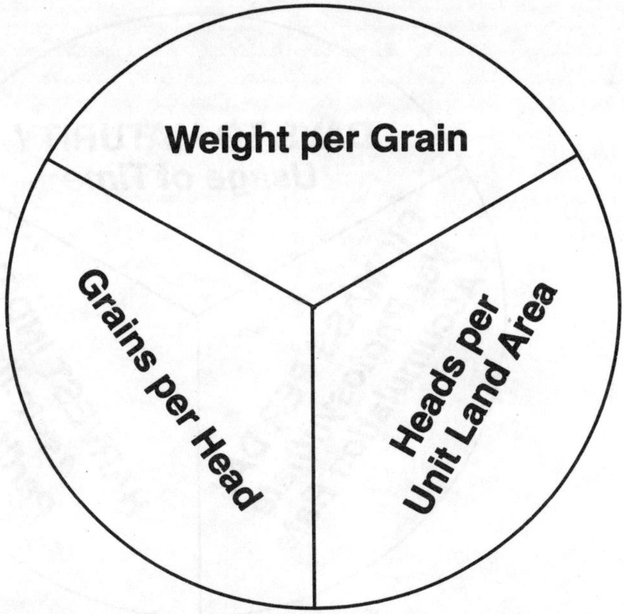

Fig. 11.3. Compensation among the morphological components of yield. Assuming that all categories in Fig. 11.1 other than these morphological yield components are fixed, then the capacity of the subsystem consisting of these components is also fixed, for the given environment. Therefore, each of these three morphological components of the yield must compete for its part of their subsystem's fixed capacity. This fixed capacity is indicated by the diameter (and area) of the circle above. Different allocations of the capacity among the three yield components may give different or the same yield. The yield is the product of all three components. But, since each component competes for part of its subsystem's constant capacity, the yield cannot be high if any one of the three components is either too small or too large relative to the other two. The yield can be at any level from zero (if any one of the components becomes zero), through large to the highest yield potential for the specific environment and genotype. Attaining the maximum potential yield of the environment requires that compromised level of each of the three morphological components that maximizes functioning of the subsystem for yield. The morphological yield components presented here are for cereal crops like wheat. Other crops will have different morphological components of their yield. (Modified from Yan and Wallace, 1995a.)

4. The capacity of the system will be enlarged by a superior environment. That is, the functioning system capacity that is represented by the radius and therefore the area of the circle will be enlarged by a more favourable environment. Nevertheless, for comparison of the multiple genotypes of the species, the enlarged system capacity is again near-constant. The same five categories, including all their components and subcomponents, must share the enlarged but again constant capacity represented by the area of an enlarged circle.

This hypothesis of a constant-capacity plant system leads to the following statements:

Fig. 11.4. Compensation among the physiological components of accumulating yield. Under a given growing environment, if levels of the categories and traits in Fig. 11.1 other than these three physiological components of the process of accumulating the yield remain unchanged, the proportion of the capacity of the holistic system represented by these three components also remains unchanged and constant. Thus, these three major physiological components of the yield compete for the fixed capacity of their subsystem, this capacity being represented by the area of the circle. Different allocations of the constant capacity of this subsystem among its three components may give different or the same yield. The yield may vary from zero (if one of the components becomes zero) through large to the highest potential yield of the given environment combined with the given genotype. The yield is the product of all three components. A compromised and thereby optimized combination of levels among the three physiological components is required to maximize the system functioning and thereby maximize the yield. The yield cannot be high if (i) any one of the three components is either too small or too large relative to the other two or if (ii) other of the components of the holistic system (shown in Fig. 11.1) have genetically directed levels that are inappropriate for the given environment. The control is by the system as a whole.

1. All five categories of traits are required for a successful cultivar. However, the constant capacity of the system renders it impossible to maximize the genetically established levels of all five categories and/or of their component traits.
2. Under each environment, improvement of one category (or trait) unavoidably forces shrinkage of one or more of the other categories (or traits). Likewise, reduction of the level of any one automatically raises the level(s) of one or more of the others.
3. The overall system capacity is established by the environment, but is nearly

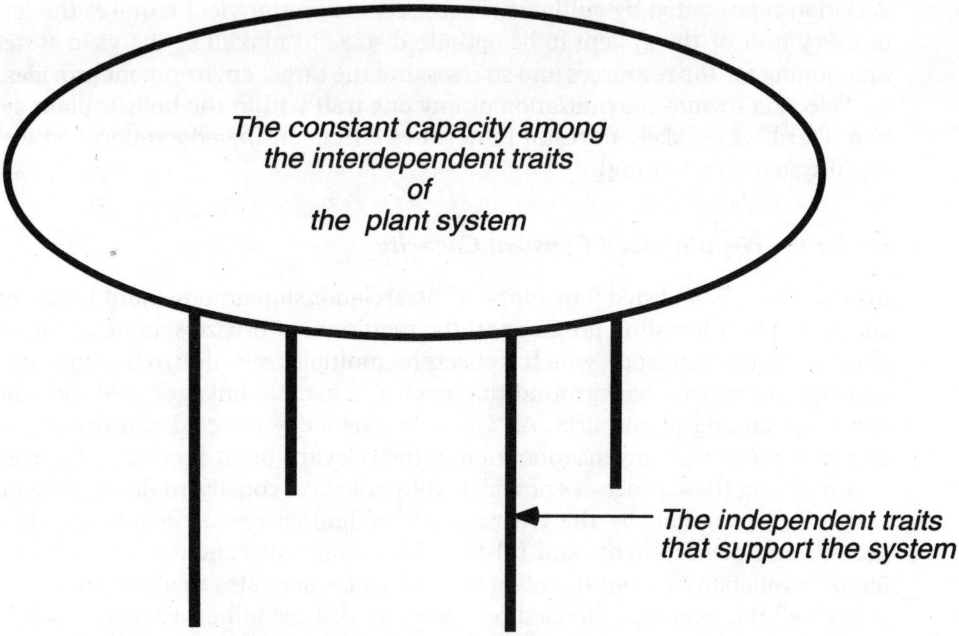

Fig. 11.5. An illustration of the essentiality to a cultivar of both the interdependent and independent traits. Each genotype is comparable to a round table. Its interdependent and quantitatively variable traits compete for the constant capacity of its plant system, which is represented by the circular table top. The independent traits that support the system functioning are represented by the legs of the table. Breeding a successful cultivar involves two virtually separate tasks: maximizing levels of the independent traits, but optimizing levels of all the interdependent traits. Both tasks are required to maximize the system output, such as the yield.

the same for every genotype grown in that environment. The different genotypes do differ in yield (and in other categories and traits), however, because the genetically directed differences in their allocation of the system capacity among the categories and traits establish superior or inferior adaptation of the genotype to the given environment, and thereby result in higher or lower (from Fig. 11.1) yield, product quality, days required to develop to the different stages and maturity, resistance to stresses, and competitive ability with other plants. These variations are the structure of their different system functionings.

4. Based on statement 2 above, the relation between the quantitatively variable levels of any two categories or their component traits can vary from being highly positive to being highly negative, depending on the levels of all of the other categories and traits. In commonly used terms, the relation between any two traits is dependent on both the genotype and the environment as detailed below in section 11.3.1. Actually, the relation between any two of the five categories (i.e. between any two quantitatively variable traits) is usually, but not always, negative.

5. The relation between yield and any trait level within the system is curvilinear rather than linear – for comparison across a wide-enough range of the genotypic variation represented by cultivars. Therefore, maximum yield requires the level of every trait of the system to be optimized so as to maximize the yield system functioning for the resources and stresses that the target environment provides.

6. Selection toward maximization of any one trait within the holistic plant system (Fig. 11.1) is likely to result in negative effects on the adaptation and yield (yield system functioning).

11.2 Bases for the Hypothesized Constant Capacity

Rasmusson (1987) listed four types of interrelationship among plant traits that constrain plant breeding progress: (i) the requirement of size symmetry among plant parts; (ii) pleiotropy, which is effects on multiple traits due to the same gene activity; (iii) genetic background (unfavourable genetic linkage); and (iv) compensation among plant parts. All four are genetically directed constraints. We add to the fourth: compensation among the relevant plant processes. Compensation among these processes provides a physiological constraint due to either or both of: (i) provision by the environment of limited resources for support of growth and development; and (ii) the consequent environmentally controlled limited availability within the plant of sufficient assimilates to allow full expression of all the genotype-directed processes needed for full expression of all the categories and traits within the plant system (Fig. 11.1). That all four constraints are genetically directed for establishment of a given level for each trait, but that that level is also environmentally modulated adds another well recognized constraint of plant breeding progress; it is the genotype × environment ($G \times E$) interaction control over the system functioning and trait levels.

Unfavourable genetic linkage slows breeding progress, but genetic linkage can be broken, given sufficient time, a large enough number of meiotic events, or molecular genetic techniques, and selection across populations of sufficient size. Once an unfavourable linkage is broken, it becomes a favourable linkage. From the long-range viewpoint, therefore, genetic linkage is not a major problem. The requirement for symmetry among plant parts, is actually a kind of pleiotropy. Pleiotropy, physiological compensation and partitioning of the available metabolites among plant parts and processes constitute the biological bases for the hypothesis of a constant-capacity plant system.

11.2.1 Examples of pleiotropy

Pleiotropy is control over several traits by the same gene(s). Genetically and environmentally caused changes in the level of activity of the gene(s) both change the levels of the other system-interconnected traits. Gene actions are integrated into biochemical pathways. Pleiotropic effects are expected, therefore, to be a common phenomenon of plant biology. The following are important examples of pleiotropism.

11.2.1.1 Pleiotropy due to photoperiod-gene activity

It has long been recognized that the insensitive versus sensitive alleles of photoperiod gene(s) control the days to flowering (DTF) and DTHM. Wallace *et al.* (1993b, see Chapters 4 and 5) reported control by just two alleles (*ppd* and *Ppd*) of a single photoperiod-gene in bean (*Phaseolus vulgaris* L.) over both the DTF and the DTHM (sections 3.6 and 3.7). This gene simultaneously controlled the quantitatively variable levels of at least 26 plant traits (Table 4.3). These many traits are controlled simultaneously, because the photoperiod insensitive allele *ppd* establishes partitioning of the available photosynthate predominantly toward support of growth of the reproductive organs (direct accumulation of the yield of seed crops). On the contrary, if functioning under a delaying *P*, the photoperiod sensitive allele *Ppd* establishes predominance of partitioning of the available photosynthate toward continued vegetative growth including shoot elongation and more leaves and branches (sections 4.7, 4.8 and 4.9). Thus, the two alternative pathways toward which the available photosynthate can be competitively partitioned must share the available intraplant metabolites (photosynthate, nitrogen, minerals, etc.) that the plant has assimilated. Therefore, receipt by the reproductive organs of any of the available metabolites is at a metabolic cost of reduced growth by the vegetative organs. Vice versa, any receipt of these metabolites by the vegetative organs is at metabolic cost of reduced growth of the reproductive organs that become the yield. Without a delaying *P*, both the homozygous insensitive genotype *ppdppd* and the sensitive genotype *Ppdppd* or *PpdPpd* partition the metabolites predominantly toward the reproductive organs. This strongly raises the yield accumulated day^{-1} and the HI, the metabolic cost of achieving this being a reduction of the net accumulated aerial biomass and an associated reduction also of the number of leaves and total leaf area (Tables 4.3, 4.4; Figs 5.1, 9.1). Alternatively, a larger biomass and associated larger leaf area will be accumulated when activity of the dominant photoperiod sensitive allele *Ppd* directs the partitioning of the photosynthate predominantly toward continued vegetative growth. But, this is at a metabolic cost that reduces the reproductive growth. A further consequence is a delayed DTF and DTHM (Figs 5.1, 9.1). Through these compensatory metabolic costs due to having to share the available photosynthate and other metabolites within the plant, and the interconnections between virtually every part of the plant system (Figs 2.1, 11.1), the DTF and DTHM become positively correlated with the net accumulated biomass, but negatively correlated with the HI, while the HI becomes negatively correlated with the net accumulated biomass. These system-established interconnections among the three major physiological–genetic components (Fig. 11.4) of the process of accumulating crop yield were first derived in Chapters 4 and 5. They are most thoroughly demonstrated in sections 9.2 and 9.3).

The 26 traits pleiotropically controlled by the photoperiod-gene activity (Table 4.3) fully demonstrate that a single gene can influence most if not all the categories of traits within the holistic system (Fig. 11.1), and the extensive interconnections among the system's components (Fig. 2.1). As illustrated in Figs 11.1 and 2.1, the level of virtually every quantitatively variable trait within the system affects the level of the other quantitative traits. Just as Trewavas (1986) stated, 'Biological systems include a great density of connections between parts of the system.' The metabolic cost upon other traits, from sharing of the available intraplant metabolites which represents the system capacity, and caused by enlarging some traits to an excessively high level is a sufficient explanation for the great density of interconnections among the parts of the yield system (Figs

2.1, 5.1, 9.1), and among all the parts of the more holistic plant system (Figs 11.1 to 11.5)

11.2.1.2 Pleiotropy due to vernalization gene activity

Pleiotropic effects due to vernalization occur commonly in cereal crops. Vernalization genes establish winter hardiness, by reducing the rate of early development, i.e. the expression of earliness. Slower and later development occurs because, the high *T* before winter enlarge the vernalization gene (*vrn*) activity. Any *vrn* activity slows the rate of reproductive development (Chapter 6 and 8), so that the plants remain in the vegetative state when winter arrives. In contrast, genotypes with the *Vrn* genes for spring habit develop rapidly during the fall while the *T* is high. They can develop to the reproductive stage well before winter arrives, thereby rendering them vulnerable to injury by the low winter *T*. Through its control of the phenological development, the vernalization gene system (*vrn*/*Vrn* genes) also partially controls the morphological components of yield (Flood and Halloran, 1986; Yan *et al.*, 1993). Consequently, in comparison with spring cultivars, winter cultivars are always more resistant to cold injury, have more tillers and produce more heads plant^{-1} and land area^{-1}. At the same time, they are always slow (late) to develop to heading, and tend to have smaller heads and shorter durations of grainfill. It is commonly concluded in the literature that vernalization genes influence not only the phenological traits, but also the agronomic traits. Stelmakh (1993) extensively studied the effects of different vernalization genes on various agronomic traits of isogenic lines of wheat. He concluded that the *Vrn* genes appeared to act as non-complementary genes (duplicate or additive, but with different effects) with classical (for *Vrn1*) or incomplete dominant epistasis (for *Vrn3*). The triple dominant genotype (*Vrn1Vrn2Vrn3*) was the earliest and the triple recessive (*vrn1vrn2vrn3*) the latest.

Between these extremes, earliness of the other six genotypes is in the following order: *Vrn1vrn2Vrn3*, *Vrn1Vrn2vrn3*, *Vrn1vrn2vrn3*, *vrn1Vrn2Vrn3*, *vrn1vrn2Vrn3* and *vrn1Vrn2vrn3*. Under the experimental conditions of Stelmakh, the largest grain weight plant^{-1} was characteristic of the genotypes carrying *Vrn1* and/or *Vrn2*. The *Vrn1vrn2vrn3* genotype had maximal grain yield plant^{-1}, resulting from its intermediate levels for all the yield components and excellent grain filling. The double dominant genotype (*Vrn1Vrn2vrn3*) had many tillers, and on average, its earliness provided high potential for grainfill. The triple dominant genotype had the lowest grain yield plant^{-1}, because it was too early. The *vrn1Vrn2vrn3* and *vrn1Vrn2Vrn3* genotypes had high spike number and grain number plant^{-1}, but their lateness prevented realization of this potential for yield, because of incomplete grain fill.

11.2.1.3 Pleiotropy due to genes for dwarfness

Usage of dwarf genes in rice and wheat, beginning in the 1950s and 1960s, led to the higher yields of the 'green revolution' of agriculture (Jordan, 1993). Dwarf habit can result from a single gene, yet each of the several dwarf genes profoundly influences many plant traits. The smaller stature due to a dwarf gene is always more lodging resistant. In addition the plant has more spike-bearing culms and a higher HI. The higher HI of modern cereal (wheat, barley and rice) cultivars, most of which possess one gene for dwarfness, is another example of a higher HI being achieved through smaller sharing of the metabolites with the vegetative growth. Other traits that are altered follow. The shorter culms are often associated with shorter coleoptiles, shallower roots and, therefore, greater vulnerability to drought, heat, nitrogen deficiency, and to soil-borne diseases. Plants that are exceptionally short, due to possession of multiple dwarf genes, are so vulnerable to drought, heat and soil-borne diseases that they produce only shrivelled kernels (Yan, 1989).

11.2.1.4 Symmetry among plant parts

An example in cereals is that the desirable large inflorescences or heads are usually, if not always, associated with large and drooping leaves and thick stems. The large and droopy leaves reduce penetrance of light into the crop canopy (Figs 3.1, 3.2) and thereby reduce the assimilation (Fig. 3.3). A similar example is association in bean of large and small leaf size with corresponding large and small seed size.

Evans (1994) pointed out that the genes controlling desired characteristics are often pleiotropic in their expression, so the trade-offs between their various effects may result in counter-intuitive outcomes. He listed three interesting examples of pleiotropy. One is that selection for low stomatal density in expectation of increasing water-use efficiency in barley actually resulted in the opposite outcome, via unconscious selection for larger stomates and leaves and delayed senescence. His second is that selection for higher photosynthetic rate per leaf area has actually diminished yield in several crops by reducing leaf size. The third is that selection for uniculm habit of barley, which was emphasized by Donald (1968) as being a high yielding characteristic, actually resulted in lower or no higher yield than the cultivars of traditional tillering habit even at high sowing densities. The examples of pleiotropy above are sufficient to show it is impossible to improve all desired traits at the same time.

11.2.2 The physiological basis is compensation among components

Based on the hypothesis of a constant-capacity plant system (compare Fig. 2.1) every trait competes with all other traits for occupancy of part of the constant capacity of the holistic plant system (Fig. 11.1). Magnitude of the expressed constant capacity is always constrained – by the limited availability of the environmental resources to be used for growth and development. The subcomponents within the categories of traits of the plant system (Figs 11.2 through 11.4; compare Figs 2.1, 5.1 and 9.1) also compete for their part of the holistic plant system. Finally, negative or positive correlations of levels (system occupancy) among the interdependent traits occur. They quantify the interconnections among the traits, because all traits must share the metabolites assimilated by the plant and/or the time required to give expression of the traits. Consequently, reception of metabolites by one trait is inevitably at a metabolic cost to the level of one or usually more of the other plant traits. The paradigm shift here is: correlations between the levels of quantitative traits do not arise just from chromosomal linkages of the genetic controls or physiological pathways that are genetically linked. Most, of the correlations arise from competition for the inevitably environment-limited intraplant metabolites that must be shared among the gene actions that implement the several quantitative traits (Figs 11.1 to 11.5, and compare Fig. 12.4).

11.2.2.1 Competition among the traits for developmental time

Usage of time and associated rates is an integral physiological basis for the expression of every trait (Figs 1.2, 4.1, 9.1, 11.2). Traits or plant organs that originate at different stages of the plant's growth and development must compete for the available and always limited calendar time of the growing season.

Based on a 4-year field study of up to 95 diverse wheat genotypes, Yan *et al.* (1993) divided the growth duration of wheat among four developmental phases (Fig. 11.2). The time span from sowing to the single-ridge stage (phase I) determined, among other characteristics, the winter hardiness. This time span is short for spring genotypes but long for genotypes with winter habit. The duration from the single-ridge to the double-ridge stage (phase II) determined the number of spikelets head^{-1} and hence the size of the heads, which constitutes a measure of the 'sink' strength of the yield organs. The proportions of the available 365 days of a year occupied by phases I and II established the ecological adaptation, maturity and yield potential of the different wheat genotypes (Yan *et al.*, 1993; Yan and Wallace, 1995a). Most of the biomass, however, was accumulated during phase III (the time span from the double-ridge stage to anthesis) and phase IV (the time from anthesis to maturation). Thus, phases III and IV facilitate, or fail to facilitate, realization of the yield potential established during phases I and II. Therefore, the so-called 'sink–source balance' is strongly determined by the relative time span allotted for the vegetative development (phases I + II) versus the time span for the subsequent reproductive development (phases III + IV). In addition, the durations of these stages and the associated traits are quantitatively modulated by the quantitative variations of the environmental factors such as the *T*, *P*, radiant energy, moisture, nutrient availability, etc.

For annual plants, the larger the calendar time between maturation of the crop and the sowing of the succeeding crop ('phase 0' of Fig. 11.2) indicates greater need for earliness of the genotype. The longer 'phase 0' means the genotype requires and uses a shorter time to develop to maturation. Earliness adapts the cultivar to a short growing season, or allows a larger part of a long growing season to be available for growing other crop(s). Thus, time to maturity is an important trait for the adaptation of a cultivar to the ecological environment and/or to the agricultural system. Usage of time by a genotype is one of the three major physiological–genetic components of the total yield land area^{-1} year^{-1} (Figs 2.1, 5.1, 9.1, 11.2, 11.4). The yield day^{-1} land area^{-1} is a measure of the efficiency of the genotype, and the yield year^{-1} is a measure of the efficiency of the overall output of the agricultural system.

The constancy of 365 days year^{-1}, and forced usage of but part of that time for growth of a crop, results in inevitable compensation among the time durations of the four phases of growth (stages I to IV of Fig. 11.2), and the earliness ('phase 0'). Alternative partitioning (sharing) of that time (Fig. 11.2) results also in compensatory changes among the levels of the traits that develop sequentially across these successive time spans. A consequence of these compensatory usages of time is the triangular relationship among winter hardiness, head size and earliness of wheat described in section 11.1 of this chapter.

11.2.2.2 Competition among organs and processes for limited resources

Competition among the plants for the limited resources for growth available from the environment (sections 10.1 and 10.3), interacting with metabolic costs spread across the organs and processes of the plant due to sharing of the environmentally limited assimilates are the direct reasons for the commonly occurring compensation among levels of the morphological components of yield (sections 11.1 and 11.2). The seed size can be enlarged, but doing so will be a metabolic cost for either or both of the grains head^{-1} and the heads land area^{-1} (Fig. 11.3). Thus, many plants land area^{-1} lead to fewer heads plant^{-1}; many heads plant^{-1} or land area^{-1} are associated with smaller heads; many grains head^{-1} lead to smaller grains, etc. Such metabolic-cost-established compensatory changes among the morphological components of the expressed yield dictate an impossibility of maximizing all of the morphological yield components at the same time for any environment. Enlarging one necessitates reduction of others, unless the assimilate or time requirements of traits of the system other than these yield components are reduced, or unless the rate of assimilation and the proportional partitioning of these assimilates can be improved (Sedgley, 1991). Other examples of compensation are: higher rates of photosynthesis leaf area^{-1} in association with smaller but thicker leaves (Bingham and Austin, 1993; Evans, 1993), and many stomata on a leaf in association with smaller stomata, etc.

The conclusions are: first, that the constancy of the calendar time of a year, the conservative similarity of the major metabolic processes of all plant species (Evans, 1993), and the limited supply of resources available within each environment provide sufficient bases to force a near constant-system capacity within which all the genotypes of a species must function when grown under the limitations of each given environment. Second, all traits for which their expression requires much time and/or much of the available intraplant metabolites are forced to be interdependent and quantitatively variable traits. All are components of the complex, many-component, constant-capacity system that is represented by a circle (Fig. 11.1, compare Fig. 2.1).

Third, the independent traits (section 11.1.1) seem not to follow the first and second conclusions given in the preceding paragraph. Nevertheless, expression of some independent traits is just as essential to the success of a cultivar in some environments (section 11.5) as any of the interdependent traits. We suggest that independent traits require only small and negligible time and/or metabolites as compared with larger requirements for all interdependent traits (compare Figs 11.5 and 12.3). The expression of these independent traits, therefore, invokes an inconsequential metabolic cost on the expression of the other traits.

To achieve higher yield in the future, two complementary approaches are evident from Fig. 11.1. One is the agronomic approach, i.e. enlargement of the system capacity as represented by an enlarged diameter of the circle (Fig. 11.1)

that represents the constant capacity of the plant system. The capacity and yield potential will be enlarged by enhanced agronomic inputs. The second is the genetic approach. It requires breeding and selection toward maximization of the yield within any given environment, and separate breeding and selection toward maximized yield within any other environment. Direct selection for yield will always be effective, in spite of yield's low heritability, because the yield is a consequence of the integration of all the components of its system. Yield advance represents progress toward the compensatory optimizations of the levels of all these components that maximizes the system functioning within the given environment. Any indirect selection for advance in yield must improve this compensatory optimization among all of the quantitatively variable and genetically directed component traits that make up the plant system, especially among the plant systems components and subcomponents (Fig. 11.1) that constitute the yield system (Figs 2.1, 5.1, 9.1, 11.4). Absolute maximization of the yield requires optimization of the morphological (Fig. 11.3), the physiological components of the yield (Fig. 11.4), and of the time components (Fig. 11.2), i.e. of all traits and subtraits of all categories of the plant system (Fig. 11.1). Even though variable from year to year, usually, the environment is the component of the system that a breeding programme will judge as being fixed. To the extent that the environment is relatively fixed, it is the genetic instruction of the genotype that must be changed to raise the yield.

Not infrequently, societal demands impose metabolic costs that limit the expressed yield, by requiring product quality that competitively reduces the achievable yield. As detailed more thoroughly in the breeding procedures presented in Chapter 12, optimization of a trait requires reducing its level when that level is higher than needed for the target environment, but raising it when it is lower than optimal for the specific genotype grown in the given environment.

11.3 System Requirements for Higher Yield

11.3.1 Increase of agronomic inputs

It is a fundamental principle that continuing to raise crop yield requires both genetic improvement and agricultural innovation. As stated by Way *et al.* (1983): 'The evolution of better crops and the development of better cultural systems are intimately related and are a mutually defining system.' Evans (1993) indicates that higher crop yield comes from synergistic interactions between plant breeding and agronomic innovations. That is, greater agronomic inputs and greater genetic potential for yield are each a prerequisite for successfully benefiting from the other. Reciprocation between plant breeding and agronomic innovation will continue to create new opportunities for synergistic interactions, just as the evolution of new functions has done in nature. Evans (1994) stated: 'Rather than being preoccupied with the ceiling to yield and with identifying processes which might be "improved", we should be more alert to the evolving opportunities for synergisms between breeding and agronomy as innovations on both sides are made.' This is in full agreement with our hypothesis of a constant-capacity system. The interrelationships between genotypes and the environment are such that, if there is no improvement in the environment and then the genetics, or vice versa, there will not be continuous improvement of the yield. To illustrate, Shroyer and Cox (1993) reported that a century of wheat breeding in the US has produced no genetic improvement in yield under low-fertility conditions. A similar situation is true for wheat breeding in central India (Fischer, 1993). Plant breeders are not assigned the role of improving the agronomic management. Nevertheless, awareness of the ecological environment and agronomic potential of the target area, and collaboration with agronomists charged with implementing potential improvement of the environment, are essential to continuing advance of yield.

Any improvement in agronomic inputs will be reflected by enlargement of the radius and hence area of the circle illustrated by Fig. 11.1. However, the components of this enlarged capacity are forced to share it. Therefore, it simply becomes an enlarged constant-capacity system.

11.3.2 *Balance among the yield components optimizes system functioning*

Change toward larger yield within an agronomically improved environment will require some of the other category(ies) within the system to be changed toward occupying a smaller part of the holistic system. Without such change among the other categories of the plant system (Fig. 11.1), the multiple compensatory genetic changes required to raise the yield can occur only among the components of the constant-capacity subsystem consisting of the morphological components of the yield (section 3.1; Fig. 11.3) and/or of the physiological components of the process of accumulating the yield (Fig. 11.4; compare Figs 2.1, 4.1, 9.1).

To breed for the appropriate level of either class of yield components, three things have to be considered. First, the yield is the product of the three morphological components (Fig. 11.3) and also of the three physiological components (Fig. 11.4). Therefore, any of these components that is smaller than optimum for the given environment will lower the yield. Second, the environment dictates that there is never an unlimited supply of all environmentally supplied resources or of the intraplant assimilates and metabolites. Third, since the three components of the yield system *per se* compete within their constant-capacity subsystem of the plant system, if any component is too large, this will force one or both of the others to be too small, so the yield will be lower than it could be. Consequently, for each given environment, there is an optimum combination among all the categories of the plant system (Fig. 11.1), including among the components of the yield system (Figs 2.1, 5.1, 9.1, 11.3, 11.4). This will often be just a range of near optimum combinations.

Rather than extreme levels of their yield components, successful cultivars have a balance among the levels of these yield components as well as among all the categories of components of the plant system that optimizes their integrated system functioning within the given environment. These interconnections explain why breeding programmes that aim to improve only one of the components fail (Donald, 1968). As frequently observed, there is little merit in selecting for but one of the yield components to assist selecting for yield *per se*, because of compensatory reduction of the other components (Whan *et al.*, 1993). Selection to an excess level for any one of the morphological components of yield (Fig.

11.3), or for the physiological components of yield (Fig. 11.4), without selection at the same time for yield *per se*, can be counter-productive (section 11.2.2; Evans, 1993). Compromised and thereby optimized levels of these components is required to maximize the yield; each component must be optimized relative to the levels of all the others that occupy both the yield subsystem and the holistic plant system.

Unfortunately, the procedure for visualizing and selecting toward optimization of multiple components is far less obvious – and more complex – than visualizing and selecting for maximization of but one component at a time. A strategy that can improve understanding about the reciprocal compensations among all the traits within the constant-capacity system for each different environment is introduced in the next paragraph.

11.3.3 Minimization of system components other than traits for yield

The concept of a constant-capacity plant system suggests that reduction of any trait(s) that occupy more of the plant-system's capacity than the environment requires will, if that system capacity is not reoccupied by excess levels of other traits, make more capacity available for the yield component category of traits within the holistic plant system. This will enlarge the crop yield potential (Fig. 11.1).

11.3.3.1 Breeding for weak competitors

Selection for weak competitive ability is a well-known suggestion of Donald (1963, 1968, 1981), as discussed throughout Chapter 10. Donald believed the following: (i) The genotypes with strongest competitive ability have an advantage during natural selection, where each genotype competes mostly with plants of other genotypes; but (ii) strong competitors are poor yielders when grown in pure stands, where each plant competes with plants of only its own genotype. (iii) High yielding cultivars are weak competitors; and breeding for high yield, therefore, should be breeding for weak competitors. Donald and Hamblin (1983) listed seven traits that make a strong competitor: annual habit, tallness, leafy canopy, much tillering or branching, small seeds, rapid germination, and a strong root system. Of these, tallness is the visually most obvious provider of competitive ability for the ten crops they discussed, including wheat, rice, bean and others. Donald's generalization that tall genotypes are strong competitors and poor yielders was confirmed by recent research on wheat (Thomas *et al.*, 1994). They concluded, in agreement with Donald and Hamblin (1983), that low interplant interference, achieved through reduced plant height (genes for dwarfness), is a positive factor for yield, and this reduced competitiveness is always present in high-yielding cultivars.

These findings, in combination with the hypothesis of a constant-capacity system, led us to a slightly altered explanation of why breeding for dwarfism resulted in the higher yields of the 'green revolution'. We agree that the lower plant height reduced the competitive ability of the plants, and more strongly stated; it reduced the interference of one plant by another. The concept we emphasize (section 11.2.1.3) is that the dwarf habit allocates a larger part of the system capacity toward higher yield, at a metabolic cost to the plant height and the attendant competitive ability. Physiologically, these weakly competitive plants allocate more of the available metabolites toward growth and development of the reproductive organs that become the yield, rather than competitively partitioning it toward the vegetative organs that establish the competitive capability against the other plants for the available light and other environmental resources needed for growth. This explains why high HI is an integral part of Donald's weak competitor ideotype. It explains also, as discussed in section 11.2.1.3 above, why higher levels of fertilizer and water availability (agronomically improved environments) are essential to the highest yields of the dwarf cultivars.

A recent study of Reynolds *et al.* (1994) reinforces conclusions by Tanaka *et al.* (1966) that lines with high yield potential are less affected by competition for light, and by below ground competition, than low yielding lines. Reynolds *et al.* concluded that the higher yield results from superior adaptation within the plant community. However, they indicated that the traits that contribute to this reduced competition with other plants, and to lack of intraplant interference, remain unidentified.

In Chapter 10, we described a conclusion about interplant competition and the crop yield potential of a genotype that is highly relevant to breeding for higher yield. It is that the crop yield potential of each genotype grown in a given environment is determined by: (i) the potential yield of its individual plants in that environment, which potential can be expressed only when the crop density is low enough to prevent all interplant competition; and (ii) the sensitivity of the genotype to plant density, measured as the reduction in yield plant^{-1} in response to increasing plant density. The latter is a more precise and quantifiable way to describe what Donald (1968) called the competitive ability of the genotype. High competitive ability can reduce the crop yield potential. The points needing emphasis here are that high yielding genotypes should have a high potential yield plant^{-1}, in combination with low reduction of that yield plant^{-1} by an increasing density of the plants of that genotype. These concepts are summarized in section 11.5.

11.3.3.2 Breeding for less resistance to environmental stresses

In harsh environments (with extreme *T*, and/or water and/or nutrient deficiencies), plants must divert much of their life span and/or assimilated metabolites toward survival. In environments where the limiting factors are less extreme or are consistently sufficient, more of the growth duration, its efficiency and its assimilates can be used for growth and development of the yield organs. For barley, wheat and other cereal crops, the most effective single trait for conferring tolerance to water and nutrient deficiencies is taller plant stature, which may be related to a deeper root system, better access to water and minerals and storage of photosynthate for remobilization to grains during grain fill (Ceccarelli and Grando, 1993). The requirements for a longer duration of vernalization and/or greater response to *P* are the major mechanisms for resistance to the cold *T* of the winter and early spring. The higher yield of modern wheat cultivars grown in productive environments is associated with reduced plant height, reduced requirement for vernalization (Yan, 1992) and reduced responsiveness of the reproductive development to the *P* (Donald and Hamblin, 1983). The latter include reduced delay by higher T of the development through the photoperiod- and vernalization-gene activities (sections 6.2 and 7.3).

Reduction of plant height is the most frequently mentioned trait believed to be causal of higher yield potential of the cereal crops. One interconnected change due to the reduced plant height can be a reduction of the resistance of the plant to water and/or nutritional deficiency. Therefore, many genotypes with reduced height can be relatively high yielding, but they lack the adaptation to be high yielding within the lower-yielding environments. The reduction in plant height and associated reduction in resistance to stresses, has led to large increases in the yield of many crops through a higher HI, but only in the higher yielding environments. The yield gains occurred through an initial breeding objective of higher resistance to crop-lodging (Evans, 1993, 1994).

A similar explanation applies for the reductions of the vernalization requirement and sensitivity to P of modern cultivars of wheat and other crops. Jointly, these two traits provide the winter hardiness of the winter cereals, in addition to control of maturity, the latter being far better understood than the former. The primary merit of cold resistance, especially the requirement for vernalization, is to reduce or totally prevent active growth and especially development before low T. For winter cereals, reduction of the vernalization requirement shortens the duration of the development phase I, which spans from planting to the single-ridge stage (Fig. 11.2). Shortening of phase I enlarges the calendar time available for phase II, during which the spike and sink size are determined, and/or enlarges the time for phases III and/or IV, during which most of the photosynthate is accumulated, and/or extends the time span of 'phase 0', which indicates (section 11.2.2.1) the earliness required for the cultivar to be adapted to the growing season. A longer time available for phases II to IV, made available by reduction in winter hardiness, explains why, in regions with mild winters, wheat cultivars with spring or semi-winter habit usually have greater yield potential than the winter cultivars, even though they flower and mature in less time than the winter genotypes. As illustrated (Fig. 11.1), reduction of the winter hardiness enlarges the opportunity to increase the yield potential of the crop (Yan, 1992). Supporting this is a recent statement by Evans (1994). He wrote:

> If the preceding analysis is on the right track, i.e., that the rise in yield potential has been due to the rise in harvest index which in turn is dependent on greater agronomic support for the crop, it follows that the further improvement of yield potential is likely to be at the expense of cultivar adaptability to adverse conditions.

It is essential to point out here that for a target environment that has been improved, breeding for higher yield may require selection for lower resistance to stresses than was essential to the adaptation of a genotype in the unimproved environment. In regions where winter is severe, winter habit is a mandatory requirement for survival of the genotype, and therefore for any yield. Breeding for adaptation will be discussed further in section 11.4.

11.3.3.3 Breeding for cultivar maturity

Early maturity is a favourable trait (Donald and Hamblin, 1983) for many environments and crops because: (i) it provides adaptation for rain-fed regions where water availability is limited or variable, and for places where the growing season is shortened by the low *T* of both spring and fall; and (ii) it makes it possible in some regions for farmers to grow other crops during part of the year, in order to increase the total yield land area^{-1} year^{-1}, and/or to enlarge the number of crop products.

If only one crop is grown within a year, the yield will be below the genotype's potential when either less or more time is required to develop to harvest maturity than is provided by the growing season duration of the environment (sections 5.5.2 and 5.9.2). If the growing season is longer than the cultivar requires, the unused days of the growing season are wasted. If the cultivar requires a longer time to develop to harvest maturity than the growing season, then the full yield potential of the cultivar cannot be realized, and at least part of that which has been realized will be wasted due to incomplete development. Therefore, the objective has to be to breed for that duration of growth and development that matches the growing season duration of the target environment.

As discussed in the immediately previous section, if the stress level of the environment allows it, reduction of the vernalization requirement and/or of photoperiod sensitivity, can improve both the yield potential and the earliness. If the growing season duration provided by the environment allows the time to maturity of the genotype to be maintained or extended, i.e. if the earliness of the current cultivars can be maintained or reduced, the gain in yield potential due to reduction of the vernalization requirement and/or of the photoperiod sensitivity would be yet larger. Much discussion in the literature suggests breeding for a longer reproductive (post-anthesis) duration; however its genetic control(s) are not at all understood (Evans, 1993). Yield system analysis based on the three traits of Fig. 11.4 (compare Figs 2.1 and 9.1) considers the DTHM to be one of the three major physiological–genetic components of the process of accumulating crop yield.

The importance of optimizing plant maturity according to environment was highlighted by a finding of Wallace *et al.* (1993a). Variation in length of the growing season from location to location and year to year resulted in correlation between the DTHM of the tested cultivars and their yield being positive within about one-third, negative within about one-third and near zero within about one-third of 51 yield trials (sections 5.3 and 5.4). The range of these 51 correlations was −0.74–0.70, and the average was 0.01. Thus, for some years and/or locations the early maturing cultivars gave the higher yields, for other years and/or locations the later maturing ones yielded the most, and for some the intermediate maturities yielded the most. This is an expected result from the hypothesis of a constant-capacity plant system. According to this hypothesis, correlation between quantitatively variable traits is dependent on the levels of the other traits and also on the environment (section 11.1.2). Therefore, growing some cultivars that are slightly earlier, some that are intermediate, and some slightly later than the average growing season duration is likely to maximize the long-term average crop yield of a region.

11.3.3.4 Breeding for acceptable product quality

Maximization of crop yield may demand compensatory minimization of the requirement for product quality (Fig. 11.1), and vice versa. No component of the constant-capacity plant system (Fig. 11.1) is more diverse than the human-society-dictated 'product quality' traits desired for different crops and also for the same crop. Protein content is a general standard of quality for grain crops. Negative correlation between grain yield and the nitrogen or protein content has been reported for wheat (Bingham and Austin, 1993; Delzer *et al.*, 1995), triticale (Fossati *et al.*, 1993), barley (Bulman *et al.*, 1993), lentil (Mamni *et al.*, 1991) and many other crops. Therefore, breeders must expect a reduction in protein concentration or other aspects of product quality, if yield is the overriding objective of their breeding effort, and vice versa. Two facts favour breeding to some extent for higher yield at the expense of protein content, at least in wheat. First, greater yield is always accompanied with greater total protein yield land area^{-1} (Bingham and Austin, 1993). Second, quality can be improved through other approaches. For example, some high molecular weight (HMW) subunits of glutenin are more important for making good bread than the total protein concentration. These HMW proteins have simple qualitative inheritance, so breeding for them allows both yield and the bread-making quality to be improved (Bingham and Austin, 1993). In this case, presence versus absence of the HMW glutenin subunits for conferring good baking quality can be regarded as an independent (section 11.1.1) trait. There is need to continue to seek vigorously for such simply inherited traits of product quality.

Except for protein concentration in most grain crops and sugar content in sugarcane and sugarbeet, product quality is very much a man-made concept. It differs greatly for different crops, uses of the same crop, and countries and cultures. For example, some cultures prefer black coloured beans, while others demand red or buff colour; some require a solid colour, while mottled is a must for others. In addition the culture may demand a large- or intermediate- or small-size bean. Such society–culture established product–quality traits can negate benefit from an extreme focus on breeding for higher yield. Thus, just as there can be too much focus on the individual traits that contribute to yield, there can be too much focus on breeding for higher yield, product quality, stress resistance etc. Another example of excessive attention to breeding for higher yield follows. Breeders from developed countries bred a higher yielding maize cultivar intended for use by subsistence farmers in a less developed country. After trying the superior 'high yielding cultivar' the farmers rejected it, and continued to use their low yielding cultivar, because their culture needed the taller stalks of their original cultivars to build and maintain the fences needed to keep the chickens confined and out of their fields. Understanding the needs of the human society is always essential to establishing the breeding objectives.

11.4 Breeding for Cultivar Adaptation to the Environment

Breeding for higher yield potential is a major objective of almost all breeders. However, high yield potential does not necessarily mean 'high' yield. The highest yield potential, is a luxury for the majority of farmers across the world, especially in developing countries. For regions where the climate is harsh, water is scarce, soil fertility is low or agronomic inputs are too expensive, the major objective should not be to breed for the highest yield potential. It should be to breed for resistance to the stresses of the environment and for better yield stability. This amounts to breeding for maximum adaptation to the adverse conditions, or in other environments, for adaptation to the excellent conditions. In statistical terms this is breeding for an adapted $G \times E$ interaction effect on the yield. The expressed yield can be considered as the product of the yield potential of the genotype times its adaptability to the environment:

$$\text{Yield} = \text{Yield potential} \times \text{Adaptability}$$

where the yield potential is determined by the proportion of the system capacity occupied by the 'yield components' (Fig. 11.1), and the adaptability is determined, within the remaining proportional capacity, by the balance between the proportion of the system capacity of the genotype allotted to the stress resistance (in a broad sense) as compared with the stress intensity of the environment in which the genotype is growing:

$$\text{Adaptability} = 1 - |\text{Stress resistance} - \text{Stress intensity}|$$

where:

1. The stress intensity is the magnitude of shortage or scarcity of resources for plant growth available from the environment, including water, nutrients, and the length of the growing season that results from the available water, nutrients, T, P, etc.

2. The stress resistance is the capability of the genotype to grow under the given stresses. This may include resistance or tolerance to cold, drought, poor soil fertility, etc. In a broader sense, it is likely to include earliness, as a mechanism to escape the stresses due to, for example, drought or hot wind during the final stage of the plant development. All values for the stress resistance and stress intensity in the above equation are defined as being between zero and 1. Therefore, the adaptability also has a value between zero and 1. Consequently, the actual yield is determined by the following formula:

$$\text{Yield} = \text{Yield potential} \times [1 - |\text{Stress resistance} - \text{Stress intensity}|]$$

The best adaptability = 1, so the expressed yield can be the potential yield possible for the cultivar, only if the stress resistance of the genotype exactly matches the stress intensity of the environment. Either a larger or smaller level of stress resistance than this will reduce the adaptation of the genotype, and therefore reduce realization of its yield potential in that environment. Thus, strong winter hardiness provides adaptability to regions where the winter is severe, but is excessive for regions with warm winters. Similarly, high resistance to drought provides adaptability to dry areas, but results in low adaptability for environments where water is plentiful. Likewise, early maturity provides adaptability to regions where the growing season is short, but not for regions where it is long. As thoroughly illustrated in Fig. 2.1, Table 4.3 and Fig. 11.1, because of the interconnections between each system component and others, a given system component often serves more than one function.

In environments with low stress intensity relative to the stress resistance of the existing cultivars, higher yield can be achieved by breeding for high yield potential and/or for less stress resistance. More specifically, the hypothesis of a constant-capacity system (Fig. 11.1) suggests usage of less of the system capacity for stress resistance will be essential to the highest yield potential possible. This hypothesis implies that breeding for higher crop yield potential and breeding for less resistance to stress are reciprocal views of the same compromised balance and optimization among the traits within the constant-capacity system. However, it is only for a circumstance where the stress intensity of the environment is low relative to the stress resistance of the existing cultivars that selection for less stress resistance will lead to high realizable yield.

In contrast, where the stress intensity of the environment is large relative to the stress resistance of the existing cultivars, achieving higher yield requires breeding for both high yield potential and greater stress resistance. Can higher yield potential and greater stress resistance be achieved simultaneously? In many and probably most cases the answer is no. The crop yield potential, the stress resistance and the earliness must each competitively occupy a proportion of the constant-capacity system. Therefore, greater stress resistance can be achieved but at the cost of crop yield potential, and vice versa (Fig. 11.1).

The hypothesis of a constant-capacity plant system provides an explanation of why the genotypes with high crop yield potential measured in superior environments do not give high yield in harsh environments (Fischer, 1993; Shroyer and Cox, 1993). They give lower rather than higher yield than the stress adapted

Table 11.1. Grain yield (kg ha^{-1}) at low- and high-yielding sites of the top 5% of barley lines selected for high grain yield in low- or else high-yielding sites. (The data are from Ceccarelli and Grando, 1993.)

	Test site and its rainfall			Mean yield of 5% top yielding lines which were selected at		
Year	Site	Rainfall (mm)	Mean yield	Low-yield sites	High-yield sites	Low *Y*-High *Y* difference
Low-yield sites						
1985	Breda	178	743	1284	779	505
1986	Bouider	180	1138	1935	1340	595
1987	Breda	164	673	1019	654	365
1988	Bouider	382	2910	4199	3383	816
1989	Bouider	184	687	1287	658	629
1990	Breda	183	471	794	429	365
1991	Bouider	207	1050	1693	953	740
High-yield sites						
1985	Tel Hadya	373	3489	3339	4210	−871
1986	Tel Hadya	316	4007	4139	4970	−831
1987	Tel Hadya	343	2668	2739	3613	−874
1988	Tel Hadya	499	4420	4672	6100	−1428
1989	Athalassa	214	5824	4874	7814	−2940
1990	Terbol	317	3346	3066	4122	−1056
1991	Terbol	559	4740	4705	6073	−1368

genotypes with lower yield potential. This is illustrated by data of Ceccarelli and Grando (1992) shown in Table 11.1. They found lines selected for high yield potential in productive environments to have lower yield in stress environments than lines selected under stress. This indicates lines with high yield under stress will be discarded if the yield is measured only in high yielding environments. Ceccarelli *et al.* concluded the alleles controlling high grain yield in low yielding environments are at least partially different from those controlling high yield in high yielding conditions.

Based on the hypothesis of a constant-capacity plant system, the 'alleles' controlling high grain yield in low yielding environments must provide higher stress resistance, which reduces the system capacity for and lowers the yield potential. The alleles which establish high grain yield in high yielding environments must provide high crop yield potential but low stress resistance. This conclusion is also supported by Ceccarelli and Grando (1993). They concluded that crop yield potential and stress tolerance (defined as yield under stress) were, by and large, mutually exclusive. Therefore, selection in high-yielding environments is expected to produce negative or no yield response for low yielding environments. This finding for barley has been more or less repeated for wheat by Ud-Din *et al.* (1992) and Shroyer and Cox (1993).

Obviously, stress resistance is a very complex trait. With the unpredictable differences in frequency, timing, and severity of various climatic stresses, each growing season (environment) is likely to require a different combination of the several traits to confer the needed 'stress resistance'. The interactions among these traits can be expected to enhance the importance of specific trait(s) for each specific stress environment. In this context Ceccarelli *et al.* (1991) conclude that it is difficult to consider 'drought resistance' as a character with its own identity in terms of inheritance. Nevertheless, genotypes that yield better in low yielding conditions are generally taller, earlier to head, better in early growth vigour, more tolerant to cold, and have greater kernel weight (Ceccarelli *et al.*, 1992; Ceccarelli and Grando, 1993).

Oosterom and Acevedo (1992) identified two barley ideotypes for environments with terminal drought. The first is for Mediterranean environments with mild winters; the ideotype includes early heading with good early vigour, erect growth habit, light plant colour, and ability to recover from cold damage. The second is for environments with harsh winters, which require medium-early heading, prostrate winter growth habit, dark winter plant colour that changes to pale-green in the spring, and cold tolerance. The first ideotype is a syndrome of traits for a spring-type barley, and the second for a winter type. Early heading is positively related to higher yields in most environments with terminal drought, especially in the lowest yielding environments. For the second ideotype, its winter habit combined with relatively early heading essentially means that a very large part of the total growing duration is not used for active plant growth. This fully illustrates that, for protection from damage by various unfavourable environmental factors, plants adapted to harsh environments have to be equipped with multiple stress-resistance mechanisms, which are achieved only at the expense of the crop yield potential. The resistance in this example is due to prevention of developmental advance until after the low winter temperatures.

Adaptation to the variable environments of year to year or place to place, is a problem for other than the environments with strong stresses. To some extent it is a problem for every breeding programme. Therefore, breeding for compromise between yield potential and adaptability to adverse environments is a universal task for breeders.

Statistically, adaptation is due to the $G \times E$ interaction effects on the yield (Kang, 1990; Gauch and Zobel, 1996; Kang and Gauch, 1996). $G \times E$ interaction effects are always evidenced in multiple location yield trials of multiple genotypes. Both the $G \times E$ interaction effect on the yield and the effect by the environment are usually larger than the effect on the yield due to differences among the genotypes (Fig. 9.7J). The implications are that environment establishes the largest constraint to selection and breeding for higher yield. Therefore, breeding for adaptation, i.e. for the $G \times E$ interaction, should be a primary goal in addition to selection for the yield of the genotype. The hypothesized constant-capacity system provides some biological explanation for the $G \times E$ interaction. Philosophically, the hypothesis differs from any statistical model used to quantify the $G \times E$ interaction, such as the AMMI analysis applied throughout Chapter 9. The difference is that the biological hypothesis precludes any independent ('main') effect by either the genotype or the environment (section 9.3.2.3). It excludes these effects, because every phenotype results directly from the interaction of the genotype with the environment and nothing else (Figs 1.1, 2.1, 5.1, 9.1, 11.1; Table 4.3). The gene actions implement the biological processes, but they are inevitably modulated quantitatively by the variation of the levels of the environmental resources.

11.5 Breeding for Independent Traits and Defect Elimination

Focus above in this chapter has been predominantly on the interdependent, quantitatively-variable traits. However, certain independent traits, controlled by one or a few genes (sections 11.1.1 and 11.2.2.2) can be equally essential to a successful cultivar. The plant system of a genotype can be compared to a round table (Fig. 11.5). The top of the table represents all the categories of the interdependent quantitative traits that constitute the constant-capacity plant system illustrated in Fig. 11.1. The legs of the table represent independent traits that constitute essential support of the system and its expressed capacity. Such independent traits include resistances to definite diseases and pests in some environments. Both the top and legs are essential for system functioning. If the diseases or pests are severe, a new and superior genotype for these independent traits may be more decisive for higher yield, than any genotype for a higher crop yield potential. Such roles of independent traits are never the main topic of this book. Nevertheless, their essentiality in some environments endows their incorporation as essential breeding goals. Effective control of diseases and pests is a prerequisite to serious study of the physiological-genetics of crop adaptation and yield, and learning how to breed for higher yield efficiently.

Apart from the large yield gains brought about by dwarf genes, the history of raising yield of most crops has largely been breeding for resistances to the diseases and pests that limit functioning of the yield system. For many crops, including high yielding environments, the main reason for replacement of older cultivars by new ones is greater resistance to the prevailing diseases or pests that limit the yield. For wheat in China, change of the predominant physiological races of yellow (stripe) rust, followed by finding and incorporating essential resistance into the existing or new cultivars have been the predominant reasons for release of new cultivars during 1950 to 1980. In irrigated areas, as a consequence of increased application of artificial fertilizers, new more severe diseases emerge. For instance, yellow rust was the major wheat disease in north China before the 1980s. Now, however, powdery mildew, head blight and several leaf blights have become equally and sometimes more limiting diseases.

Breeding for disease- and pest-resistances will continue to be a major objective in most breeding programmes, especially for the highest yielding environments. It is fortunate that many needed resistances are independent traits that are comparatively easy to incorporate into the already adapted germplasm. The difficult part is to find the needed gene(s) and determine their inheritance. Modern molecular biological techniques will certainly accelerate finding and incorporating these genetic resources that are otherwise not easily available to plant breeders.

11.6 Determining the Optimal Level of the Components Based on the Constant-capacity Hypothesis

The constant and limited capacity of the yield subsystem of the plant system has long been recognized by the compensatory changes among the morphological components of yield (Fig. 11.3). More recently, it has been illustrated by negative or positive correlations among the levels of the physiological components of the process of accumulating the yield (Figs 2.1, 5.1, 9.1 and 11.4). These figures illustrate the subsystem for yield that exists within the constant capacity of the holistic plant system (Fig. 11.1). Improved understanding of the compensatory trade-offs among the components of the constant capacity of holistic systems can accelerate breeding for higher yield. Largely, the control is by the system as a whole, rather than the limitations of individual components. The compensatory sharing of the constant-capacity system among its components that is optimal changes with each different environment. The major need, therefore, is understanding of the genetically-established level of each component trait that is optimal (best adapted) for each different environment. Each environment must be considered separately, because each alters: (i) the constant capacity of the system, and (ii) the compensated and thereby optimized levels among all the multiple traits that are needed to maximize the whole-system functioning.

Whan *et al.* (1993) pointed out the following. Comparison of the historical cultivars of wheat can identify the traits that have contributed to yield improvements of the past. But, it cannot be assumed that continued selection for these same traits will lead to yield improvement of future cultivars. To illustrate, comparing a set of modern wheat cultivars with the older cultivars indicates that dwarf habit and higher HI account for the recent dramatic increases in crop yield potential. Can we conclude that future improvement of wheat yield will lie in continued reduction of plant height? Obviously the answer is no;

further reduction in plant height can bring serious micro-climatic problems. The double-dwarfism genotypes of wheat (60 cm tall or less) result in much higher *T* within the crop canopy, and on the soil surface, so soil-borne diseases become much more severe as compared with the taller mono-dwarf genotypes (Yan, 1989). Even in the absence of this greater biotic stress, excessive dwarfness significantly reduces the total accumulated biomass, so the grain yield land area^{-1} is lower, and the grain quality becomes unacceptable. The described effects due to excessive dwarfism illustrate the general principle. Selection for every trait can cause either positive or negative effects, depending on its level relative to the level of the other traits that contribute to the yield. Another illustration is that continuing to raise the HI will ultimately bring back lodging, which will reduce the radiant energy distribution within the leaf canopy, and lower the biomass and the yield (Austin, 1993).

Every trait that contributes to yield can cause both positive and negative effects on the yield. This can occur because all the traits are essential to a successful cultivar, but each competes with the others for its part of the limited capacity of the system. The sharing of the assimilates adapts the system to the resources versus stresses provided by the environment. Any trait with a genotype-established excessively low level will be detrimental to the yield of a cultivar. Less obviously, any trait with a genotype-established excessively high level will also be detrimental. Based on optimization of system functioning, its excessive part of the constant-capacity system forces one or more of the other traits to be at a sub-optimal low level. Therefore, determining which trait will enhance yield requires knowing the required level of all the traits for each environment. The need is for the genetically directed level of every component trait of the yield system to be optimized for the given environment, so that its positive effects on the yield are maximized, while its negative effects are minimized. Yield enhancement always requires a superior balance and optimization among the levels of all the genetic and environmental factors that make up the constant-capacity level of the system functioning within the target environment.

Selection toward the extreme level of any component can lead to failure of a breeding programme. Failures of the uniculm ideotype (Balkema-Boomstra and Masterbroek, 1993; Dofing and Knight, 1994) and of the narrow and erect-leaf ideotype of barley (Rasmusson, 1987) are examples.

Loomis (1993, see section 2.4) indicated the theory of optimization should be applied to yield improvement. He described optimization in the sense of alternative uses of scarce resources, with the largest net response to a resource occurring when there are no other limiting factors. This concept is in full agreement with the hypothesis of this chapter.

Chapter 11 Summary

1. The hypothesis of a constant-capacity plant system is that all the interdependent and quantitatively variable traits of a plant species constitute its biological system. The fixed capacity of this system is forced to be essentially constant across all the genotypes of the species by the limited resources for growth available within each given environment, and by the stresses of that environment. The needed level for each component trait of the system is determined by the environmentally provided resources to be acted upon by the genetic directions, and also by the stresses that limit the needed gene actions. But, the genetic direction for the level of the component traits is established by the genotype. Under each given environment, there is no way to maximize the genetically-established functioning level of one quantitative component trait, without causing the level of one or more other quantitative traits to become limiting. Maximization of the yield requires all the component traits of the yield system to be compensatorily and simultaneously optimized at that intermediate level that maximizes the system functioning. The required optimal levels of the components change with each different environment.

2. The constant-capacity hypothesis indicates three possible strategies for increasing crop yield. They are: (i) enlargement of the system capacity by improving the environment in which the crops are growing. (ii) Optimization of the system for the specific environment, i.e. to reduce the level of the interdependent traits that reduce the yield because their levels are excessive for the given environment, while enlarging the genetically established levels of traits that reduce the yield by being suboptimal. And, (iii) improvement of some essential independent traits (disease and/or pest resistance(s) as an example). These traits are independent of all other traits, because their requirements for time and intraplant metabolites are small, as compared with the large and competing requirements for these metabolites by all the interdependent traits.

3. Every breeding programme for higher yield should use all yield trials conducted by the breeder, and by agronomists and horticulturists, to quantify the levels of the several interdependent and quantitatively variable near-fully integrated (major) components of the process of accumulating the crop yield. This will optimize functioning of the system within the target environment(s). This will advance physiological–genetic understanding of why some of the genotypes are high yielding, while others are unadapted and have low yield. This information is essential to a rational set of breeding objectives and selection standards. To achieve system understanding, it is recommended that the yield system analysis (YSA/AMMI) procedure (Chapter 9; Figs 9.2, 9.3 and 9.7; Table 9.1) be applied to every yield trial. Application of YSA/AMMI to ongoing yield trials to assist identification of the component trait levels that will optimize the system functioning for each different environment will be presented in Chapter 12 as an integral aspect of breeding for higher yield.

4. Not infrequently, societal demands impose metabolic costs that limit the expressed yield, such as requiring product quality that, within the holistic plant system, competitively reduces the achievable yield.

12 Maximizing Efficiency of Breeding for Higher Crop Yield

Paradigm shifts 28 to 31 all relate to selection during plant breeding. Shift 28 is that selection for a physiological–genetic component trait, which is intended to give indirect selection toward higher level of a system output such as yield, must always be toward whichever of a lower or a higher level of that trait that will establish a more optimal system functioning within the given environment. It is as likely this selection should be toward a lower as toward a higher level for most traits considered to control the yield, or other system outputs. Paradigm shift 29 is that, during the early segregating generations, selection for yield can be achieved by selection for the maximum potential yield plant^{-1}, and also for the only other component from the viewpoint of interplant competition, which is selection for a smaller reduction of the yield plant^{-1} as the plant density is increased to that needed to maximize the crop yield land area^{-1}. Paradigm 30 is that it is unlikely a shift will ever be needed from the paradigm that the most effective selection for yield is selection for yield *per se* in yield trials. Paradigm shift 31 is that a way to find and effectively use genetic diversity is to choose parents, even closely related parents, that collectively span the range of the observed levels of that smallest number of (the major) component processes that account in full for an expressed system output such as the yield.

Sedgley (1991) indicated three approaches to breeding for yield applied in the past are defect removal, direct selection for yield, and breeding for ideotype. Defect removal eliminates yield limiting traits by replacing them with positive traits. Examples are breeding for disease resistance, and breeding for the lodging resistance that led to the high harvest index (HI) and higher yields that attend the dwarf plant habit of cereal crops.

Selection for yield *per se* is an empirical approach. It has been successful

(Evans, 1993) because of the 'great integrating power' due to yield being the system output. Being a system output, the yield integrates all effects by all component and subcomponent processes, including all modulations of the thousands of gene actions by the varied levels of many environmental factors. Because of this integration, selection for yield *per se* within a given environment becomes selection toward the compensated and near-optimal levels of every component of the system that collectively result in superior functioning of the constant-capacity plant system.

The difficulty of indirect selection for yield by selection for components within the yield system is explained by the hypothesis of a constant-capacity plant system illustrated in Figs 11.1 to 11.5. This difficulty will continue into the future because of the system-established interconnections (negative or positive correlations) among all the components of the yield system illustrated in Figs 2.1, 5.1, 9.1, 11.4, and Tables 4.3 and 4.4. The yield system (Fig. 11.4) consists directly of only two of the five components of the plant system (Fig. 11.1). Nevertheless, because of necessity to share the available metabolites, the yield level is affected by the level of all five components, and vice versa.

12.1 High Yield Results from Compensatory Optimization among the Levels of all Traits of the Plant System

Acceleration of gains in yield through plant breeding requires replacing an old paradigm for plant selection with a new paradigm based on systems thinking. The old paradigm is: if a trait contributes toward higher yield, then selection for a higher level of that trait should raise the yield. Many breeding programmes have applied this old paradigm to one or a few traits at a time. Limited thought was given to the possibility that a yet higher level of each trait may become so high that the yield will be lowered. The latter is the new paradigm shift 28 derived by systems thinking.

Most quantitative traits are essential to yield expression. The level of each may be either higher or lower than the level that is optimal for high yield. This optimum level of each trait is dependent upon and compensatorily related to simultaneous optimization of the expressed levels of all other component traits of the functioning plant system (Fig. 11.1, sections 11.2.2 and 11.3.2). Therefore, the optimal level of every trait depends on (i) each genotype's genetically established potential level for all the plant system's gene actions, and on (ii) the environmental-factor implemented modulations of the levels of all these gene actions, including (iii) reductions of these gene actions by the inevitably limited resources for growth that will be provided by the environment.

12.2 Examples of Selection for Excessive Levels of Traits

Chapters 4, 5 and 9 describe the interconnectedness among the three major physiological components of the process of accumulating crop yield. The interconnections are such that the expressed level of the HI is negatively correlated with both the days to maturity (DTHM) and the aerial biomass, while the biomass and DTHM are correlated positively (Figs 2.1, 5.1, 9.1 and 9.7). The DTHM can be either too long or too short to exactly match the duration of the growing season. Either will reduce the yield.

Kenworthy and Brim (1979) showed that selection within a segregating population of soybean for higher HI lowered yield, because it shortened the DTHM and reduced the plant biomass below levels that can support high yield. Similarly, selection for largest biomass will lower the yield, because it will result in a DTHM that is too long in combination with a HI that is too low to give high yield.

Systems thinking (particularly *the hypothesis of a constant-capacity plant system* described in Chapter 11) was not in place as the above interconnections were originally reported and discussed. Nevertheless, an obvious conclusion was that a compromised intermediate level of each of the negatively correlated traits will give the optimal functioning of the subsystem of the plant system being studied. But, it was not adequately emphasized that the optimal level of all quantitative traits will shift: (i) with altered levels of both the resources for plant growth provided by the environment and the stresses it provides, and (ii) with any alteration of the genetically established potential level(s) of the quantitatively variable trait(s), even traits which seemingly should be genetically independent. Shifts of the optimal trait level through both (i) and (ii) will arise – through the forced sharing by all traits of a large portion of the always limited supply of intraplant metabolites, which in turn will arise through the always limited resources for growth available from every environment and the resultant stresses to growth that differ for every environment.

The above examples indicate negative effects on yield from selection for above-optimum levels of a component trait for which it was thought selection toward a higher level should/could contribute to higher yield. The should/could is to emphasize that often in the past breeder's expectations have not been met. The reverse of the breeder's expected gain often occurs. Such non-intuitive results indicate need for greater understanding of the complex plant system.

12.3 Supplemental Supports of Empirical Selection for Higher Yield

12.3.1 Yield system analysis and optimization theory

Evans (1993, 1994; section 4.1) indicated past yield gains were not achieved through analytical understanding of the yield system. We suggest (section 11.3.2) gains occurred because the selection for yield resulted in indirect selection for the compensatorily optimized levels among all the physiological–genetic components integrated into the process of accumulating the yield. Acceleration toward yet higher yields through future breeding will require application of this systems thinking, and improved understanding of the interconnections among the system components. This must include understanding of the alterations of the interconnected levels of the component traits that result from different genotypes, and from modulations of the implementing gene actions by the variable factors of the environment.

Better understanding of the plant system will never reduce dependency of breeding for higher yield on empirical and phenotypic selection for yield. Empirical selection for yield *per se* will remain integral to effective use of any genetic variability of plant traits for which the gene actions and expressions occur early to intermediately within the system, i.e. occur before and become integrated into the penultimate outputs of the yield system. These gene actions include most enzyme activities that result from a single or a few genes, and most networks of gene actions that are integrated into such processes as photosynthesis, respiration, nitrogen uptake and its fixation, biosynthesis of proteins, etc. There is an inability to measure these intrasystem traits cheaply, and a demonstrated lack of correlation (section 3.4.3) of the levels of photosynthesis, nitrogen fixation, ribulose bisphosphatase activity, etc. with the expressed yield. This leaves selection for yield *per se* as the only way to benefit from the genetic variability for these early-functioning intra-system traits. All early positioned traits are integrated into at least one of the three penultimate outputs of the yield system. The further integration of the three penultimate system-output traits gives the ultimate system output, which is the yield.

Appreciable support for yield gains beyond the gains due to empirical selection for yield *per se* can arise by application of the yield system analysis (YSA) procedure. The YSA will quantify the penultimate and antepenultimate outputs (Table 9.1; Fig. 9.1) of the plant's subsystem (Fig. 11.4, compare Fig. 11.1) that accumulates the crop yield. The level of each of these component traits should be compared with the yield levels expressed across multiple genotypes grown in multiple environments (as detailed in Chapter 9). These comparisons can be achieved cheaply by applying the YSA to the multiple location or multiple year yield trials already being conducted. The only added expense is measurement of the biomass accumulated on each trial plot, and attention to recording the exact days to flowering (DTF) and DTHM. Support from YSA toward better selection for yield *per se*, will be increased further if the measured yield and each of the penultimate and antepenultimate system output levels are subjected also to an Additive Main effects and Multiplicative Interaction (AMMI) statistical analysis (sections 9.1.4, 9.2 and 9.3). The YSA/AMMI analysis will quantify the genotype × environment ($G \times E$) interaction effects that are caused by each genotype and are caused by each environment on the levels of the yield and its major components. The supplemental support to selection for yield *per se* will arise primarily through improved understanding of the system, especially of the physiological–genetic bases for the interconnections among the levels of the components of the system.

Loomis (1993) suggested crop yields can be raised by optimization of crop structure and physiology. He indicated this will require improved understanding of optimization theory and of how the system's network is interconnected. Understanding of the system established interconnections will provide the capability to recognize and understand the trade-offs between traits, and to select for the optimal system functioning. Limited description of optimization theory is given in Loomis and Connor (1992). Givinish (1986) presented arguments for optimization theory. The YSA/AMMI quantification of the $G \times E$ interaction effects on the yield, and on each of yield's major physiologically and genetically controlled components can lead to the needed system understanding. Quantifications by YSA of the penultimate system output levels (biomass, HI, and DTHM) and the antepenultimate system outputs (rates of the biomass and yield accumulations) make it possible to recognize and then select toward that relative level of each that is optimal for each different environment. This will supplement selection for yield *per se*, particularly by improved insight about selecting the parents for crosses.

12.3.2 Selection for optimized rate of biomass accumulation and compensatory optimized rate of yield accumulation

Hay (1995) indicated that for maize, a cross-pollinated crop, selection of hybrid cultivars adapted to high population density, high soil fertility, and use of pest and disease control has resulted in large increases in grain yield. These yield gains are due mainly to increased biomass production, rather than increased HI which has remained relatively stable for maize. Relatively few pairs of inbreds have been used as parents of the hybrids. The inbreds used have been selected from among tens to hundreds of thousands of tested inbreds, and through reciprocal recurrent selection to maximize the hybrid vigour. The high hybrid vigour (heterosis) results in the yield gain being through an enlarged biomass. On the contrary, for most self-pollinated crops, the biomass has remained relatively constant, while the yield gains have come through increase of the HI. Understanding of the (i) system-established interconnections among the biomass, HI and DTHM (Chapters 4, 5, 9), and of (ii) the genetic control over the biomass being by almost all genes, in contrast to (iii) the strongest control over the HI being by a few genes (Figs 2.1, 5.1, 9.1), and of (iv) a much higher heritability of the HI than of the biomass because of the control by few genes, explains (Chapters 4 and 5) why past selection for yield *per se* has readily raised the HI of self-pollinated crops, but has only infrequently given genotypes with yield gain through an enlarged biomass. For self-pollinated crops, we are not aware of studies that fully evaluate whether hybrid varieties will have larger biomass than the standard cultivars.

We do suggest there should be selection toward enlarging the accumulated biomass. *Raising the biomass should be based predominantly on selection for the rate of biomass accumulation per day*, however. This should be attended by simultaneous selection against increase of the DTHM, reduction of the rate of yield accumulation per DTHM or per day of seedfill, and reduction of the HI. The simultaneous selections are required because YSA/AMMI has shown the biomass to be negatively correlated with both the DTHM and the HI (section 9.3.4, compare Chapters 4 and 5). Selection for all three traits requires application of YSA/AMMI to the ongoing yield trials. The YSA/AMMI should begin as the segregates from crosses advance beyond the F_4 to F_6 generations. These yield trials of the intermediate to late generation segregates should be across multiple environments (locations, years or seasons) and should always include commercial checks. Testing for magnitude of reduction in yield plant^{-1} in response to increasing plant density (see Chapter 10, and section 12.4.5.2 below) could be incorporated into the yield trials.

12.3.3 Continuous advance of biological understanding of the plant system

Within this section 12.3, YSA/AMMI has been emphasized for its supplemental support of selection for yield *per se*. In the long run, the major benefit from YSA will be continuous advance of understanding about how the plant system varies in functioning and the levels of its components across the range of existing genotypes grown across a wide range of environments. The quantification by the YSA/AMMI of the levels of the accumulated biomass, HI and DTHM that give the highest yields in each environment will reveal the compensated levels among these traits that optimize functioning of the plant system in the several environments. Just as cultivar yields differ within each tested environment, the optimal levels of each of the physiological–genetic components of the process of accumulating the yield will differ for each environment. The advanced-generation genotypes that give the highest yield and merit release as new and superior cultivars will be identified simultaneously with the gain in understanding of the yield system. Simultaneously also, the YSA/AMMI will identify the broad range of genetic diversity that should be crossed to continue breeding toward yet more superior cultivars.

12.3.4 Selection for high potential yield of individual plants and tolerance to planting density

Selection for crop yield *per se* has been effective only during the advanced generations after the segregates become relatively homozygous and there is enough seed for yield trials. Effective selection during the early segregating generations would greatly reduce the number of lines that need comparison in yield trials. This

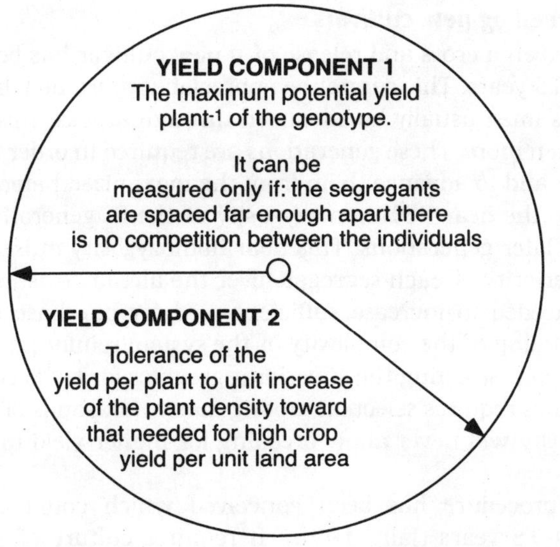

$$Z_{max} = (a + b)^2/(4b)$$

where:

a = yield component 1

b^{-1} = yield component 2

Fig. 12.1. Two negatively correlated components of the crop yield based on the expressible yield plant^{-1}. Evidence is mounting that: (i) It is possible to select effectively for the crop yield potential of F_2 and other early generation segregates, by selecting for the yield plant^{-1} of the segregates spaced far enough apart there is no competition between the individual plants. (ii) Selection for a high harvest index of these same individuals will be selection for low competitive ability, so these segregates will have a lower reduction of their yield plant^{-1} as the plant density is increased. (iii) Selection for high yield plant^{-1} among plants grown at the density required for high crop yield is likely to be selection for competitive ability rather than selection for high crop yield of the genotype. (iv) Compare: Yan and Wallace, 1995a; Fasoulas and Fasoula, 1995.

would improve breeding efficiency. Chapter 10 (compare Fig. 12.1) describes a viewpoint from which the crop yield potential (Y) of a genotype is determined by just two components as follows:

$$Y = \frac{1}{4}a^2b^{-1}$$

where a is the potential yield of an individual plant of the genotype; and b^{-1} is that genotype's tolerance to increasing plant density. Therefore, during the early generations, crop yield potential can be indirectly selected for, if there are associated selections for both components (Fig. 12.1). The maximum yield potential of the individual plants can be selected for directly, using plants spaced so far apart there is no or at least limited interplant competition. Tolerance to density cannot be directly selected for at such a single widely spaced plant density. But, in addition to selection for high yield plant^{-1}, selection for high HI and other traits associated with weak interplant competitiveness and described in section 10.4.1.2 could help achieve gain in crop yield.

12.4 Breeding of Higher Yielding Cultivars

12.4.1 The time span of breeding new cultivars

The usual time between a cross and release of a new cultivar has been and will continue to be 10–15 years. This cannot be appreciably shortened, because segregates from crosses must usually be advanced through at least this number of generations, and often more. These generations are required in order to select the superior segregates, and to advance them from the maximized heterozygosity of the F_2 generation to the near 98% homozygosity of the F_8 generation, and the nearer 100% of all later generations. This near homozygosity must be followed by proving the superiority of each segregate over the already existing cultivars. Yet more time is needed to increase sufficient seed for purchase by farmers. Enhanced understanding of the complexity of the system reinforces the requirement for this time span. Selecting the superior segregates for traits controlled by more than a few genes requires selection among many thousands of segregates. The system complexity will never allow breeding for higher yield to be an easy task.

One breeding procedure has been conceived which could shorten the requirement for 10–15 years (Jain, 1996). It requires culture of plants from many haploid pollen grains of F_1 plants, followed by doubling the chromosome number, thereby generating homozygous diploid plants. This cannot yet be achieved with the efficiency needed to obtain many thousands of such homozygous segregates.

12.4.2 The cyclic steps of plant breeding programmes

The major advantage to be gained from improved understanding of the plant system, especially of its subsystem for yield, is greater yield gain at the end of the 10–15 years required to complete one cycle of plant breeding. In an on-going plant breeding programme, a new 10–15 year breeding cycle begins each year. And, each cycle started previously is advanced one or more generations. A new cycle has to anticipate the cultivar traits required 10–15 years into the future.

Simmonds (1989) described two tasks of plant breeding. Stoskopf *et al.* (1993) described five steps that span Simmonds' two tasks. We subdivided the first step of Stoskopf *et al.* into two steps which lie within the first task of Simmonds.

1. The first task of breeding is the creation of superior genetic segregates. Its two steps are:
   - identification of genetic variability for use as parents;
   - making crosses to create new genotypes.
2. The second task of breeding is selection and testing of the superior segregates. Its four steps are:
   - selection of the superior genetic segregates;
   - comparison among the selected segregates;
   - comparison of the potential new cultivars with the existing cultivars;
   - final selection of the most superior genotype.

If the breeder wants the shortest possible time to cultivar release, final selection of the most superior selection must be preceded by increase of seed of a number of candidate superior genotypes, only one of which will ultimately be identified as the most superior. Longer than 10–15 years results from waiting until selection of the superior genotype is finalized, before beginning the several generations of seed increase needed to make the new cultivar commercially available to farmers.

12.4.3 Selection of parents for creating higher yielding segregates

12.4.3.1 Choose parental germplasm based on genetic diversity for the levels of all traits that affect acceptance of a cultivar

Creation of many superior segregates requires goal-oriented selection of the genotypes to be used as parents of crosses that is based on both system understanding and genetic diversity. This parental germplasm should span a wide range of genetic diversity. A broadly based non-goal-focused source of diversity is simply plants from diverse geographical locations, or use of related species. *The parents should represent at least some selection applied toward traits that seemingly should/could support the goal of higher yield, especially traits or trait levels not previously observed within the breeding programme. Whether from exotic or closely related germplasm, this genetic diversity should span most of the full range of variability of each major physiological–genetic component of the process of accumulating yield.* Selections based on both the highest yield plant^{-1} grown at wide interplant spacings to reduce interplant competition and also selected for the negatively correlated tolerance to plant density (section 12.3.4 and Chapter 10) provide another kind of variability of major components and of genetic diversity. Genetic diversity among the parents can come from all levels of observations and selection within the breeding programme and from other breeding programmes. Intuitively, from the phenotypic viewpoint, it seems that plants with major component levels that are high are more appropriately used as parents than, plants that have low levels. But, low levels do add to genetic diversity as does any variation in the background genes. In the past, it has seemed advantageous for the levels of all the component traits to be associated with at least a medium or higher expressed yield within the target environment.

Sections 12.3.1 to 12.3.3 of this chapter presented tandem application of the YSA/AMMI analyses detailed in Chapter 9 as a superior way to supplement the empirical power of selection for yield *per se*. Here, the very same YSA/AMMI is presented as the superior way to facilitate selection for the genetic diversity of the parental germplasm needed to maximize the number of segregates from crosses that will have a new, and more frequently superior, pathway and capability for high crop yield. Multiple-genotype/multiple-environment yield trials are already being conducted, especially by the international crop research centres. With minimal cost, YSA/AMMI can be applied to all ongoing trials to simultaneously achieve all three goals of: (i) selection based on yield *per se* of the highest yielding genotypes for release as new cultivars; (ii) identifying parents to provide the wide range of genetic diversity represented by diverse levels of the several major physiological–genetic components of the process of accumulating the yield; and (iii) continuous advance in understanding of the system-established biological bases for the interconnections among the components of the plant system, and especially of its subsystem that accumulates the crop yield and its interconnections with other components of the yet more holistic plant system.

12.4.3.2 Parental germplasm for genetic diversity of the components of traits essential to product quality

The five components of the plant system (Fig. 11.1) each compete for a portion of that system's capacity, that capacity being near-constant across all genotypes of the given species. Four of the five components, including the physiological–genetic components of yield, time to harvest maturity, resistance to abiotic and biotic environmental stresses, and appropriate interplant competitiveness have all been discussed above as relevant to breeding for higher yield.

A genotype that expresses the optimum level of all the component traits for accumulating yield, and therefore expresses the highest of all yields – yet lacks the product quality demanded by the human society of the target area (environment) – will not become a successful cultivar. A superior and successful cultivar requires a level of genetic instruction for all traits such that the expressed level of all gene activities including their modulations by the environment are optimal (or near optimal) for the given environment's resources and stresses and the demands of the human society.

12.4.4 Breeding procedures based on systems thinking

12.4.4.1 Continued breeding for independent traits with qualitative inheritance

Independent traits are controlled by one or two (only a few) genes (Fig. 11.5). Presence versus absence of the gene is obvious through expression of the dominant versus recessive phenotype. Based on systems thinking (section 11.1.1), expression of independent traits requires such a small part of the available intraplant metabolites, i.e. of the constant capacity of the plant system, that the trait's expression is not obviously altered quantitatively by the larger proportions used by all quantitative traits (Fig. 11.1). Because such a small part is used, presence versus absence of the gene does not obviously alter the levels of other traits. Therefore, most 'qualitatively inherited' traits are largely physiologically independent of the levels of all the quantitative traits. Nevertheless, the yield system functioning often requires the expression of some independent traits (Fig. 11.5). For example, resistance to or chemical prevention of a disease is essential to high, or even minuscule yield in some environments. Breeding for such qualitative traits has occupied 50–80% of prior plant breeding efforts. Need to breed for them will continue. It will often have higher priority than breeding specifically for higher yield. This is because selection for yield *per se*, and quantifications of its physiological components using YSA/AMMI, cannot be achieved if disease or insect damage prevents their full expression as allowed by the other resources and stresses of the environment.

A few physiological traits with qualitative genetic control can alter the functions of the plant to use the environmental resources more effectively and counter its stresses. Tolerance of roots to aluminium in highly acid soils is an example. Bushamuka and Zobel (1988) report qualitative genetic control over aluminium tolerance. Depending on the cultivar, aluminium tolerance may be expressed by only the basal, or the tap, or the lateral roots. The tolerance of some genotypes results because one of these categories of roots grows more effectively in the top soil layer that has been limed to reduce its acidity. For other genotypes the resistance occurs because the roots are truly aluminium tolerant, and can grow into the uncultivated soil below the ploughed layer where the soil acidity has not been reduced. That mode of aluminium tolerance enlarges the soil volume from which the plant can acquire mineral and water resources. A conclusion is that both qualitatively controlled disease resistance and qualitatively controlled aluminium tolerance can enlarge the functioning yield system. The aluminium resistance is doing so by enlarging the environmental resource base that the roots can use. Such qualitative genetic directions enlarge the constant capacity of the resistant genotype within the given stress environment.

12.4.4.2 Expanded biological explanation for quantitatively variable traits

Genetic and plant breeding terminologies have defined a quantitatively variable trait as being controlled by many genes, and strongly influenced by the environment. Our categorization as 'interdependent traits' has the following basis. Typically, *but with definite exceptions of strong control by one or few genes*, the interdependent traits are controlled by many genes. Most of these genes supply indirect control only. Indirect control may be through (i) the direct interconnections between the genetically directed biochemical pathways. But, the quantitative variability and interdependency results primarily, because (ii) the quantitatively variable system components obligatorily share the constant capacity of the plant system (Fig. 11.1) that is established by the limited available environmental resources needed for plant growth. An interdependent trait requires a large enough proportion of this near constant capacity (of the intraplant metabolites) that its level is quantitatively altered by any change of the proportion(s) required by the genetic direction for the other components that also require a large proportion. Change of the genetically directed level for one or more additional of the interdependent traits is required for a genotype that has been changed for one component, if it is to be adapted to the unchanged available environmental resources. These interdependencies arise through the network of interconnections (Fig. 2.1) among the component traits within the system. The many components are all essential to functioning of the plant system. Therefore (Fig. 11.1), enlargement of any component trait forces reduction of one or more of the other component traits. Likewise, genetically instructed reduction of any component that uses a large proportion of the resources allows enlargement of one or more of the other component traits as illustrated in the next paragraph.

A few interconnections among the major components of the system arise from the activity of the different alleles of a single gene with near qualitative control. As an example, it was shown (Tables 4.3, 4.4; Figs 5.1, 9.1) that the photoperiod gene *Ppd* controls as many as 26 traits in bean. Many may seem to be unrelated traits, but only until the biology of their interconnections is understood. This pleiotropic control over 26 or more traits by a single gene results from a negative control (section 6.4.1) by the dominant photoperiod gene *Ppd* over partitioning (Chapters 4, 5) of the available photosynthate to continued support of growth of the reproductive organs. This negative control inhibits partitioning predominantly toward the reproductive organs, and thereby inhibits the implementation of the plant's autonomous capability for rapid development to flowering (Chapter 5). The insensitive (*ppd*) and photoperiod sensitive genotypes (*Ppd*) initiate flower buds at the same time. But, the autonomous capability to develop to flowering is quantitatively inhibited by the photoperiod-temperature modulated level of the post-flower-bud-initiation activity of the photoperiod gene *Ppd* (Chapters 6, 7, 9).

In the absence of allele *Ppd*, i.e. presence of the recessive allele *ppd*, without the negative control over partitioning to the already initiated reproductive organs, sufficient photosynthate is received by them that they grow rapidly. Their rapid growth causes them to develop rapidly to flowering, and continue to grow and develop rapidly to harvest maturity (sections 7.3.2 and 7.3.3). This control over partitioning of the photosynthate to growth of the reproductive organs versus to growth of the vegetative organs becomes a simultaneous control over the interdependent levels of the accumulated biomass, HI and the DTHM (Figs 2.1, 5.1, 9.1). Additional traits for which the expressed level changes as a consequence of this differential in partitioning include (from Tables 4.3 and 4.4) the number of leaves, leaf area, node to flower, number of branches, plant height, first flowers on branches versus at the apex of the main stem, clusters of flowers versus individual flowers scattered across time, and rates of biomass accumulation and of yield accumulation.

Early in the 20th century, the DTF and DTHM were classified as quantitative genetic traits, the interpretation being control by many genes and strong control by the environment. Subsequent research with many crops has demonstrated by far the strongest control over the DTF and DTHM is due to one to a few photoperiod genes, with consequences described in the preceding paragraph and illustrated in Figs 2.1, 5.1 and 9.1. In the last 75 years, this strong control by one to a few photoperiod genes has been demonstrated for many crop species (Murfet, 1977a, b). In addition to the strong control by one or a few genes, much of the quantitative variation of the DTF and DTHM is due to the smaller control by an enhancer gene, and more of it is due to the genetic variability represented by the permutations of the many background genes that consist of all the other genes of the plant (Mmopi *et al.*, 1995).

12.4.4.3 The whole system, diverse component, recurrent selection breeding method

Jensen (1994) described recurrent selection as follows:

> Recurrent selection is controlled-breeding management for quantitatively inherited characters. Its basic modus operandi involves a repetitive sequence of selecting and crossing desirable genotypes to increase the frequency of favorable genes in the population. . . . In its simplest form it may be seen as selecting and breeding individuals from the high-end tail of the normal curve of a desired character's frequency distribution. Recurrent selection is uniquely suitable for cross-pollinated crops but with family resemblances to the backcross.

Figure 12.2 illustrates phenotypic recurrent selection from the high-end tail of the distribution of yield using empirical selection for yield *per se*. This is recurrent selection for a single trait.

A computer database search for recurrent selection combined with plant breeding brought up hundreds of research reports. They involved self-pollinated crops as often as cross-pollinated crops. Recurrent selection breeding is used for any trait that is controlled by more than a few genes and therefore has low heritability. Recurrent selection is the preferred method of breeding for higher yield because, definitely, yield is controlled by many genes with most causing such small effects that the effect due to these individual genes cannot be detected. The *whole system, diverse component recurrent selection breeding method* described below provides for the parents of crosses to introduce physiological–genetic diversity, to thereby maximize the genetic segregation and number of superior genotypes. The caption of Fig. 12.2 indicates recurrent selection maintains needed genetic diversity. The whole-system, diverse-component, recurrent-selection

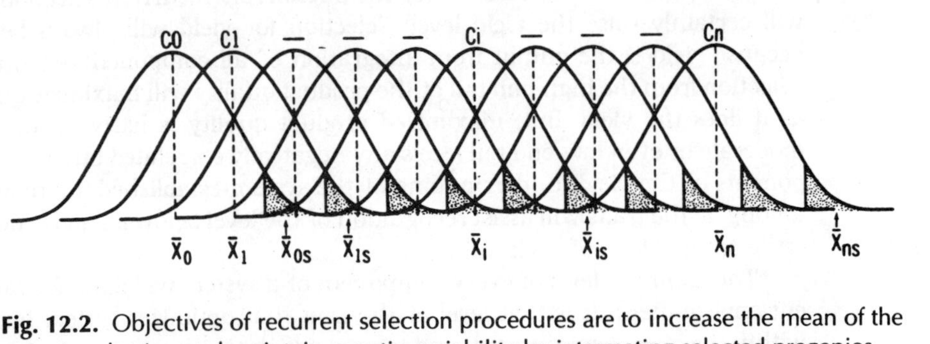

Fig. 12.2. Objectives of recurrent selection procedures are to increase the mean of the trait via selection and maintain genetic variability by intermating selected progenies. Each cycle originated from the C0 and Cn increases the mean while maintaining variability. (Reprinted from Hallauer, 1991.)

breeding method enlarges the genetic diversity beyond just that selected from the high-end tail of the yield *per se*. It does so by using as parents, for each successive recurrent intercrossing, germplasm that spans a wide range of the levels of each of the eight physiological components of the accumulation of the yield measured by a YSA (Table 9.1).

12.4.5 Breeding methodology based on systems thinking

Systems thinking introduces the following as a first change in the methodology of breeding for higher yield.

12.4.5.1 Phenotypic recurrent selection for genetic diversity of all the physiological–genetic components of the process of accumulating yield combined with product quality

We suggest the genotypes to be crossed should be selected to maximize the range of the genetic diversity they collectively represent. This diversity should span most of the recognized variability for all major physiological components that collectively give expression of the ultimate system output. This system output from the breeding programme should be higher yield combined with acceptable product quality. Recurrent selection using maximized genetic diversity departs to a compromised extent from Jensen's statement that recurrent selection involves recurrent intercrossing of individuals selected from the high-end tail of the normal distribution curve of the desired character's frequencies as illustrated in Fig. 12.2.

Whatever the traits required for product quality, they are as essential to success of a cultivar as high yield. As far as the system-established interconnections among all the plant traits and their components allow, therefore, selection for product quality should be simultaneous with selection for yield. Like for yield, the component trait levels for quality should be as diverse *as needed and as can be identified*. The italic is to recognize the germplasm used to incorporate superior yielding capability often introduces negative traits for product quality. And, the major (penultimate) system components of the product quality may not have been recognized. Certainly, some parents should have high product quality.

Selection from the high-end tail of one trait (yield) as Jensen (1994) and Hallauer (1990, 1991) describe for the phenotypic recurrent selection method, will certainly raise the yield level. Selection for yield will always be effective because yield is the output from integration of all components of the system. Selection from the high-end tail of the product quality will maximize quality just as it does the yield. But, maximized product quality is likely to force toward moderate or even low-end tail levels any negatively correlated but essential components of the yield. Understanding of the system-established interconnections among all the traits will assist recognition of the level below its high-end tail that is the optimal level of each trait.

The optimum level of every component of a system will differ for each different environment. It is that level of the trait that optimizes system functioning within the given environment, and toward which the successive parents used for the recurrent intercrossing might advantageously advance. As the trait levels of the successive parents of the recurrent crosses change toward being closer to their optimal level, each recurrent intercrossing cycle constitutes a backcross toward a more nearly optimized functioning of the plant or yield system. Rapid progress requires either (i) having already learned through YSA/AMMI the optimal levels of all the traits, or (ii) empirical selection for optimal system functioning which does require selection from the high-end level of the system output, i.e. for yield *per se* and for product quality *per se*.

The selected parents should not all come from the high-end tail of the yield *per se*. They should encompass enough genetic diversity toward low yields to include some segregates with intermediate to high levels of all the major components of the process of accumulating the yield, the HI, biomass and the optimum DTHM. Trade-offs among the component levels and the yield are needed for the parents, because the HI is almost always negatively correlated with both the biomass and the DTHM, while the DTHM and biomass are usually correlated positively. And, the effects on these traits by the genes that most strongly control them are altered by the permutation of all the background genes.

Need to select toward an optimal level is obvious for the DTHM. It is far less obvious for the HI and biomass. The time to maturity needs to be neither longer nor shorter than the growing season duration. However, a complexity arises, because the growing season duration of most locations varies irregularly from year to year. Pfeiffer (1996) reported selection of a soybean cultivar with the highest yield for each of several maturity groups resulted in more stable yield than selecting just the cultivars giving the highest yield across years. This is supported by the average of the correlation within each of 51 bean yield trials between yield and the DTHM of the cultivars being 0.01 (Table 9.2). The near zero average correlation is the average (section 9.2.4) from significant negative correlations as high as −0.74 within about one-third of the 51 yield trials, significant positive correlations as high as 0.70 within another third, but low positive or negative correlations for the remaining third. The continuum among the levels of the HI, DTHM and biomass constitutes the range of variability of the physiological pathway to yield. Pfeiffer (1996) found two soybean cultivars differed in maturity by 26 days, yet gave the same yield. The 0.01 average correlation and the quantitative continuum of the pathway to yield indicates the stability of crop yield across years will be highest from growing near-optimally adapted cultivars of a range of maturities, rather than only the highest yielding cultivar quantified across several years.

There is need for the parental germplasm collectively to span the range of levels of all the major components of the physiological-pathway to yield. All genes that direct expression of these alternative genotypes cannot be specifically recognized. The background genes collectively but partially control the DTF (Mmopi *et al.*, 1995) and simultaneously also partially control the biomass, HI and DTHM. Recurrent selection and intercrossing involving this wide range of germplasm variability may initially slow breeding progress. However, use of this wide range of physiological–genetic diversity is integral to creating the largest possible number of new segregated genotypes to be selected among for a maximized capability for crop yield combined with product quality for the target environment.

Germplasm diversity can be enlarged at the start of any recurrent cycle by introducing as parents lines known to enlarge the yield or its components in other environments, and by incorporating parents possessing any other trait(s) that seemingly should/could raise the yield. Only infrequently can it be known from which parents (and genes) the highest yielding segregates will arise. This is especially so because of the many unrecognized modifying genes and background genes and their combined genotypes. We can recognize only the very few qualitatively acting photoperiod-, dwarf- or vernalization-genes that most strongly confer adaptation versus non-adaptation and high yield versus low yield within the target environment. Some modifying- and background-genes may strongly alter the usual mode of action of these near-qualitatively acting genes. For plant breeding objectives, the segregation should not be limited to just recognizable genes.

A segregating population within a *whole-system, diverse-component recurrent selection breeding programme* will consist of progenies from many crosses. Attempts to follow the pedigree of every segregate will require near-impossible record keeping. Usual lack of pedigree records does not negate changing to use of pedigree selection for more focused breeding, inheritance and physiology research objectives.

12.4.5.2 Selection of early-generation segregates for maximal yield plant^{-1} and its tolerance to increasing plant density

This suggested change in breeding and selection methodology attains feasibility through a largely unrecognized and therefore seldom used potential for indirect selection for crop yield, which is detailed in Chapter 10 and summarized briefly in section 12.3.2. The potential for indirect selection of crop yield land area^{-1} during the F_2 to F_4 generations is based on yield being fully accounted for through two negatively correlated physiological–genetic components (Fig. 12.1). One is the maximum potential yield plant^{-1} of each genotype. The other is the tolerance to plant density of this yield plant^{-1}.

Low reduction of a genotype's yield plant^{-1} with increasing plant density indicates a genotype-established tolerance to the strong interplant competition due to the high plant density used and required for high crop yield land area^{-1} (section 10.2). Jointly, the maximum potential yield plant^{-1} and its tolerance to increase of plant density account in full for the genotype's potential yield per land area (sections 10.1 and 10.3). These two negatively correlated components of the yield (Figs 11.1, 12.2) are always functioning. Both always partially control the crop yield. Neither is accounted for by the quantifications by YSA/AMMI of the major physiological components of yield unless spacing trials are conducted. It is a compromised optimal level for each of these two components that will maximize the yield. The needed compromized levels will be established largely by the resources the given environment can provide. But, the instructions of the genotype must implement the needed compromised optimal levels of the maximum potential yield plant^{-1} and of the sensitivity of that yield plant^{-1} to the increase in plant density needed for maximum crop yield.

The only way to select simultaneously among the early-generation segregates for both their maximum potential yield plant^{-1} and lack of its reduction per unit increase of plant density is to: (i) select for the highest yield plant^{-1} of

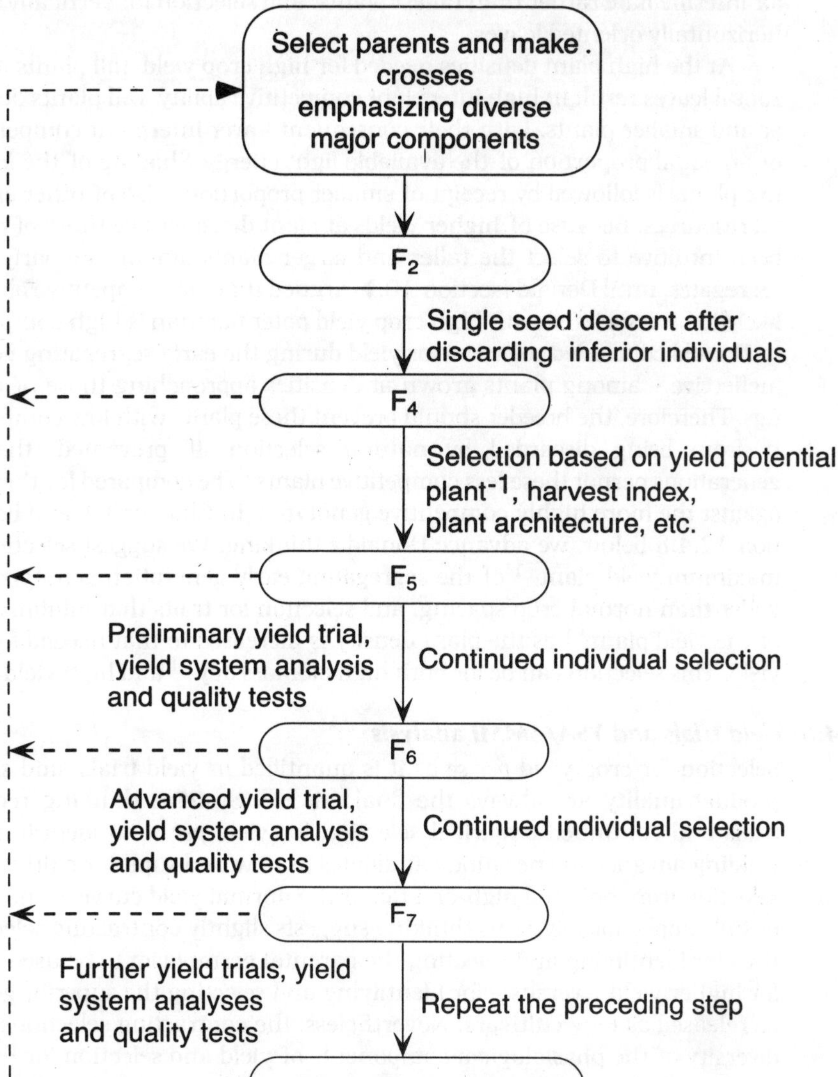

Fig. 12.3. Suggested procedures for the whole-system, diverse-component, recurrent-selection breeding method.

segregates spaced far enough apart there is no interplant competition; while (ii) indirectly selecting simultaneously against traits known to confer large interplant competitive ability. High HI is one trait known to be associated positively with low interplant competitive ability (sections 10.1 to 10.4). The HI is suggested as the best trait to use for strong indirect selection for high tolerance of the yield plant^{-1} to increasing plant density (Fig. 12.3). Other traits include selection for intermediate rather than tallest plants, and selection for vertically rather than horizontally oriented leaves.

At the high plant densities needed for high crop yield, tall plants and/or horizontal leaves result in high interplant competitive ability. Tall plants deprive shorter and smaller plants, with their consequent lower interplant competitive ability, of an equal proportion of the available light energy. Shading of the less competitive plants is followed by receipt of smaller proportions also of other environmental resources. Because of higher yields at plant densities like those of crops, it has been intuitive to select the taller and larger plants among the early-generation segregates, until Donald (section 10.1) argued that low competitive ability is more likely to be associated with high crop yield potential than is high competitive ability. Donald concluded selection for yield during the early segregating generation is ineffective – among plants grown at densities approaching those of crop plantings. Therefore, the breeder should prevent these plants with low competitive ability from being discarded by natural selection. If prevented, the advanced generations permit these less competitive plants to be compared for their crop yield against the more highly competitive genotypes. In Chapter 10, and briefly in section 12.4.8 below, we advance Donald's thinking. We suggest selection for both: maximum yield plant^{-1} of the segregating early-generation genotypes grown at wider than normal crop spacing, and selection for traits that minimize reduction of the yield plant^{-1} as the plant density is increased to that needed for high crop yield. This selection can be for both high biomass day^{-1} and high yield plant^{-1}.

12.4.6 Yield trials and YSA/AMMI analysis

Selection for crop yield *per se* as it is quantified in yield trials, and selection for product quality are always the final criteria. Systems thinking requires yield trials and continued empirical selection for yield *per se* to identify the highest yielding advanced-generation candidates as new and superior cultivars. For this, selection from only the high-end tail of the normal yield curve is and always will be fully applicable. Systems thinking suggests slightly contrasting selection objectives for identifying and selecting the parental germplasm to be used in breeding for higher yield – versus – for identifying and selecting the superior genotypes to be released as new cultivars. Nevertheless, the contrasting selections for genetic diversity of the physiological components of yield and selection for highest yield *per se* can be achieved by the same applications (Chapter 9) of YSA/AMMI analysis to the ongoing yield trials.

Most yield trials are conducted to identify the highest yielding genotypes within a target environment. The genotypes compared will already have been selected for adaptation and high yield in that environment. They are highly likely to span a truly wide range of levels of all the components of the process of accumulating the yield. Maximizing this diversity requires multilocation yield trials such as those often conducted by the international crop research centres. Gauch and Zobel (1997) describe use of AMMI analysis to identify mega environments, which may but need not be geographically adjacent. The AMMI analysis can identify one or more given cultivars that will give high yield across an entire mega environment. This provides high yield, while lowering the need to select a different cultivar for every different environment within the mega environment.

12.4.6.1 Quantification of the genotype × environment interaction controls over the expressed crop yield

Selection for yield among the superior advanced-generation selections will always require several growing seasons or years because ever-present environmental variation from one season to the next causes $G \times E$ interactions that result in different genotypes giving the highest yields from one season to the next comparable season. A single yield trial allows application of the YSA procedure, and its supplemental assistance to selection for yield *per se*. Three or more trials grown at different locations and/or during successive seasons or years facilitate, in addition, application of the AMMI analysis described in section 9.1.4. AMMI can statistically quantify the $G \times E$ interaction effects on the yield, and on each of the physiological components of its accumulation. This includes quantifying the relatively negative or positive control due to each genotype and to each environment through the level of the $G \times E$ interaction effect. The AMMI-quantified levels of these multiplicative controls combined with the breeder's knowledge about known genes and alleles of the cultivars and about the *P* and *T* of the environments is causally related to the largest $G \times E$ interaction controls over the DTF. This $G \times E$ control over the DTF carries through as system-established controls over the major components of the yield accumulation, including the DTHM, net accumulated biomass, HI and rate of yield accumulation (sections 9.2, 9.3).

12.4.6.2 Deriving system understanding from YSA/AMMI analyses of ongoing yield trials

Most yield trials of the past have wasted the biological information that resides in each yield trial, because the research focused only on the economic goal of identifying the cultivar(s) that give the highest yield, to enable the agronomist and extensionist to recommend them for planting by farmers. Sections 12.3.1 to 12.3.4 described tandem application to yield trials of the YSA/AMMI analysis procedures (see also sections 9.2 and 9.3). Advance in understanding of the subsystem of the plant system that accumulates the crop yield can be achieved most cheaply by applying YSA/AMMI to the yield trials that are already being conducted.

12.4.7 The role of ideotype in breeding for higher yield

Donald (1968, 1981) introduced the concept of breeding for yield based on an ideotype. His ideotype involved selection for morphological traits thought to maximize distribution of light interception across all the leaves and thereby its efficiency of use for photosynthesis. The ideotype also involved prevention of lodging using dwarf habit, and selection for a higher HI. He assumed selection for each of these traits results in indirect selection for yield. A 1991 symposium concluded that breeding for ideotype has not been successful (Belford and Sedgley, 1991; see also Balkema-Boomstra and Masterbroek, 1993). Rasmussen (1991) suggested the ideotype should incorporate more physiological genetic traits, in addition to selection for just morphological traits.

Systems thinking indicates that the ideotype was too rigid. It did not allow adequately for the different ideotype each different environment requires.

Application of YSA/AMMI to all yield trials will facilitate recognition of the ideotype with superiority for each different environment. From section 12.4.6.1, this superior ideotype will consist of the compensated levels among the physiological–genetic components of the process of accumulating the yield that give optimal functioning of the plant system within the given environment.

12.4.8 *Yield gains that support the suggested methods of breeding for higher yield*

12.4.8.1 Support for the diverse component recurrent selection breeding method

The *hypothesis of a constant-capacity plant system* is detailed in full for the first time in Chapter 11 of this book. This hypothesis began with observations by Yan (Chapter 11; Yan, 1993a; Yan *et al.*, 1991a, b; Yan and Wallace, 1995a) of compensatory interconnections among the quantitatively variable levels of three seemingly unrelated traits of wheat: winter hardiness, head size and DTHM. Yan found selection for a higher level of each of these three traits always lowered the level of one or both of the other two. Wallace (Chapters 4 and 5) reported similar compensatory changes in bean among the levels of the aerial biomass, HI and DTHM. Literature review indicated negative correlation between the HI and the DTHM of many crops, and negative correlation between the accumulated biomass and the HI, but positive correlation between the DTHM and the biomass.

These interconnections among the three major physiological–genetic components of the process of accumulating crop yield have not been the major focus of published reports by others, even though review of the literature (Wallace, 1985 and Chapters 4 and 5) indicated these interconnections among the major physiological components of yield exist in many crops. Corroborating demonstrations of the effect on yield through these compensatory interconnections among its components are evidenced most by the YSA/AMMI analyses of bean and pigeonpea yield trials presented in sections 9.2 and 9.3, respectively.

The first bean cross between a genotype with high HI and early maturity times a late genotype with large biomass was made in 1967. The cross resulted in selection of the early cultivar Redkloud, which has the photoperiod insensitive genotype *ppdppd*). Redkloud gave the same yield in 85 days as its late parent (cv. Redkote) with the photoperiod sensitive genotype *PpdPpd*, which gave the larger biomass in 105 days (section 4.2.2; Wallace *et al.*, 1993b). The equal yield of Redkloud was due its higher HI, and 20% higher yield day^{-1} to maturity. After these observations the first steps toward a complete YSA were to measure at harvest maturity the range of the biomass, HI, DTHM and yield expressed by bean in yield trials. Wallace recognized that the early high yielding plant was one of the small, weak, apparently low yielding F_2 segregates he had observed but always discarded rather than saved in prior segregating progenies. The early high yielding segregate giving Redkloud had been saved this time, simply because the cross was made and all the F_2 segregates were advanced to the F_4 generation while he was at a tropical location. He had recognized the need for ultimate selection of the progenies within his target temperate target environment. Therefore, all segregates having acceptable seed colour (a product quality factor) were saved and advanced in generation until Wallace returned to the USA.

Return to the USA has been followed by recurrent yearly intercrosses among lines found to have the highest HI, highest biomass, and highest yield. The F_2 and F_3 generations were advanced by single seed descent. It is obvious superiority will not be represented by either the earliest or latest DTHM. The optimum and intermediate DTHM obviously has to match the growing season duration. This duration varied sufficiently from year to year and across locations that the correlation between yield and DTHM within each of 51 yield trials averaged 0.01 while ranging from 0.70 to −0.74 (section 9.2.4). In general, inclusion of the highest level of biomass among the intercrossed parental germplasm involved the latest DTHM; inclusion of a near highest level of HI involved the earliest acceptable level of DTHM, and; the highest yielding germplasm was usually intermediate for these traits. Several yearly applications of this *diverse component recurrent selection* and intercrossing gave a cultivar with 20% higher yield than Redkloud. Unfortunately, it did not possess the progressively higher quality of the canned bean product the bean canning industry was demanding, so it did not become a successful cultivar. The breeding effort had focused too much on yield and insufficient on the product quality component (Fig. 11.1) of the plant system. Subsequent selection focused on canning quality in addition to yield has resulted in a cultivar that has the best available quality for canning and a 15% gain in yield.

A Cooperative Dry Bean Nursery has been distributed by the University of Idaho annually for 40 years. It is grown at 15–20 sites in the USA and Canada. Several locations recently applied YSA as an adjunct to these yield trials. These YSA showed that, as compared with all other classes of dry bean, the pink and pinto classes from the Northwestern United States have the: (i) highest rate of biomass accumulation day^{-1} to harvest maturity, (ii) highest HI and (iii) shortest DTHM, but (iv) the highest yields. Therefore, the pink and pinto classes were crossed with red kidney bean, the major dry bean class in New York State. The YSA showed the red kidney class has the lowest rank for all four traits. This breeding effort is still in progress.

The selection and recurrent intercrossing has proceeded as illustrated (Fig. 12.3) for the whole-system, diverse-component, recurrent-selection breeding method. Use of pink and pinto beans as parents introduced the high-end tail of four traits measured by YSA and specified within the previous paragraph. The red kidney class of interest had the low-end tail levels of these traits. Some red kidney lines used as parents introduced the best canning capability in combination with the early maturity of the photoperiod insensitive genotype. Others introduced the highest canning acceptability, which has been expressed by some of the late maturing photoperiod sensitive lines. Another subset of progenies involves crosses between just the best canning early and best canning late lines, with the objective of raising acceptability of the canned product for the early maturing lines preferred by farmers. The need is for less seed coat splitting after canning, with consequent reduction of intact beans, and starch caused gelatinization of the water around the beans within the can. These industry demands make product quality the trait with highest priority for focused selection. This application of the whole-system, diverse-component recurrent-selection breeding programme has as yet advanced only to the F_4 and F_5 generations.

The above several paragraphs present the only reports we are aware of in which, deliberately, the *genetic diversity* of the parents spanned wide levels of the biomass, HI and yield and the associated range of the DTHM and involved recurrent intercrossing. Therefore, *experience with the whole-system, diverse component, recurrent selection breeding method is too limited to present strong experimental evidence it is a superior way of breeding for higher yield.* The literature abounds, however, with reports that recurrent selection for yield *per se* leads to steady yield gains, even within germplasm populations for which the parents have not been selected to incorporate the *genetic diversity* represented by different component levels. Plant breeding literature abounds with arguments for need of *genetic diversity* (Tanksley and Nelson, 1996; Tanksley and McCouch, 1997). Genetic diversity will be selected for, even within closely related germplasm, by simultaneous selections for biomass, HI, and DTHM. The genetic diversity can be enlarged further by incorporation into the original population through crosses with any germplasm having high levels of other traits that should/could lead to higher yield. This method of selecting parents combined with recurrent intercrossing progressively assembles germplasm inclusive of all the genetic diversity required for expression, within the target environment of the breeding programme, of the entire range of the adapted quantitatively variable physiological– genetic and morphological-continuum of accumulating the yield. This holistic continuum consists of the collective continua of the high to intermediate or for some low levels of at least all eight of components of the process of accumulating the yield that are measured by the YSA (Table 9.1).

12.4.8.2 Support for early-generation selection for maximum potential yield plant^{-1}

Additional recommendations for the whole-system, diverse-component recurrent breeding follow. Systems thinking suggests the F_2 and/or other early-generation segregates be grown at spacing far enough apart there is no (or at least reduced) competition between the individual plants. There should then be selection among these segregates for: (i) highest yield plant^{-1} combined with (ii) high HI. The selection for high HI is intended to be indirect selection for minimal reduction of the high yield plant^{-1}, as the plant density is raised to that needed for a high crop yield. In section 12.4.5.2 above and Fig. 12.3, rather than beginning such selection with the F_2, it is suggested that the selection for high biomass plant^{-1} and tolerance to increasing plant density be done using F_4 generation plants developed by single seed descent from the F_2 plants. It is suggested the F_4 plants be spaced further apart than usual crop spacing, but not far enough to eliminate all interplant competition. Spacing plants so far enough apart would require an excessively large land area. And, each such widely spaced plant will be stressed by visits from that number of insects or herbivores that would initially share many more plants with normal spacing. Each widely spaced individual plant stands alone in facing all environmental stresses due to wind, rain, hail, etc. Some segregates may lodge even without these stresses, because the plant weight bends and may break the supporting stem(s). In dense stands of crop plantings individuals mutually support each other. Because each plant stands alone, the weed control required, and any viruses or diseases brought in by these pests (and damage by wind, etc.) result in reduction to zero of the interplant competition being impractical.

Fasoulas and Fasoula (1995) and Fasoula and Fasoula (1996), as stated without reference to the honeycomb selection method in sections 10.1 and 10.4.3 of Chapter 10, presented data from thesis research on 15 crops: wheat, maize, sunflower, rice, barley, rye, cotton, sugar beet, faba bean, tomato, potato, melon, pepper, carrot and asparagus. They cited data of other plant scientists for three additional crops (oat, the aromatic plant *Artemisia pallens*, and Pensacola Bahia grass). The results for all 18 crops indicate selection for yield plant^{-1} among wide and equally spaced F_2 segregates using the honeycomb selection method will result in selection toward higher crop yield. The honeycomb selection method is that the central plant within one or more hexagons formed by the equally spaced plants surrounding it, is selected only if its yield is larger than that of any plant within the surrounding hexagon.

A computer database search revealed three additional crops for which selection of widely spaced F_2 plants gave superior yield gains. These were mungbean (Gill *et al.*, 1995), the herb crop *Cymbopogon martinii* (Kulkarni, 1990) and lupin (Ratinam and Thurling, 1989). Another study with rye (Littlejohn *et al.*, 1994) reported an 84% yield gain after eight successive generations. Using the honeycomb selection procedure as described, the first and second generations each gave a 17.5% yield gain. Only one study, of faba bean (Lulsdorf and McVetty, 1986), has failed to find the gain in yield increased by selection for high yield plant^{-1} among the equally spaced early-generation plants selected using the honeycomb design. The goal of measuring yield plant^{-1} with the honeycomb method is to prevent the selected high yield plant^{-1} from being due to heterogeneity of soil rather than due to a genetic instruction (sections 12.4.5 and 12.4.9).

12.4.9 Elimination of effects on yield due to environmentally caused differential interplant competition

In Greece, Fasoulas (1973) published his first recommendation of selection for higher yield of equally spaced early-generation individual segregates. His nearly 30 years of research on this have been summarized (Fasoulas, 1988; Fasoulas and Fasoula, 1995; Fasoula and Fasoula, 1996). These reviews compared the yield plant^{-1} of individual segregates with that of individuals of check cultivars when all plants are spaced with a triangularly positioned equal-distance between each plant. This is the *honeycomb design*. The triangular and equal interplant-spacing places one plant at each of the six points of a hexagonal, honeycomb-shaped arrangement surrounding each plant. Within the hexagon formed by the six surrounding plants, the seventh, being central within the six-plant hexagon, is compared with the six plants that encircle it. Alternatively, it can be compared with the 13 plants that make up the two closest circles around it, or the 19 plants of the three closest circles, the 31 that include the fourth circle, or with the 43, 55 or 61 plants included in the fifth to seventh circle of plants around the central plant. These authors suggest the central plant be selected only if visual inspection indicates it has higher yield than any of the plants within the surrounding circle. They used a circular frame made of wire and moved it from plant to plant, so that every plant was considered once to be the central plant. When a central plant is selected only if its yield is greater than that of the six plants in the surrounding circle, the selection intensity is 14.3%. With inclusion of the second to seventh surrounding circles, the selection intensity becomes 7.7% followed by 5.3%, 3.2%, 2.7%, 2.3% and 1.8% (Fasoulas and Fasoula, 1996).

Honeycomb selection provides equal spacing between all plants but does not eliminate all interplant competition. It does equalize the interplant competition, to maximize the selection being for instruction by the genotype of the yield, and it reduces the environmental influence on the selection decision. A sufficiently wide spacing eliminates all interplant competition but does not equalize all the other environmental influences.

12.4.10 The economics of selection for yield plant^{-1} at wide enough spacings to eliminate interplant competition

Mitchell *et al.* (1982) reported selection of one F_2 wheat plant out of seven using the honeycomb design with 60-cm spacing between the plants gave larger yield gain than selection using the honeycomb design at 30-cm spacing. Selection at 60-cm spacing gave an average 4% gain in yield as measured in the F_4 generation, after F_3 seed increase without selection. Mitchell concluded this 4% gain from one generation of honeycomb selection did not merit the large time and effort required for the honeycomb procedure. He did recognize the economic advantage a 4% yield gain through each of several generations could provide. He suggested genetic variability to permit further gain may still reside within the F_4 generation. Mitchell did not consider the possibility that any continued segregation during the F_4 generation would likely have lowered the average expressed yield of the F_4 progenies grown at the density used for crop plantings. This is likely, because loss in yield due to a weakly competing genotype is usually larger than the yield gain of the strongly competing genotype. Knott (1972) also reported yield gain from honeycomb selection in the F_2, yet concluded the gain was not large enough to compensate for the labour required. On the contrary, Valentine (1979) found heritability of the yield of single plants of barley to be only 5.3%, yet argued no opportunity to select for yield during the early generations should remain unused.

Existing evidence indicates effective yield gain from selection for yield among the early generation segregates grown at equal wide spacing for at least 21 crops. This implies need to explore, more seriously than in the past, both the long-term biological and economic merits of selection for maximum potential yield plant^{-1} among the early generation segregates. A graded set of alternative selection procedures spans (i) selection for yield plant^{-1} among the widely spaced F_2 and the several subsequent generations of segregates, (ii) such selection for yield plant^{-1} among only the F_2 segregates, (iii) such selection for only the F_3 and/or F_4 and/or F_5 generation progenies, (iv) selection for the yield plot^{-1} of these F_3 through F_5 generation progenies grown at conventional crop spacings, and (v) no actual selection for yield by the breeder until the yield trials of the advanced near-homozygous F_7 or F_8 generations.

Lack of breeder selection for yield among the early generation segregates leaves all selection for yield during these generations in the hands of nature. If the segregates are grown at close crop-like spacings, the unequal competitive abilities of the genetic segregates will enforce natural selection by the interplant competition. The 'hands of nature' will select for any high competitive ability established by the genotypes of the segregates, or established by environmental differences. This constitutes selection for low rather than high crop yield potential, and may eliminate many segregates with high potential crop yield from being available in the succeeding generations for selection by the hands of the breeder. Is selection of widely spaced segregates really too expensive? Or, alternatively, is it the long-term cost from natural selection among the segregates toward lower crop yield potential that is too costly? The answers require more research.

12.4.11 Labour reductions for the honeycomb selection approach

One way to reduce labour of the honeycomb selection procedure is to select each plant only if its yield is visually superior to that of all plants within a moving square grid surrounding each individual plant. Two straight rods or sticks of the chosen length, each equally distant from the central plant, can be moved along just two sides of the individual plants to mark the moving square. These straight reference rods for a grid can be moved much more rapidly than a circular frame. Selection within this moving square (between the straight rods or sticks) will also facilitate selection of the central plant only if its yield is visually judged as superior to that of all the adjacent plants. This procedure largely guarantees, like the honeycomb design, that the superiority of the selected plant is due to its genotype, rather than due to plant variation caused by any soil-established or other environmentally caused superior growth and yield. Bos and Van Boxtel (1990) concluded grid selection and honeycomb selection are more effective than truncation selection for yield such as selecting the highest yielding 10% of all the segregates. They tentatively concluded honeycomb selection is more reliable than grid selection. The biological and economic merits of these alternative ways of selecting for crop yield potential among early-generation segregates merit intensive evaluation by the breeding programmes of many crops.

12.4.12 Single-seed descent from F_2 until selection at wide spacing

What is essential to applying the whole-system, diverse-component recurrent-selection breeding method toward breeding for higher yield is prevention of removal by natural selection of the weakly competitive and consequently lower yielding genotypes within near crop-density spaced segregating progenies. Such removal will be largest among segregates of the widely diverse genotypes. Two generations of single-seed descent from the F_2 generation spaced far enough apart to guarantee survival and seed collection from the weak competitors will advance all the F_2 segregates to the F_4 generation. Three will give the F_5. The single-seed descent to the respective 75% or 87.5% homozygosity of the F_4 and F_5 generations can be an alternative to using the very wide spacing needed to select among the segregates for their maximum potential yield plant^{-1}. This single-seed descent, the major physiological–genetic yield components, the hypothesis of a constant-capacity plant system, and recognition of the but two components of yield through interplant competition are the four theoretical biological bases for the whole-system, diverse-component recurrent-selection breeding method. Figure 12.3 summarizes a minimally time-consuming and inexpensive breeding procedure. A variation of single-seed descent from each F_2 segregate is descent through an equal number of multiple segregates from each F_2 plant.

12.5 Breeding for Cultivar Release Versus Research Intended for Publication

Using the honeycomb method and moving a wire frame to precisely, time-consumingly and accurately encircle each plant, the same land area, plus actual measurement of the yield plant^{-1} of the central and surrounding plants will necessarily delay the selection decisions. Relative to a breeding programme, this will usually provide only a retroactive decision. The next generation will likely be growing in the greenhouse or warmer-geographic area, before the data analysis is completed. Dependence on measurement of such extensive data will certainly retard breeding progress. Such time-consuming methodology is often required for peer reviewed and accepted publication, and for demonstrated advance in biological understanding. It is not required for applied plant breeding, the peer review of which is by acceptance versus rejection of the new cultivar.

Plant breeding is an art, the effectiveness of which depends on correct understanding of plant biology. It is an art because breeders make most of their selections by observing and comparing the expressed phenotype(s) with their eyes, although they base the selection decisions on their understanding of the relevant plant biology. Decisions based on excessively time-consuming measurements will not allow selection across the large number of segregates required for rapid progress toward cultivar improvement. Breeders must make as many selection decisions as possible in the limited time the developmental stages of the plant materials and calendar time allow. This requires selection based on rapid visual comparisons.

Making correct decisions visually and rapidly requires true understanding of the effect on yield through the levels of interconnection between all the quantitatively variable phenotypic traits, and true understanding of both the genetic and environment factors that cause the different levels of these phenotypic traits. There is need for systems thinking, and recognition of how the optimal level of each trait which will be altered by change of the environment but depends also on the potential gene-action levels established by the genotype. There is need to recognize that selection for different levels of the individual system components also alters the levels of the other components. These concepts have been emphasized throughout this book. These are the biological understandings needed to arrive more consistently at correct selection decisions within the art of plant breeding.

When need exists for more biological understanding to support or reject selection for a quantitatively variable trait, or a procedure for selection, a breeder may apply his/her time to precise measurement of levels of the trait and their consequence. More likely, the breeder will continue to apply most of his/her effort toward selection, evaluation and release of new cultivars, while encouraging plant physiologists and/or graduate students to do this meticulous research.

The interpretations of such research data must meet the criteria of peer review and acceptance for publication. The corresponding review for plant breeding is acceptance and continued use by farmers and industry of superior cultivar(s).

12.6 Breeding Methods for Self- Versus Cross-pollinated Crops

Both of this book's authors breed a self-pollinated crop. Predominantly, therefore, the concepts presented have been derived and illustrated using experiences with bean and wheat. Nevertheless, data have been cited from many crops. Three cross-pollinated plant species (maize, sorghum, *Arabidopsis*) have been used to illustrate effects on the DTF and other physiological–genetic traits that affect yield. The major difference between breeding self versus cross-pollinated crops follow. First, natural cross-pollination provides an inexpensive mechanism to facilitate the crossing of genetically different lines needed to produce commercial quantities of seed of hybrid cultivars. Second, rather than being selected for their yield *per se*, inbreds used as parents of hybrid cultivars are evaluated for their combining ability, measured as the gain in yield they cause through heterosis. Third, the inbred parents have low yields due to the inbreeding depression described below.

12.6.1 *Breeding hybrid cultivars of cross-pollinated crops*

Transfer of pollen by wind or insects in cross-pollinated crops eliminates having precise knowledge of the male parent, and of the alleles this parent has contributed to its offspring. The wind or insects bring pollen from many plants, thereby maintaining large genetic variability through each successive generation of open-pollinated cultivars. This large genetic variability of open-pollinated cultivars contrasts with breeder-developed cultivars of self-pollinated crops. The latter become more homozygous with each successive naturally occurring self-pollination followed by selection. It is because of the larger genetic variability of the cross-pollinated crops, and inability to know the identity of the alleles, that Jensen (1994) described phenotypic recurrent selection as a breeding method uniquely suitable for cross-pollinated crops. Jensen described (section 12.4.4.3) recurrent selection and intercrossing among only the segregates from each successive generation that express the high-end tail of the yield curve (Fig. 12.1). Recurrent selection is considered to be the superior way to accumulate the favourable alleles for yield or other targeted traits (section 12.4.4.3). Nevertheless, Hallauer (1990, (and compare Duvick, 1996) states that since its suggestion in 1909, the pure-line (enforced self-pollination) breeding method has continued to be the basic breeding method used to develop maize inbreds for use as parents of hybrid cultivars. The enforced selfing is to maximize the genetic uniformity of the inbreds to be used as parents of the hybrids. Also important is that the enforced self-pollination and subsequent testing for combining ability eliminates harmful alleles present in the naturally cross-pollinated population. Genetic uniformity of the inbreds is needed to maximize genetic uniformity of the hybrid. As described in section 10.4.3, lack of genetic uniformity lowers the yield.

The pure-line method consists of controlled self-pollination of the successive generations until genetic uniformity of the inbred line has been achieved, or balanced against the strong inbreeding depression (reduction of yield) that accompanies selfing of cross-pollinated crops. Mild if any inbreeding depression occurs with self-pollinated crops. Thus, developing the inbred lines to be used as parents of hybrid cultivars requires manual transfer of self pollen whereas self-pollination occurs naturally for self-pollinated crops. Controlled self-pollination leads to near-complete genetic homozygosity after about eight generations of selfing (Table 4.2). The physiological and genetic bases for breeding of the inbred lines of cross- and self-pollinated crops are identical. Maize and sorghum are the two cross-pollinated crops for which hybrid cultivars are most extensively used. Onion, cabbage, broccoli, carrots and squash are cross-pollinated vegetable crops for which hybrid cultivars are important.

There is minimal similarity in the method used to select a superior advanced-generation self-pollinated line as compared with selecting an advanced-generation cross-pollinated inbred. The naturally self-pollinated line will be selected and released as a cultivar – if its yield *per se* and other economically valuable traits are superior to those of the compared lines and existing cultivars. In contrast, the advanced-generation, manually selfed inbred of a cross-pollinated crop will be considered only for use as a parent – and will be used as a parent of a hybrid cultivar – only if the yield is superior for one or more of its hybrid(s). This is determined using test crosses with other elite inbreds. Only one or few of an inbred's hybrids will have superior yield *per se* and/or other commercially important traits as compared with the already existing hybrid cultivars (Hallauer, 1990, 1991). Comparison of the relative yields of hybrids between new inbreds and elite testers is referred to as testing the combining ability of the inbred. Selection based on yield *per se* versus based on combining ability is the only major difference in the breeding methodology for self-pollinated versus for the cross-pollinated crops for which hybrid vigor (heterosis) is used to achieve yield gains. For some cross-pollinated crops, including vegetable crops, the uniformity needed for once-over mechanical harvesting that the hybrid cultivars have is as or more important than the higher yield through heterosis.

12.6.2 The role of statistical models of gene action in breeding hybrid cultivars

Comparisons of the phenotypes due to alleles of the same gene brought recognition of the dominant, recessive, additive and heterosis of gene action. Gene action is interpreted as additive if the F_1 of the cross has a phenotype intermediate to the phenotypes of the two parents. The gene action is negative heterosis if the phenotype level (e.g. yield) of the F_1 is lower than that of the parent with the lower phenotypic level. It is positive heterosis if the yield is higher than that of the parent with the higher phenotypic level. The gene action is dominance if the F_1 phenotype equals or nearly equals that of one of the parents. Statistical geneticists have applied these gene action types for single genes to the multiple gene actions controlling the quantitatively variable traits such as yield (Griffing, 1994). Selection for additive effects on a trait is effective. On the contrary, direct selection for high levels of dominance and especially selection for heterosis effects is ineffective. High heterosis requires selection for superior combining ability of the inbred parents. An inbred line has specific combining ability if it gives high heterosis in crosses with only one or a few inbreds. It has high general combining ability if it gives high heterosis for the cross with many inbreds.

Application of the statistical genetic method has provided the major scientific guidance used in the art of breeding hybrid cultivars of cross-pollinated crops. This is especially so for maize and sorghum. The biological information used in making selection decisions is based predominantly on statistical genetic concepts and measures of additivity, dominance and heterosis. The plant breeding remains an art, because a large part of the breeder's selection for combining ability depends on visual selection (Hallauer, 1990; Duvick, 1996). Because of the numerous genes and environmental variabilities involved, all breeders select the parents to be used and make crosses with incomplete knowledge. Breeders always have but partial ability to foresee the end product of their crosses and selection.

12.6.3 Breeding non F_1-hybrid cultivars of cross-pollinated crops

There are two reasons hybrid cultivars may not be used for a cross-pollinated crop. The first is lack of an economical and biologically efficient method to develop the required genetically uniform inbreds, and/or accomplish cross-pollination between the inbred parents to obtain adequate seed of the hybrid. When either occurs, breeders can make use of every plant of a cross-pollinated cultivar being a hybrid, while recognizing each plant is a different hybrid so the cultivar is far from being genetically and therefore phenotypically uniform. The lack of plant-to-plant uniformity can be reduced, by selecting a small number of lines with superior vigour, yield and combining ability, and then isolating them. The cross-pollination occurring between just this smaller number of individual lines will be less variable. Cultivars developed in this way are called polycross or synthetic varieties.

Cultural and economic factors can result in hybrids not being used, even when they are biologically more efficient. For some areas in developing countries transportation, communication and finances of those who grow the crops will not support the cost of buying hybrid seed every year. Such farmers prefer to select open-pollinated seeds from the plants they visually judge as having superior yield and product quality. For them, in addition to the unaffordable cost of purchasing a hybrid cultivar, seed from a hybrid results in a lower yield when planted the next year, because the heterosis is reduced, and extensive segregation and variability of the F_2 generation results from the cross that created the F_1 hybrid.

12.6.4 Heterosis and the hypothesized constant-capacity plant system

Heterosis is a phenomenon whereby yield (or any trait) expressed by the F_1 hybrid between two parents is higher than that of the superior parent. Heterosis is used worldwide to get higher yield of major cross-pollinated crops. Use of heterosis for self-pollinated rice is becoming common in many countries. Heterosis of wheat is being vigorously investigated in China.

Breeding hybrid cultivars is a specialized aspect of breeding for higher crop yield. Duvick (1996) stated: 'With all our advances in maize genetics and technology, we still do not know what causes hybrid vigor, nor do we know the genetics of yield or other important agronomic traits such as drought tolerance or resistance to root lodging.'

Table 12.1. Yield and its components of two rice parents and their F_1 progeny. (The data are from Xiao *et al.*, 1995, 1996.)

Traits	P1 Japonica	P2 Indica	F_1 hybrid
Grain yield (t ha^{-1})	6.02	6.53	7.88
1000-grain weight (g)	22.2	24.6	27.1
Grains plant^{-1}	911.0	960.0	1005.0
Grains panicle^{-1}	105.9	84.2	93.1
Panicles plant^{-1}	8.6	11.4	10.8
Plant height (cm)	104.0	94.0	114.0
Days to heading	86.0	83.0	86.0
Days to maturity	125.0	118.0	129.0

Genetic dominance between alleles of the same locus, combined with additive effects due to different loci, constitute the accepted interpretation for heterosis (e.g. Stuber, 1995; Xiao *et al.*, 1995). The constant-capacity plant-system hypothesis (Chapter 11) provides a physiological–genetic explanation that extends and supports this. The hybrid vigour can result from compromised optimization among levels of the components (Fig. 11.1) that compete for part of the constant-capacity plant system. These compromised component levels and the consequent optimization of system functioning will maximize the yield. This will be examined using data for rice of Xiao *et al.* (1995, 1996). Our interpretation follows.

First, the F_1 hybrid between a japonica (parent P1) and an indica (P2) rice gave heterosis for grain yield per unit land area (Table 12.1). Second, heterosis for yield resulted from heterosis for both of yield's two morphological components: the 1000-grain weight and grains plant^{-1}. Third, the heterosis for grains plant^{-1} resulted from compromised, intermediate levels of its two subcomponents: the grains panicle^{-1} and panicles plant^{-1}. For both, the F_1 has levels intermediate to those of the two parents. In conventional statistical–genetic terms, the genetic control over both is additive gene action. Fourth, the heterosis for 1000-grain weight resulted from an extended grain-filling duration. The DTHM of the F_1 was longer than that of both parents, while its days to heading was that of the later parent (P1). Thus, the enlarged 1000-grain weight of the F_1 is attended by reduction of the early maturity. Fifth, heterosis was expressed for plant height, which we interpret as a substitute measure of the biomass. Based on the hypothesis of a constant-capacity plant system, the conclusion is: compromised optimization among levels of all the physiological components of the process of accumulating the yield resulted in heterosis of the biomass and the yield. Larger biomass accounts for most of the yield gain of hybrid cultivars (Hay, 1995), whereas larger HI does so for most self-pollinated crops.

Yan (1994) suggested two major requirements for selecting a parent for breeding hybrid wheat. First, both parents should have an intermediate to high level of expressed yield in the target environments. This will ensure reasonable adaptation. Second, on this basis, differences between parents in their yield components should be as large as possible. This will increase likelihood the parents span sufficient genetic diversity with which to express hybrid vigour. Common to all breeding programmes is that the parents should complementarily provide the required resistances to diseases and pests. These conclusions hold from the viewpoint of the constant-capacity plant system. Yan's suggestions for selecting the parents for crosses rely on phenotypic observations, while emphasizing genetic variability for the major components of accumulating yield.

Tanksley and McCouch (1997) suggest genes for higher yield can come from wild germplasm that has no phenotypic indication of high yield. These genes can come from selecting as parents the lines that that have quantitative trait loci (QTLs) that contribute to high yield and account for a large percentage of the genotypic variation in the yield.

12.7 Molecular Biology and the Hypothesis of a Constant-capacity Plant System

Biotechnology is currently a very active research area. Law (1995) indicated biotechnology offers two remarkable opportunities for plant breeding. One is gene transformation, i.e. insertion of genes from other species. The second is DNA marker assistance to selection. He suggested already genetically engineered traits and those likely to be in the near future are controlled by only one or a few genes. In this book most but not all traits with such simple genetic control are classified as independent traits for which, the expressed level does not obviously alter the level of other traits.

Law concluded little opportunity exists that the traditional characters of plant breeding such as plant height, leaf width and length, grain numbers and weight, and yield itself can be genetically engineered. These are complex traits, classified as interdependent traits in this book, traits for which the level of each does affect the level of the other quantitatively variable traits. Law (1995) made the following statement.

> Little is known about their detailed genetics, or about the physiological processes underpinning their expression. . . . The location and full evaluation of QTLs affecting these traits would be the first step down the road to their exploitation, followed by their directed subsequent manipulation, either through gene tagging or, having isolated the genes responsible, by gene transfer. To pursue a genetic engineering approach with such characters is a formidable undertaking because apart from the difficulties of identifying and then isolating the gene, it is very likely that the gene involved is only one of the many affecting the character concerned. Before embarking on this course it is essential to know the importance of the gene within the overall breeding programme, hence the need for evaluation, not just of the gene itself, but of other genes influencing the character. Almost certainly these are the genes which are directly involved in plant productivity and are, therefore the genes of greatest importance to plant breeding in the long term when yield requirements from agriculture are likely to be high on the list.

Buiati and Bogani (1995) described another aspect of this problem. 'Present knowledge of the regulation of the dynamic interacting networks of life is still not sufficient to predict the limits within which single components can be changed and yet allow the general metabolic architecture of the network to be maintained.'

Most QTL studies have targeted only individual traits, without attention to relations among the traits, even if they identified QTLs for multiple traits. A few recent studies have identified QTLs for the components of complex traits. Those that have, have found some QTLs to control multiple traits, which the researchers attributed to pleiotropy (Xiao *et al.*, 1996), or linkage (Pan *et al.*, 1994). Others have shown negative or positive correlations to occur between levels of the morphological components of yield (Dirkwanger, *et al.* 1994; de Koning and Weller, 1994; Veldboom and Lee, 1995).

Breeders have recognized negative or positive correlations between many quantitative traits, mostly without study of the biological bases thereof. Molecular geneticists generally apply Mendelian-gene actions to QTLs, while recognizing the QTL may represent one gene or multiple genes. Some studies recognized genotype × year or genotype × season ($G \times E$) interaction effects, i.e. that the significant QTLs identified for one environment do not significantly affect the trait level in another environment (Asins, 1995; Champoux *et al.*, 1995; Hyne *et al.*, 1995).

Backes *et al.* (1995) suggested: 'Instead of looking for QTLs for yield, QTLs for yield components should be determined.' We suggest the major shift in viewpoint needed for both plant breeding and for QTL and other marker assistance to selection is systems thinking. It is only for system outputs – like the yield *per se* – that selection by breeders should pursue the highest possible trait levels. For all component traits of a system output, systems thinking indicates the selection should be toward that set of compromised levels among them that will optimize the system functioning, i.e. maximize the system output (yield) within the given environment. Quantitatively variable traits must share the available metabolites. Therefore, few if any quantitative traits can be enlarged without reducing the level of one or more of the other quantitative traits. The complex of controls by many genes, and by the holistic plant system which includes the environment, are illustrated in Figs 11.1 and 11.4. Figure 12.4 illustrates the complexity of just the genetic control over the components of the plant system needed for a successful cultivar. This figure shows how it comes to be that every quantitative trait is controlled by many genes, each of which partially controls the level of many to all other quantitative traits. This control over each quantitative trait by close to if not all genes arises through forced sharing by each gene action of part of the available interplant metabolites. Adding to the complexity, this control through the partitioning and sharing of the intraplant metabolites due to control by almost all genes is through tandem, parallel, feedback and feedforward interconnections among the gene actions and their role in recognized metabolic pathways.

Here, we suggest use of YSA/AMMI to quantify the genetic, environmental and $G \times E$ interaction controls will complement and accelerate QTL research (Tanksley and McCouch, 1997) in the same way it will accelerate breeding for higher yield. Section 12.4.4 indicates YSA/AMMI will progressively increase understanding of the plant and yield systems. From YSA/AMMI of yield trials, the molecular geneticists can identify the QTLs for yield, including both the

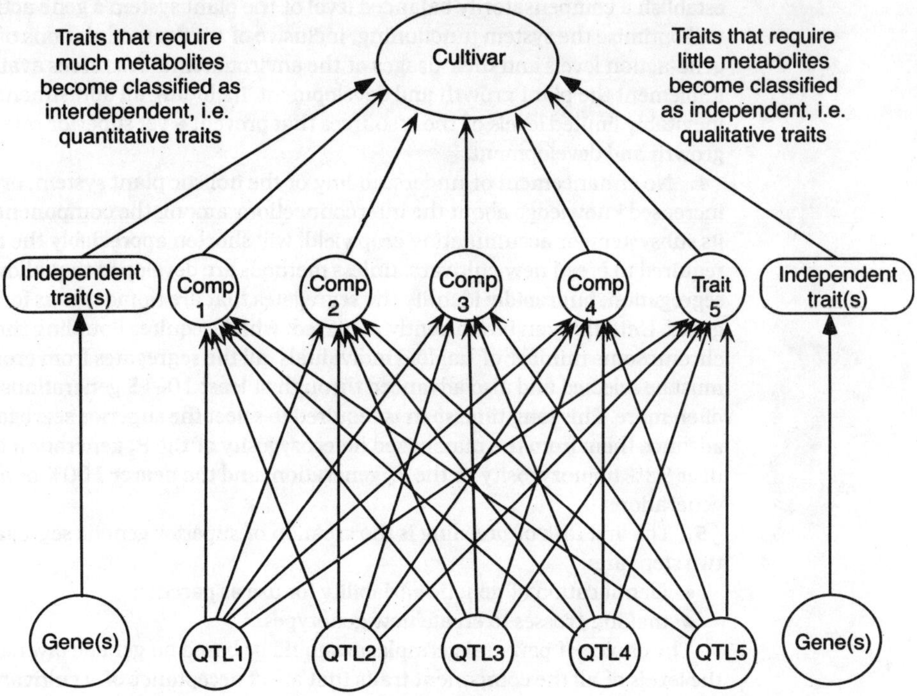

Fig. 12.4. The compared consequences of: (i) the small requirement for intraplant metabolites needed to implement a strong control by one or a few genes over the independent traits, in contrast to (ii) the competitive requirement for a large part of the intraplant metabolites needed to implement the control by many genes (including QTLs) over the interdependent traits. The genetically directed level of every biochemical and physiological activity is modulated by the environment to give the actually expressed trait level. Each interdependent trait competitively requires a large share of the available intraplant metabolites. Therefore, the level of every quantitative trait is quantitatively influenced by each of the many genes that control all the quantitative traits. And, each of these genes quantitatively influences the level of many if not all quantitatively variable traits.

genetically directed and the environmentally modulated controls over the processes of accumulating the yield. The consumer, farmer, plant breeder, crop and plant physiologist and the molecular geneticist will all benefit from such collaboration.

Chapter 12 Summary

1. Selection for yield *per se* will never be replaced as the final criterion for breeding for higher yield. This is because yield is an output from a complex holistic system.

2. Negative or positive correlations (system-established interconnections) occur among all the major component processes of accumulating the yield.

3. The yield of every cultivar depends: (i) on how closely its genetic directions establish a compensatorily balanced level of tne plant system's gene actions that will optimize the system functioning, inclusive of (ii) the modulations of the gene-action levels and their usages of the environmental resources available to implement the plant growth and development, including all adjustments for the inevitably limited levels of the resources that prevent a yet superior rate of growth and development.

4. No enhancement of understanding of the holistic plant system, or increased knowledge about the interconnections among the component parts of its subsystem for accumulating crop yield, will shorten appreciably the time required to breed new cultivars, unless methods are developed that allow segregation, but rapidly identify the segregates that are homozygous for most genes. Until this can be efficiently achieved, which requires doubling the chromosome number of haploid individuals, all the segregates from crosses must be selected and also advanced through at least 10–15 generations, and often more. This long time span is required to select the superior segregates, advance them from the maximized heterozygosity of the F_2 generation to the near 98% homozygosity of the F_8 generation, and the nearer 100% of all later generations.

5. The first task of breeding is the creation of superior genetic segregates. Its two steps are:

- identification of genetic variability for use as parents;
- making crosses to create new genotypes.

The choice of parental germplasm should be based on genetic diversity for the levels of all the component traits that affect acceptance of a cultivar. Likewise for yield, the genetic diversity of the parents should span most of the full range of variability of each of the major physiological–genetic components of the holistic process of accumulating the yield. Tandem application of the YSA/AMMI analyses detailed in Chapter 9 is suggested as the superior way to select for the genetic diversity of the parental germplasm needed to maximize the number of segregates from crosses that will have a superior pathway and capability for higher crop yield.

6. The second task of breeding is selection and testing of the superior segregates. Its four steps are:

- selection of the superior genetic segregates;
- comparison among the selected segregates;
- comparison of the potential new cultivars with the existing cultivars;
- final selection of the most superior genotype.

It is suggested that the *concepts*: (i) the potential maximum yield plant^{-1}; and (ii) the sensitivity of the yield plant^{-1} to increasing plant density be jointly applied to select early-generation segregates for their potential for high crop yield. The term concept is italicized to indicate the maximum yield plant^{-1} will not actually be determined, because spacing the plants far enough apart there is no interplant competition is impractical. Likewise, the sensitivity of the yield plant^{-1} to increasing plant density will not be measured. Selection for both will be applied, however.

All F_2 generation segregates that are not discarded because of obvious faults should be advanced by single-seed descent to the F_3 generation. This provides a single F_4 progeny for each selected F_2 segregate. Any progeny with obvious faults can be discarded. Each F_4 progeny should be grown with larger spacing between the plants than the normal crop spacing. Within and among the F_4 progenies, the individuals should be selected based on having high yield plant^{-1}

combined with a high HI. High HI is known to be associated with low biomass accumulation and consequent interplant competitiveness at standard crop densities. Yield trials and YSA/AMMI analyses thereof should be used to compare the F_5 and all later generations. In addition, there should be continued selection of individuals toward (i) belonging to progenies showing the higher yields, and (ii) greater uniformity for all traits, i.e. toward greater homozygosity. The combined plant breeding first- and second-tasks of conclusions 4 and 5 are called the *whole-system, diverse-component recurrent-selection breeding method* (Fig. 12.3).

7. That selection for yield *per se* of self-pollinated crops has raised yield through higher HI, with little if any gain in biomass, suggests need to select deliberately for higher biomass day^{-1}, because selection for higher biomass alone will enlarge the DTHM and lower the yield.

8. Control over the crop yield is by too many genes, component physiological processes and the modulations of them are by too many environmental factors to ever allow reliable prediction of the yield performance by a genotype the breeder can describe in full. The diverse-component, whole system, recurrent selection breeding method is designed to gradually recombine the several tens of thousands of genes to bring together the genetic directions that optimally use the available environmental resources and account for the stresses to thereby optimize the system functioning and enlarge the yield in the target environment.

9. The eight above conclusions all relate to improving yield through optimization of the level of each of the component quantitative traits that is directly involved in the process of accumulating the yield.

10. Breeding for independent traits with qualitative inheritance and gene actions that protect the yield system from limiting biotic and abiotic stresses will always be an integral to achieving and maintaining high yields.

13 Systems Thinking Requires Multidisciplinary Expertise and Collaboration

Of the 32 paradigm shifts presented in Chapter 1, the only one derived within this final chapter is shift 32. It is: The societal and cultural community as a whole will no longer allow the plant scientist and agricultural communities to independently establish agricultural research agendas. Those of the 31 other shifts that contribute the greatest insight to whole-system research are reviewed succinctly as concepts and conclusions rather than as the paradigm shifts as they are presented in Chapter 1.

13.1 The Hypothesis of a Constant-capacity Plant System Enhances Understanding of Quantitative Traits

Trewavas (1986) indicated that thinking in terms of systems emerged in the late 1960s as a revolution in biological research and its interpretation. He wrote:

> In its way the effects of system attitudes on biology may be as profound as those of relativity in physics. There are some parallels. Just as relativity points to the lack of absolutes, that all measurements are relative to each other and that the observer is an equal component in physical measurements, systems thinking analogously does the same for biology. Effects occur relative only to the context in which they are experienced, and experimental conditions are an essential part of the context. . . . Metabolic systems are networks with great density of connections and counter-intuitive properties. . . . The counter-intuitive character of network behaviour is that properties are shared throughout the network by virtue of the connections through the various parts. Therefore, perturbation of one part leads to a response by the whole network. Parts do not function in isolation, but only in a context or framework of the whole of metabolism. In networks individual pieces cannot be meaningfully studied without reference to the whole since, self-evidently, isolation of the pieces destroys much of the essential character of the system we wish to detect.

Counter-intuitive results commonly occur during breeding for higher yield. The yield system consists of its components. The interconnections among them result in the control being by the holistic system. A *successfully accepted cultivar* is one focus of plant breeders. A cultivar does not consist of only its multiple traits and physiological processes; the pathways and mechanisms of the interconnections among the components are integral to the control by the system as a whole.

One unique contribution of this book toward improved understanding of the interconnections among the parts of the plant system that accumulates crop yield is: *the hypothesis of a constant-capacity plant system*. The constant-capacity plant system, some of the components and subcomponents of which constitute the subsystem for accumulating crop yield (Table 9.1), can be represented by a circle (Fig. 11.1). The circle's radius is fixed differently by each given environment, i.e. the circle has a fixed area, which represents the near constant capacity of all the genotypes of a crop species grown in the given environment. The limited resources and attendant stresses for plant growth of every environment establish the size of the constant capacity of the plant system. It is competitively shared among the five categories of traits of the plant system (Fig. 11.1). All five categories are essential to a successful cultivar. Quantitative traits express a continuum of their level because each requires and competes with the others for a large proportion of the quantitatively variable level of available intraplant metabolites that the environment has made possible. Therefore, genetic instruction that enlarges one quantitative trait forces reduction of the level of one or more of the other quantitative traits. The system capacity represented by the quantity of the intraplant metabolites will be enlarged by an improved environment. But, this becomes another near constant capacity, because no environment allows infinite growth.

13.2 The Theory of Optimization

The potential level of functioning of the constant-capacity plant system is established by the environmental resources for growth, and the stresses provided by that environment. The genotype establishes the fitness (adaptation) or lack of adaptation to these available resources and stresses. Maximized system functioning and yield requires a compensated optimization among the levels of all the system components. A linear relationship cannot exist between the crop yield and any individual component of the system. Consequently, the relation between any plant trait and the yield is curvilinear. To give the highest yield a genotype must have the genetic instructions that account most precisely for the positive and negative modulations of its gene activities by the given environmental factors. Balance between the genetic instructions and the environmental resources and stresses results in an optimal system functioning. Because of the constant capacity of the system, an excess level of any essential system component (plant trait) lowers instead of increases the yield. Therefore, continued enhancement of the plant traits responsible for past yield gains will not lead forever to yet higher yield (Whan *et al.*, 1993).

Trewavas (1986) suggested systems thinking makes the concept of 'limiting factors' invalid; its fallacy is exposed by simply asking 'what does not affect growth or development?' We do believe change toward the optimal level of some components can improve yield more than change of other components. But, there are no absolute and universal limiting system components. Seeking the optimum level of system functioning for each given environment and its variability is of greatest importance. This optimum is a compensatory balance among the levels of all the system components. These optimal component levels are indicated best by the highest level of the desired system output such as the yield. The limiting environmental resources are far easier to identify than are the optimal levels of the genetic instructions the genotype needs to optimize the system functioning inclusive of usages of the environment's resources and countering of its stresses.

13.3 New Concepts of Photothermal Response of Plant Development Derived Through Whole-system Research

Each geographical location has day–night duration (a photoperiod) that results from its latitude, a temperature regime due largely to its latitude and elevation, and a growing season duration controlled by the seasonal temperature (and daylength) or availability of water. Photoperiod (P) and temperature (T) are both environmental resource factors that cannot be altered on a large scale. When possible, fertilization and irrigation can solve an inadequate availability of nutrients and water. The P and T must interact with the genotype to establish jointly a developmental pattern that adapts the cultivar to the environmentally provided growth duration and its variations.

Chapters 4, 5, 6, 7 and 9 detail the modulations by T over the integrated activities of just two subsets of all the genes of the plant: (i) the integrated activity of the majority of all the genes; and (ii) any photoperiod gene(s). This majority of the genes collectively and simultaneously implement: (i) plant growth; and (ii) an autonomous capability of the plant to develop through all its stages. Growth and the autonomous development are both accelerated by each higher T, provided this T remains: (i) above the base (lowest) T that allows growth and development; but (ii) not higher that the optimal T for the collective functioning of this majority of all the genes. The acceleration of the autonomous development implemented by a higher temperature through the combined activities of the majority of all the plant's genes has been called the thermal-time response. Contrary to advancement of the autonomous development by the majority of all the genes, the activity of any photoperiod gene(s) inhibits (delays) the autonomous development. And, a more delaying P and/or a higher T will each and synergistically enlarge this delay of the development. Therefore, a higher T accelerates development through the majority of all the genes, but simultaneously delays it through any photoperiod gene(s). Consequently, only that part of the larger of the two opposite controls over the development that is not cancelled by the smaller is observable. Which is the larger control depends on the genotype, the P and the T.

While it is controlling the time needed to complete development to flowering and harvest maturity, to thereby adapt or maladapt a cultivar to the growing season duration of the site, the photoperiod-gene activity causes reverberations and adjustments throughout the network of the plant system. The indirect control over the days to flowering results from the photoperiod-gene activity that simultaneously controls the levels of many traits. It does so by controlling the proportion of the available intraplant metabolites partitioned to continued vegetative growth – versus – competitively being partitioned to the reproductive organs or tubers etc. that become the yield. Partitioning the larger proportion of the available metabolites to the organs that become the yield enlarges their growth rate. Consequently, they grow rapidly, thereby resulting in earlier flowering and days to harvest maturity, combined with a high harvest index and a higher yield day^{-1}, but a low rate of accumulation of and small total plant biomass (Figs 2.1, 5.1, 9.1). On the contrary, partitioning the larger proportion to continued vegetative growth results in a later days to maturity, with a low harvest index and lower yield day^{-1}, but larger rate of accumulation of and total accumulated biomass. Photoperiod and temperature interactively and synergistically control this partitioning to thereby pleiotropically control the levels of at least 26 plant traits (Table 4.3).

13.4 Balance is Needed Between Whole-system and Reductive Research

An intent of the research of all plant science disciplines (including breeding, genetics, agronomy, horticulture, physiology, pathology, molecular genetics and biotechnology) is to advance understanding of plant biology. An ultimate goal is to support the increases in crop yield needed to provide food, fibre, fuel and timber for the ever increasing human population of the world. These long-term intents apply whether the research is at the molecular, cell, organ, tissue, or whole-plant level.

Whole-system research becomes basic research, in addition to being applied research with only economic goals, if the researcher also asks how the plant traits are implemented by the genotype, how the trait levels are modulated by the environment, and how the system functions to interconnect the levels of its components. Wallace and Zobel (1995; section 2.5) defined whole-system research with the following statement about the plant's subsystem that accumulates the crop yield: 'Basic and truly whole-system research requires measurement of the ultimate output from the system, plus of at least the penultimate outputs from the system under study, and there must be measurements across both multiple genotypes and across multiple environments.' The system output most focused on, that is of economic value to man, is the crop yield. A penultimate output from a system is one of that smallest number of plant processes for which their further integration accounts in full for the ultimate system output. For the subsystem of the plant system that accumulates the crop yield, the penultimate outputs are the: (i) accumulated plant biomass, (ii) percentage of this biomass that is the yield (the harvest index), and (iii) the usage of time to complete (i) and (ii) as measured by the days to harvest maturity. Each penultimate output is the integrated result of many (even most) of the tens of thousands of gene actions and resultant physiological processes of the plant. Collectively, these three penultimate outputs account for all the gene actions and processes that collectively give expression of the ultimate system output, which is the yield. Chapter 9 describes and illustrates a yield system analysis. It can be applied to all yield trails to measure the system output (yield), the yield system's three penultimate outputs, and their subcomponent rates and/or durations. When measured across multiple genotypes and multiple environments, the genotype × environment ($G \times E$) interaction controls over these major components of the process of accumulating the yield can be quantified.

Any research that does not quantify the system output and relate its quantitative variability to the quantitatively variable levels of the several penultimate outputs that arise from multiple genotypes grown in multiple environments is reductive rather than whole-system research. Any research about a plant trait that deals exclusively with one crop species without comparison with comparable research about that trait for multiple other plant species is also reductive research. It is reductive, because only a limited range of genotypes and/or environments is considered. The coverage is not adequately expanded toward focus on the plant system as a whole (section 2.5). A progressively more holistic understanding can arise from fulfilling the criteria for whole-system research.

13.5 Conceptual Advancement of Whole-system Plant Research

Multiple papers in a report entitled *US Agricultural Research: Strategic Challenges and Options* (Weaver, 1993) include essentially the following statement. 'The complexity of agricultural problems appears to be much greater today than in the past. This requires a system and a cooperative approach to seeking solutions for many problems. The agricultural establishment is still learning how to develop good systems research.' A second statement is also presented in Weaver (1993). 'In 1970, Thomas Kuhn suggested that science is best understood as going through alternative phases marked by puzzle-solving and conceptual rebuilding'.

The following conceptual arguments for whole-system research of plant biology are presented within this book.

1. *The hypothesis of a constant-capacity plant system.*
2. The system that accumulates crop yield consists of but two of five major components of the plant system that compete for a part of the plant system's constant capacity.
3. Basic-research data that advance understanding of plant biology can be inexpensively derived from every applied yield trial.
4. Application of yield system analysis (Table 9.1, Fig. 9.1) to multiple genotypes grown in yield trials across multiple environments, followed by appropriate statistical analyses, quantifies the genotype, environment and $G \times E$ interaction controls over the changes in levels among the several components of the plant system's subsystem that accumulates the crop yields. These quantifications will enhance understanding of the system-established interconnected levels among the physiological–genetic components of yield accumulation.
5. Also, the quantitative controls by the genotype and the photoperiod and the temperature over the days to flowering (DTF) and days to maturity, which either adapt or fail to adapt the genotype to the environment, can be quantified by a simple physiological–genetic based mathematical model. This model is

shown in Chapter 7 to simulate with often 90–99% accuracy the expressed DTFs and the relative days to harvest maturity, and even the net accumulated biomass.

13.6 Levels of Multidisciplinary Guided Research Systems

Reports in Weaver (1993), Alston *et al.* (1995) and Huffman and Evenson (1993) point out that the present greater complexities of agricultural problems to be investigated are due in part to recognition by society that technology advances result in negative consequences, such as environmental degradation, in addition to positive benefits such as higher crop yield. Due to environmental degradation and other heightened societal problems, and reduced availability of funding, society is now demanding the central role for determining the objectives of the world's research systems. Society will no longer leave research goals to be determined by only the scientists that conduct the research as has occurred in the past. Also indicated (Weaver, 1993) is that costs and time for transportation and communication around the world have declined, so commerce occurs on an international rather than just a national scale. Therefore, world society and not just the country society is beginning to guide research. These social requirements must be reflected in breeder's considerations of the specific goals for each component of the plant system.

The *competitive-proposal and grant-funding system* of the past 50 years has de-emphasized whole-system biological research, while emphasizing basic research (Weaver, 1993; Alston *et al.*, 1995). The research has been reductive, and mostly at the molecular and individual gene-action (enzymatic) level. Whole-system research has been judged as applied rather than basic research. It has been neglected because some of its objectives are applied. And, achieving its objectives requires more than the 2–3 year duration of research grants. According to Weaver (1993), a paradox of relevance has arisen. Societal attitudes in developed countries accept that crop yields are already sufficiently high. On the contrary, societal attitudes and needs in developing countries require yet higher crop yields. Whole-system research on the plant system and its subsystem that accumulates crop yield must continue, if plant science is to benefit the largest proportion of the people of the world's society as a whole. Bunting (1995) concludes that environmental and technological resources are sufficient to provide for the largest ever need for increased food, fibre, fuel and timber. Brown and Kane (1994) argue it will not be possible to supply these needs for an ever growing world population. Cohen (1995) and Tribe (1994) summarize and compare these opposing views as argued by many authors. Answering these questions will require application of whole-system understanding of plant biology and of the societal concerns.

A major source of the arguments for need of greater whole-system research is various chapters from the book *International Crop Science I* (Buxton *et al.*, 1993). A review of this first International Crop Science Congress by Ford (1995) states: 'The first ICSC demonstrates our ability to successfully bring together people across national borders to discuss and work toward the solution of these problems. But, how and when will we bring together the disciplines?' Teaching a whole-system overview of plant biology is an urgent need. Its application presents a goal that must be accomplished in order to achieve the potentials for plant breeding and crop physiology outlined within the text of this book. Achievement of these potentials will be needed to feed the ever growing population of the world. The humanitarian viewpoint requires a focus by plant science on the inevitable need for more food, fibre, fuel and timber.

13.7 Physiological Genetics and Future Breeding for Higher Yield

13.7.1 Origins of a philosophy and methodology for whole-system research

Jackson *et al.* (1996) drew the following conclusions from responses to a questionnaire about the use of physiological understanding in plant breeding. First, there has been less use of physiological understanding in plant breeding than expected. Second, both the starting and end points of physiological research must be active breeding programmes. Third, at the start, close consideration of the factors affecting performance of the programme and use of relevant genetic populations and environments in conducting physiological research needs to occur. Fourth, at the end, the suggestions for better approaches to selection or breeding need to be evaluated against existing procedures, and then changed or developed within the context of continuing breeding programmes. Fifth, physiological research is likely to be used more in the next 20 years, because: (i) many plant scientists are interested; and (ii) crop physiology has intellectual appeal for scientists as a means of complementing the numbers game approach in plant breeding, particularly when the limited resources for research are declining. To these five conclusions we add a sixth, based on the concepts summarized in this book. The research should be holistic in the sense of being focused on (i) quantification of a specified subset or subsets of gene action and their system output and (ii) quantification of the modulations of these gene actions and of the system output by specified factors of the environment. The research should not be simply physiological research or only reductive research. The research should incorporate the philosophy and methodology of whole-system research summarized in section 13.7.3 below.

Summerfield, Roberts and colleagues (Watkinson *et al.*, 1994) progressively developed their three equations of the three-plane additive linear rate model. They based them on the recognition of the DTF being a major control over the adaptation of a cultivar to the environment in which they are grown, thereby being a control over the yield. The goal of these plant physiologists was to predict the DTF by quantifying the strong controls over it by both P and T. They recognized that photoperiod and temperature interactively control the DTF through the photoperiod and thermal-time responses (section 7.2). To avoid complexities due to the interactions between these two responses, however, and to expedite effective prediction of the DTF, they modelled the photoperiod and temperature controls separately and additively. Modelling the holistic control required greater understanding of the interaction. As they developed their three-plane model, the second control by temperature over the DTF through its modulation of the photoperiod gene activity was not fully recognized. It was not considered that the temperature is an always functioning modulator of every specific gene function and every integrated physiological process. They never personally experienced plant breeding *per se*, or the breeder's thinking about genetic heritabilities and selection procedures as related to $G \times E$ interaction.

Both authors of this book are active in plant breeding. Wallace's initial focus on the physiological genetics of yield is detailed in Chapters 3, 4 and 5. His goal has remained to acquire physiological–genetic information that explains the variation in yield among different genotypes and environments, and to use it to breed more efficiently for higher yield. Wallace's study of photoperiodism began only after finding a higher percentage of the accumulated aerial biomass to be partitioned to the yield of bean cultivars that flower and mature early. This percentage is the harvest index. The finding led to study of the holistic control by the genotype, the applied *P* and the applied *T* over the DTF, yield, harvest index and both the accumulated yield and aerial biomass day^{-1} to maturity. These were compared for the photoperiod-sensitive versus the photoperiod-insensitive genotypes. Results from the focus on photoperiodism and its relation to yield were not broadly available to plant scientists prior to Wallace *et al.* (1993a, b, c; 1995a, b) and Mmopi *et al.* (1995). These results are integrated for the first time within this text. Wallace's studies were initially in collaboration with a plant physiologist.

In section 7.7 it is shown that the three-plane model of Summerfield and Roberts (1993) and the GPTmodel of Yan and Wallace (1995) predict the DTF with equally high fits, which are superior to fits by other models. Nevertheless, section 7.4 describes that plant scientists other than the originators of both models become frustrated from applying either model. Researchers that apply the three-plane model are troubled by shifts of the *P* and/or *T* indicated to be the model's parameters: the critical photoperiod (*Pc*), ceiling photoperiod (*Pce*) and the sensitivity to temperature (*St*). Changing levels of these parameters occur because the ones seemingly related to only the photoperiod are altered by differences in temperature, and vice versa. This renders all observable (and measurable) changes of *Pc*, *Pce* and *St* as only apparent parameters. Researchers who apply the GPTmodel also have to deal with the changes of the DTF attending only apparent parameters for *Pc*, *Pce* and *St*. Plant scientists have long used these terms as the criteria for interpreting responses to the applied photoperiod and applied temperature.

As for the three-plane model, the parameters derived from the GPTmodel arise from the DTF observed and measured across a range of photoperiod–temperature combinations. For the GPTmodel, each parameter is based on the holistic control, which includes: (i) a modulation by the temperature of the level of the also photoperiod-modulated photoperiod gene activity, the consequence being the photoperiod response of the DTF; and (ii) a separate but simultaneous modulation by the same temperature of the collective activity of the majority of all the genes of the plant, which results in the thermal-time response of the DTF. The GPTmodel recognizes that the simultaneous modulations 1 and 2 by the same temperature always cause opposite changes of the DTF. The GPTmodel's parameters are fixed by the genotype, but they relate specifically to either the photoperiod response or to the thermal response. The frustrating inevitability is that these parameters can never be seen and measured directly, because both the photoperiod response and the thermal response always and simultaneously control the DTF with opposing controls in response to the temperature. Because both responses simultaneously control the DTF, the change of the DTF between different photoperiod–temperature combinations that can be directly seen and measured is just that part of the larger of the opposite changes attributable to the temperature modulations (i) and (ii) described above, which is not cancelled by the smaller opposite change of the DTF through the other modulation. This advanced knowledge of whole-system understanding assists all plant scientists to interpret the modelling results and parameters of both the three-plane and the GPTmodel.

Yan is a wheat breeder from China. The DTF of wheat is controlled jointly by low-temperature vernalization, the thermal-time response, and the photoperiod response. Yan planted wheat in greenhouses during the summer, after wheat harvest in order to advance two generations per year. This required vernalizing the seeds before planting, and controlling both *P* and *T* during growth. Yan pondered: How do vernalization, photoperiodism and thermal time interact to jointly control the DTF of wheat and thereby affect the selection for yield? What are the relations among the vernalization, thermal-time, and photoperiod responses and their effect on the DTF and the cultivar adaptation and yield?

While searching the literature for answers, Yan read the verbal description of the genotype × photoperiod × temperature (GPT) interaction control over the DTF in Wallace (1985), as then summarized (Fig. 5.4). In that review, Yan read that photoperiod-sensitive genotypes respond to low temperatures by flowering earlier while the insensitive genotypes flower later. The sensitive genotypes respond to high temperature by flowering later while the insensitive ones flower earlier (Figs 4.1, 4.2, 5.3). The earlier flowering of the sensitive genotype was interpreted as resulting from the Q_{10} effect due to the low temperature preventing the photoperiod-gene activity that delays the DTF that would occur at higher and normal growing temperatures. Likewise, the later flowering of the insensitive genotype was interpreted as resulting from the Q_{10} effect due to the low temperature preventing or slowing the gene activity that implements their development to flowering. Yan (1993b) began to think this Q_{10} effect should apply also to the activity of the vernalization genes. As a first step in developing a general model to interpret both vernalization and photoperiodism, Yan substituted simple mathematical equations to replace the verbal descriptions of the GPT interaction effects described in Wallace (1985). Yan came to Cornell in 1993. The models for holistic control over the DTF by the *P* and *T* in Chapters 6, 7 and 8 involve the Q_{10} effects on just three subsets of genes. They are the (i) photoperiod genes, (ii) vernalization genes, and (iii) all the other genes of the plant. Our conclusions are: (i) the majority of all the genes of the plant implement the autonomous capability possessed by all plants to develop as rapidly as allowed by the applied *P* and *T*. The Q_{10} of a higher *T* enlarges the rate of advance of the autonomous development and this is the well recognized thermal-time response. (ii) The activities of both the vernalization genes and of the photoperiod genes are increased also by the higher *T*. This increases the inhibition (delay) of the autonomous development caused by both the photoperiod and the vernalization gene(s). Vernalization is prevention by low-T of the delay of the autonomous development that would occur if the seed was planted at a higher T. These models and this book have resulted from the synergism of 4 years of joint effort by two plant breeders seeking understanding of the physiological genetics of crop yield.

13.7.2 The time scale for whole-system research

The linear rate model, including the three equations of its three planes, is the best known of all models for predicting the DTF. This is due to its application to many crops reported in over 260 publications (Summerfield and Roberts, 1993). These reports have resulted from 20 years of financial backing, principally by Britain's Overseas Development Administration (now the Department for International Development). This funding supported worldwide collaboration with many universities, and the international crop research institutes. The latter have been funded by many nations through the Consultative Group on International Agricultural Research.

The contributions by Wallace to the GPTmodel have arisen through continuous research since 1953 to quantify and understand the physiological–genetic controls over crop yield, and their application toward more efficient breeding for higher yield. They were derived through 40 years of support by Cornell University, with crucial supplemental support for 3 years by the Rockefeller Foundation, and continued crucial support from 1981 to 1992 by funding from the United States Agency for International Development, through its Bean/Cowpea Collaborative Research Support Program. Respectively, these supplemental funds supported joint research with the bean programme of the Centro Internacional de Agricultura Tropical in Colombia, and the bean research programme of Guatemala. Thus, the GPTmodel and the yield system analysis procedures have resulted from small but continuous funding of applied bean breeding associated with physiological–genetic research about yield over 40 years, and from 12 years of wheat breeding and physiological–genetic research in China by Yan. Cornell University funded Yan at Cornell from 1993 to 1996.

It must be accepted that whole-system research will require longer times than reductive research to acquire unequivocal answers to research hypotheses. Greater recognition is needed that whole-system research that has applied the following philosophy and methodology will answer complex research hypotheses that will not be revealed by reductive research. The answers through whole-system research will complement those derived by reductive research.

13.7.3 A philosophy and methodology for whole-system research

In this chapter and book, the system output from the plant system studied has been either the crop yield, DTF or both. For both, the philosophy and methodology of whole-system research has divided the biological expression of the system output into the smallest number of near-fully integrated physiological–genetic components that can account in full for the expression of the system output. For yield, the component processes were the net accumulated aerial biomass, its partitioning as measured by the harvest index, and the time used to complete both.

For the DTF the components were three subsets of genetic control: the majority of all the genes of the plant and their advancement of the development, and any photoperiod genes and/or vernalization gene(s) and their inhibition of advance of the development. The correlations between levels of the individual major components across genotypes and also across environments were used to quantify the system-established connections among these major components of the system. The correlations between each component's level and the level of the system output (the DTF or the yield) across the genotypes quantifies the system-established interconnections that result from the variation in genetic control due to the different genotypes. The correlations between each component part and the system output (the DTF or the yield) across the environments quantify the modulations of the genetic control by the different environments. This philosophy and methodology is essentially identical to the whole-system approach to quantification of the system-established controls over complex biochemical pathways described by Brand (1966) and Kacser *et al.* (1995). Respectively, these authors describe their methods as top-down metabolic control analysis, and analysis of the control of flux.

Our whole-system results show the control over any quantitative trait is always by the system as a whole. This holistic control consists of just (i) the holistic genetic directions of the genotype and (ii) the holistic modulations by the environment of the levels of implementation of the genetic directions. Control over the DTF by the system as a whole was shown to be through the three subsets of genes. But, the three subsets jointly involve all the genes. The accumulation of the biomass is implemented by the same majority of all the genes of the plant that implements the autonomous development to the DTF (Fig. 2.1). Likewise, the time used to accumulate the biomass and to partition it is strongly controlled by the same photoperiod and vernalization genes that inhibit the autonomous development. Actually, the photoperiod-gene activity controls whether the accumulated biomass is partitioned toward the organs that become the yield or is competitively partitioned to the other organs, and thereby strongly enlarges the DTF and the days to harvest maturity. If the genotype and P and T cause photoperiod-gene activity that partitions the biomass predominantly to the organs that become the yield, whether the yield is a reproductive or a vegetative organ, these organs grow rapidly and thereby bring the genotype to harvest maturity in a short time.

All traits discussed in the preceding paragraph are quantitative traits. The discussion shows that a system connects the expressed level of every trait with the expressed level of all the other traits of the system. The interconnections are due to both the integrated genetic directions and the modulations of their implementation by the varied levels of factors of the environment. The interconnections and control by many genes are due also to the requirement for all quantitative traits competitively to share the available intraplant metabolites. The whole-system philosophy and methodology requires elucidation of the $G \times E$ interaction controls that establish the interconnections (measurable correlations) among the levels of all the quantitatively variable traits.

Chapter 13 Summary

A philosophy and a methodology are summarized for conducting the whole-system research plant scientists and society are calling for to complement reductive research. Focus is required on the smallest number of subsets of gene actions and/or of near-fully integrated physiological processes that account in full for the system output that is the trait of economic or scientific interest. This philosophy can be applied to the yield trials that are already being conducted to elucidate, inexpensively and synergistically, the holistic system functioning and quantify the system-established interconnections among the component parts of the system.

In this final chapter the whole-system research concepts of plant science are considered jointly with recent concerns of human society. Although complex, the holistic plant system, cultivar system and yield system are but a small part of the agricultural, economic, industrial, social and technological systems and concerns of human societies and cultures. Society is demanding greater control over the research agenda of plant and other scientific research. Learning a whole-system research approach, teaching it to society as a whole, in addition to plant scientists, and conducting whole-system research of plant biology are all urgently needed. All are required if the supplies of food, fibre, fuel and timber are to be increased sufficiently to provide the needs of the expanding human population of the world.

References

Citations within quotes from other authors are not listed here unless they are highly pertinent or used elsewhere within the text.

Aitken, Y. (1974) *Flowering Time, Climate and Genotype.* Melbourne University Press, Australia, 190 pp.

Akinola, J.O. and Oyejola, B.A. (1994) Planting date and density effects on six pigeonpea (*Cajanus cajan*) cultivars at three Nigerian savanna locations. *Journal of Agricultural Research, Cambridge* 123, 233–246.

Alston, J.M, Norton, G.W. and Pardey, P.G. (1955) *Science Under Scarcity.* Cornell University Press, Ithaca, NY.

Amthor, J.S. (1989) *Respiration and Crop Productivity.* Springer-Verlag, New York, 215 pp.

Angus, J. F., Mackenzie, D.H., Morton, R. and Schafer, C.A. (1981) Phasic development in field crops. II. Thermal and photoperiodic responses of spring wheat. *Field Crops Research* 4, 269–283

Annicchiarico, P. and Perenzin, M. (1994) Adaptation patterns and definition of macro-environments for selection and recommendation of common wheat genotypes in Italy. *Plant Breeding* 113, 197–205.

Arumingtyas, E.L. and Murfet, I.C. (1994) Flowering in *Pisum*: A further gene controlling response to photoperiod. *Journal of Heredity* 85, 12–17.

Asins, M.J., Mestre, P. Garcia, J.G. Dicenta, F. Carbonell, E. A. (1995) Genotype × environment interaction in QTL analysis of an intervarietal almond cross by means of genetic markers. *Theoretical and Applied Genetics* 89, 358–364.

Atherton, J.G., Craigon J. and Basher, E.A. (1990) Flowering and bolting in carrot. I. Juvenility, cardinal temperatures and thermal time for vernalization. *Journal of Horticultural Science* 65, 423–429.

Austin, R.B. (1993) Crop photosynthesis: Can we improve on nature. In: Buxton, D.R., Shibles, R., Forsberg, R.A., Blad, B.L., Assay, K.H., Paulsen, G.M. and Wilson, R.F. (eds) *International Crop Science I.* Crop Science Society of America, Madison, WI. pp. 697–701.

Austin, R.B. and Ford, M.A. (1976) *Annual Reports, Plant Breeding Institute,* Cambridge, UK, pp. 141–142.

Backes, G., Graner, A., Foroughi-Wehr, B., Fishbeck, G., Wenzel, G and Jahoor A. (1995) Localization of quantitative trail loci (QTL) for agronomic important characters by the use of RLFP map in barley (*Horderum vulgare* L.). *Theoretical and Applied Genetics* 90, 294–302.

Bagnall, D.J. (1993) Light quality and vernalization interaction in controlling late flowering of *Arabidopsis* ecotypes and mutants. *Annals of Botany* 71, 75–83.

Baker, J.T., Gallagher J N. and Monteith, J.L. (1980) Daylength change and leaf appearance in winter wheat. *Plant, Cell and Environment* 3, 285–287.

Balkema-Boomstra, A.G. and Masterbroek, H.D. (1993) The grain yield of uniculm barley in two contrasting environments. *Euphytica* 66, 103–110.

Barber, H.N. (1959) The genetic basis of early flowering in mutant line 1/178. *Pisum Genetics* 24, 32–36.

Baumann, T.E., Eaton, G.W. and Spaner, D. (1993) Yield components of day-neutral and short-day strawberry varieties on raised beds in British Columbia. *HortScience* 28, 891–894.

Belford, R.K. and Sedgley, R.H. (1991) Conclusions: Ideotypes and physiology: Tailoring plants for increased production. *Field Crops Research* 26, 221–226.

Bernier, G. (1988) The control of floral evocation and morphogenesis. *Annual Review of Plant Physiology and Plant Molecular Biology* 39, 175–219.

Berry, G.J. and Aitken, Y. (1979) Effect of photoperiod and temperature on flowering in pea (*Pisum sativum* L.). *Australian Journal of Plant Physiology* 6, 573–587.

Berry, G.J.; Salisbury P.A. and Halloran, G.M. (1980) Expression of vernalization genes in near-isogenic wheat lines: duration of vernalization period. *Annals of Botany* 46, 235–241.

Beveridge, CA, Ross, J.J., and Murfet, IC. (1992) Mutant *dn* influences dry matter distribution, assimilate partitioning and flowering in *Lathyrus odoratus* L. *Journal of Experimental Botany* 43, 55–62.

Bingham, J. and Austin, R.B. (1993) Achievements and limitations of wheat breeding in the United Kingdom. In: Buxton, D.R., Shibles, R., Forsberg, R.A., Blad, B.L., Assay, K.H., Paulsen, G.M. and Wilson, R.F. (eds) *International Crop Science I.* Crop Science Society of America, Madison, WI, pp. 181–187.

Bleken, M.A. and Skjelvag, A.O. (1986) The phenological development of oat (*Avena sativa* L.) cultivars as affected by temperature and photoperiod. *Acta Agriculturae Scandinavica* 36, 353–365.

Blondon, F. (1972) Facteurs externes et determinisme floral d'un clone de *Dactylis glomerata* L. In: Chouard, P. and de Bilderling, N. (eds) *Phytotronique et Prospective Horticole.* Gauthier-Villars, Paris, pp. 135–181.

Blum, A., Golan, G., Mayer, J., Sinmena, B. and Obilana, T. (1992). Comparative productivity and drought response of semi-tropical hybrids and open pollinated varieties of sorghum. *Journal of Agricultural Science* 118, 29–36.

Bonhomme, R., Derieux, M. and Edmeades, G.O. (1994) Flowering of diverse maize cultivars in relation to temperature and photoperiod in multilocation field trials. *Crop Science* 34, 156–164.

Boote, K.J., Bennett, J.J., Sinclair, T.R. and Paulsen, G.M. (1994) *Physiology and Determination of Crop Yield.* American Society of Agronomy, Madison, Wisconsin.

Bos, I. and Van Boxtel, J.H.J. (1990) Comparison of several procedures for mass selection in winter rye. I. The relative quality of the procedures. *Euphytica* 48, 9–16.

Boyer, E.L. (1990) Scholarship reconsidered. Priorities for the professional. *A Special Report of the Carnegie Foundation for the Advancement of Teaching.* Princeton University Press, Lawrenceville, New Jersey, 147 pp.

Brand, M.D. (1996) Top down metabolic control analysis. *Journal of Theoretical Biology* 182, 351–360.

Bravo, A. (1975) Characterization of high yielding dry bean (*Phaseolus vulgaris* L.) cultivars by simplified growth analysis. PhD thesis, Cornell University, Ithaca, NY.

Brondum, J.J. and Heins, R.D. (1993) Modeling temperature and photoperiod effects on growth and development of dahlia. *Journal of the American Society for Horticultural Science* 118, 36–42.

Brooking, I.R., Jamieson, P.D. and Porter, J.R. (1995) The influence of daylength on final leaf number in spring wheat. *Field Crops Research* 41, 155–165.

Brown, L.R. and Kane, H. (1994) *Full House: Reassessing the Earth's Population Carrying Capacity*. W.W. Norton, New York, 261 pp.

Bulman, P., Mather, D.E. and Smith, D.L. (1993) Genetic improvement of spring barley cultivars grown in eastern Canada from 1910 to 1988. *Euphytica* 71, 35–48.

Buiatti, M. and Bogani, P. (1995) Physiological complexity and plant genetic manipulation. *Euphytica* 85, 135–147

Bunting, H.G. (1995) Feeding the world in the future. In: Spedding, C.R.W. (ed.) *Fream's Principles of Food and Agriculture.* Blackwell Scientific Publications, London, pp. 256–290.

Bushamuka, V.N. and Zobel. R.W. (1988) Differential response of maize (*Zea mays* L.) and soybean (*Glycine max* (L.) Merr.) cultivars and tap, basal and lateral roots to an acid, Al-toxic soil. *Crop Science.*

Buxton, D.R., Shibles, R. Forsberg, R.A., Blad, B.L., Asay, K.H., Paulsen, G.M. and Wilson, R.F. (1993) *International Crop Science I.* Crop Science Society of America. Madison, WI, 895 pp.

Cao, W. and Moss, D.N. (1989) Temperature effect on leaf emergence and phyllochron in wheat and barley. *Crop Science* 29, 1018–1021.

Ceccarelli, S. and Grando, S. (1993) From conventional breeding to molecular biology. In: Buxton, D.R., Shibles, R., Forsberg, R.A., Blad, B.L., Asay, K.H., Paulsen, G.M. and Wilson, R.F. (eds). *International Crop Science I.* Crop Science Society of America. Madison, WI, Crop Science Society of America. Madison, WI, pp. 533–537.

Ceccarelli, S., Acęvedo, E. and Grando, S. (1991) Breeding stability in unpredictable environments: Single traits, interaction between traits, and architecture of genotypes. *Euphytica* 56, 169–185.

Ceccarelli, S., Grando, S. and Hambin, J. (1992) Relationship between barley grain yield measured in low- and high-yielding environments. *Euphytica* 64, 49–58.

Champoux, M.C., Wang, G., Sarkarung, S., Mackill, D.J., O'Toole, J.C., Huang, N. and McCouch, S.R. (1995) Locating genes associated with root morphology and drought avoidance in rice via linkage to molecular markers. *Theoretical and Applied Genetics* 90, 969–981.

Chujo, H. (1966) Differences in vernalization effect in wheat under various temperatures. *Proceedings of the Crop Science Society of Japan* 35, 177–186.

Chun, J.U. (1993) Variation in rate of leaf emergence, initiation of primordia, stem elongation and heading time as affected by vernalization duration of barley with differing growth habits. *Field Crops Research* 32, 159–172.

Clarkson, N.M. and Russell, J.S. (1975) Flowering responses to vernalization and photoperiod in annual medics (*Medicago* spp.). *Australian Journal of Agricultural Research* 26, 831–838.

Cohen J.E. (1995) *How Many People Can the Earth Support?* W.W. Norton & Company, New York and London, 532 pp.

Connell, J.H. (1990) Apparent versus 'real' competition in plants. In: Grace, J.B. and Tillman, D. (eds) *Perspectives in Plant Competition.* Academic Press, Inc. New York, pp. 9–26.

Cooper, J.P. (1956) Developmental analysis of populations in the cereals and herbage grasses. 1. Methods and techniques. *Journal of Agricultural Science, Cambridge* 47, 262–279.

Cooper, J.P. (1960) Short-day and low temperature induction in *Lolium*. *Annals of Botany* 24, 232–246.

Cooper, M. and DeLacy, I.H. (1994) Relationships among analytic methods used to study genotypic variation and genotype-by-environment interaction in plant breeding multi-environment experiments. *Theoretical and Applied Genetics* 88, 561–572.

Cooper, M. and Hammer, G.L. (eds) (1996) *Plant Adaptation and Crop Improvement*. CAB International, Wallingford, UK, 636 pp.

Craigon, J., Atherton, J.G. and Sweet, N. (1995) Modeling the effects of vernalization on progress to final leaf appearance in winter wheat. *Journal of Agricultural Science* 124, 369–377.

Cram, J. (1980) The higher plant as a whole. In: Spanswick, R.M., Lucas, W.J. and Dainty, J. *Plant Membrane Transport: Current Conceptual Issues*. Elsevier/North-Holland Biomedical Press, New York, pp. 3–13.

Crofts, H.J. (1989) On defining a winter wheat. *Euphytica* 44, 225–234.

Cutforth, H.W., Jame, Y.W. and Jefferson, P.G. (1992) Effect of temperature, vernalization and water stress on phyllochron and final main stem leaf number of HY320 and Neepawa spring wheats. *Canadian Journal of Plant Science* 72, 1141–1151.

Damisch, W. and Wiberg, A. (1991) Biomass yield – A topical issue in modern wheat breeding programs. *Plant Breeding* 107, 11–17.

Darwin, C. (1869) *On the Origin of Species by Means of Natural Selection: or the preservation of favoured races in the struggle for life.* 5th edn. J. Murray, London, 596 pp.

Davidson, J.L., Christan, K.R., Jones, D.B. and Bremner, P.M. (1985) Response of wheat to vernalization and photoperiod. *Australian Journal of Agricultural Research* 36, 347–359.

De Jong, J. (1989) Analysis of components controlling early flowering of chrysanthemum at low temperature. *Euphytica* 40, 121–126.

Delzer, B.W., Busch, R.H. and Hareland, G.A. (1995) Recurrent selection for grain protein in hard red spring wheat. *Crop Science* 35, 730–734.

Devay, M. (1967) Biochemical processes of vernalization. VI. The change of the phytochrome content in the course of vernalization. *Acta Agronomica Academiae Scientiarum Hungaricae* 16, 289–295.

Dirkwanger, E., Isaac, P.G., Ranade, S., Belajuouza, M., Cousin, R. and de Vienne, D. (1994) Restriction fragment length polymorphism analysis of loci associated with disease resistance genes and developmental traits in *Pisum sativum* L. *Theoretical and Applied Genetics* 88, 17–27.

Dofing, S.M. and Knight, C.W. (1994) Yield component compensation in uniculm barley lines. *Agronomy Journal* 86, 273–276.

Donald, C.M. (1963) Competition among crop and pasture plants. *Advances in Agronomy* 15, 1–118.

Donald, C.M. (1968) The breeding of crop ideotypes. *Euphytica* 17, 385–403.

Donald, C.M. (1981) Competitive plants, communal plants, and yield. In, Evans, L.T. and Peacock, W.J. (eds) *Wheat Science – Today and Tomorrow*. Cambridge University Press, UK.

Donald, C.M. and Hamblin, J. (1976) The biological yield and harvest index of yield as agronomic and plant breeding criteria. *Advances in Agronomy* 28, 361–405.

Donald, C.M. and Hamblin, J. (1983) The convergent evolution of annual seed crops in agriculture. *Advances in Agronomy* 36, 97–143.

Duvick, D.N. (1996) Personal perspective: Plant breeding, an evolutionary concept. *Crop Science* 36, 539–548.

El-Mådina, I.M.D. and Hall, A.E. (1986) Flowering of contrasting cowpea (*Vigna unguiculata* (L.) Walp.) genotypes under different temperatures and photoperiods. *Field Crops Research* 14, 87–104.

Ellis, R.H., Roberts, E.H., and Summerfield, R.J. (1988) Photothermal time for flowering in faba bean and the analysis of potential vernalization responses. *Annals of Botany* 61, 73–82.

Ellis, R.H., Collison, S.T., Hudson, D. and Patefield, W.M. (1992a) The analysis of reciprocal transfer experiments to estimate the durations of photoperiod-sensitive and photoperiod-insensitive phases of plant development: an example in soya bean. *Annals of Botany* 70, 87–92.

Ellis, R.H., Summerfield, R.J., Edmeades, G.O. and Roberts, E.H. (1992b) Photoperiod, temperature, and the interval from sowing to tassel initiation in diverse cultivars of maize. *Crop Science* 32, 1225–1232.

Evans, L.T. (1960) The influence of temperature in species of *Lolium* and in *Poa pratensis*, *Journal of Agricultural Science* 54, 410–416.

Evans, L.T. (1987) Short day induction of inflorescence initiation in some winter wheat varieties. *Australian Journal of Plant Physiology* 14, 277–286.

Evans, L.T. (1993) Processes, genes and yield potential. In: Buxton, D.R., Shibles, R., Forsberg, R.A., Blad, B.L., Asay, K.H., Paulsen, G.M., and Wilson, R.F. (eds) *International Crop Science I.* Crop Science Society of America. Madison, WI, pp. 687–696.

Evans, L.T. (1994) *Crop Evolution, Adaptation and Yield.* Cambridge University Press, UK, 500 pp.

Evans, L.T. and Blundell, C. (1994) Some aspects of photoperiodism in wheat and its wild relatives. *Australian Journal of Plant Physiology* 21, 551–562.

Fasoula, D.A. and Fasoula, V.A. (1996) Competitive ability in plant breeding and agronomy. *Plant Breeding Reviews* 14, 89–138.

Fasoulas, A.C. (1973) *A New Approach to Breeding Superior Yielding Varieties.* A.C. Fasoulas, P. O. Box 1555, GR-54006, Thessaloniki, Greece, 42 pp.

Fasoulas, A.C. (1988) *The Honeycomb Methodology of Plant Breeding.* Athanasios Altidjis, Thessalonika, Greece, 167 pp.

Fasoulas, A.C. and Fasoula, V.A. (1995) Honeycomb selection designs. *Plant Breeding Reviews* 13, 87–139.

Fernandez, C.J. and Chen, H.K. (1989) Temperature and photoperiod influence reproductive development of reduced-photoperiod-sensitive mungbean genotypes. *Journal of the American Society for Horticultural Science* 114, 204–209.

Finlay, K.W. and Wilkinson, G.N. (1963) The analysis of adaptation in a plant breeding programme. *Australian Journal of Agricultural Research* 14, 742–754.

Fischer, R.A. (1993) Irrigated spring wheat and timing and amount of nitrogen fertilizer. II. Physiology of grain yield response. *Field Crops Research* 33, 57–80.

Fischer, R.A. and Kertesz, Z. (1976) Harvest index in spaced populations and grain yield in microplots as indicators of grain yielding ability in spring wheat. *Crop Science* 16, 55–59.

Flood, R.G. and Halloran, G.H. (1984) The influence of genes for vernalization response on development and growth in wheat. *Annals of Botany* 58, 505–513.

Flood, R.G. and Halloran, G.M. (1986) Genetics and physiology of vernalization response in wheat. *Advances in Agronomy* 39, 87–125.

Flower, D.J. (1996) Physiological and morphological features determining the performance of the sorghum landraces of northern Nigeria. *Experimental Agriculture* 32, 129–141.

Ford, D.M. (1995) Book review of International Crop Science I. *Field Crops Research* 41, 61–64.

Fossati, D., Fossati, A. and Feil, B. (1993) Relationship between grain yield and grain nitrogen concentration in winter triticale. *Euphytica* 71, 115–123.

Friend, D.J.C. (1965) Interaction of red and far-red radiations with the vernalization processes in winter rye. *Canadian Journal of Botany* 43, 161–170.

Gauch, H.G., Jr (1990) Using interaction to improve yield estimates. In: Kang, M.S. (ed.) *Genotype-by-environment Interaction and Plant Breeding*. Louisiana University. Baton Rouge. LA, pp. 141–140.

Gauch, H.G. Jr (1992) *Statistical Analysis of Regional Yield Trials: AMMI analysis of Factorial Designs*. Elsevier, New York.

Gauch, H.G. Jr and Zobel, R.W. (1990) Imputing missing yield trial data. *Theoretical and Applied Genetics* 79, 753–761.

Gauch, H.G., Jr and Zobel, R.W. (1996) AMMI analysis of yield trials. In: Kang, M.S. and Gauch, H.G., Jr (eds) *Genotype-by-Environment Interaction*. CRC Press, Boca Raton, FL, pp. 85–122.

Gauch, H.G., Jr and Zobel, R.W. (1997) Identifying mega-environments and targeting genotypes. *Crop Science* 37, 311–322.

Gifford, R.M. (1987) Barriers to increasing crop productivity by genetic improvement of crop yield in photosynthesis. In: Biggins, J. (ed.) *Progress in Photosynthesis*. Martinus Nijhoff, Dordrecht, The Netherlands, pp. 377–384.

Gill, J.S., Verma, M.M., Gumber, R.K. and Brar, J.S. (1995) Comparative efficiency of four selection methods for deriving high-yielding lines in mungbean (*Vigna radiata* (L.) Wilczek). *Theoretical and Applied Genetics* 90, 554–560.

Givinish, T.J. (1986) *On the Economy of Plant Form and Function*. Cambridge University Press, New York.

Gniffke, P.A. (1985) Studies of phenological variation in the common bean (*Phaseolus vulgaris* L.) as modulated by mean temperature and photoperiod. PhD thesis, Cornell University, Ithaca, NY.

Goldberg, D.E. (1990) Components of resource competition in plant communities. In: Grace, J.B. and Tillman, D. (eds) *Perspectives on Plant Competition*. Academic Press, New York, pp. 27–49.

Goldberg, R.B., Hoschek, G. and Kamalay, J.C. (1978) Sequence complexity of nuclear and polysomal RNA in leaves of the tobacco plant. *Cell* 14, 123–131.

Gotoh, T. (1976) Studies on varietal differences in vernalization requirement in wheat. *Japanese Journal of Breeding* 26, 307–327.

Gotoh, T. (1979) Genetic studies on growth habit of some important spring wheat cultivars in Japan, with special reference to the identification of spring genes involved. *Japanese Journal of Breeding* 29, 133–145.

Grashoff, C. (1990) Effect of pattern of water supply on *Vicia faba* L. I. Dry matter partitioning and yield variability. *Netherlands Journal of Agricultural Science* 38, 21–44.

Greenham, J. (1982) Vegetative and reproductive development in a photoperiod sensitive and an insensitive bean cultivar, as influenced by environment. MS thesis, Cornell University, Ithaca, NY

Grierson, W. (1995) The role of temperature in the physiology of crop plants: Pre- and postharvest. In: Pessarakli, M. (ed.) *Handbook of Plant and Crop Physiology*. Marcel Dekker, New York, pp. 61–83.

Griffing, B. (1994) Historical perspectives on contribution of quantitative genetics to plant breeding. In: Frey, K.F. (ed.) *Historical Perspectives in Plant Science*. Iowa State University Press, Ames, IA, pp. 43–86.

Griffiths, F.E.W., Lyndon, R.F. and Bennett, M.D. (1985) The effects of vernalization on the growth of the wheat shoot apex. *Annals of Botany* 56, 501–511.

Grimm, S.S., Jones, J.W., Boote, K.J. and Hesketh, J.D. (1993) Parameter estimation for predicting flowering date of soybean cultivars. *Crop Science* 33, 137–144.

Hadley, P., Roberts, E.H., Summerfield, R.J., and Minchin, F.R. (1983) A quantitative model of reproductive development in cowpea in relation to photoperiod and temperature, and implications for screening germplasm. *Annals of Botany* 51, 531–543.

Hadley, P., Roberts, E.H., Summerfield, R.J. and Minchin, F.R. (1984) Effects of temperature and photoperiod on flowering in soybean: A quantitative model. *Annals of Botany* 53, 669–681.

Halevy, A.H. (1985) Gladiolus. In: Halevy, A.H. (ed.) *CRC Handbook of Flowering Volume 3*. CRC Press, Boca Raton, FL, pp. 63–70.

Hallauer, A.R. (1990) Methods used in developing maize inbreds. *Maydica* 35, 1–16.

Hallauer, A. R. (1991) Recurrent selection in maize. *Plant Breeding Reviews* 9, 115–179.

Halloran, G.M. (1975) Genotype differences in photoperiodic sensitivity and vernalization response in wheat. *Annals of Botany* 39, 845–851.

Halse, N.J. and Weir, R.N. (1970) Effects of vernalization, photoperiod, and temperature on phenological development and spikelet number of Australian wheat. *Australian Journal of Agricultural Research* 21, 383–93.

Hamblin, J. (1993) The ideotype concept: Useful or outdated? In: Buxton, D.R., Shibles, R., Forsberg, R.A., Blad, B.L., Asay, K.H., Paulsen, G.M. and Wilson, R.F. (eds) *International Crop Science I*. Crop Science Society of America, Madison, WI, pp. 589–597.

Hamilton, M.A. (1991) Model validations. An annotated bibliography: communications in statistics. *Theory and Methods* 10, 2207–2266.

Hammer, G.L., Goyne, P.J. and Woodruff, D.R. (1982) Phenology of sunflower cultivars. III. Models for prediction in field experiments. *Australian Journal of Agricultural Research* 33, 263–274.

Hammer, G.L., Vanderlip, R.L., Gibson, G., Wade, L.J., Henzell, R.G., Younger, D.R., Warren, J. and Dale, A.B. (1989) Genotype by environment interaction in grain sorghum. II. Effects of temperature and photoperiod on ontogeny. *Crop Science* 29, 376–384.

Hartman, Th.A. (1964) Influence of daylength on vernalization of winter rye. *Netherlands Journal of Agricultural Science* 12, 132–155.

Hartman, Th. A. (1966) Influence of daylength on vernalization of winter rye. II. Continued experiments with short-day vernalization in winter rye. *Netherlands Journal of Agricultural Science* 14(2), 89–102.

Hay, R.K.M. (1995) Harvest index: a review of its use in plant breeding and crop physiology. *Annals of Applied Biology* 126, 197–216.

Hay, R.K.M. and Kirby, E.J.M. (1991) Convergence and synchrony – a review of the coordination of development in wheat. *Australian Journal of Agricultural Research* 42, 661–700.

Hayashi, K-I. (1966) Efficiencies of solar energy conversion in rice varieties as affected by plant density. *Proceeding of the Crop Science Society of Japan* 35, 205–211.

Hayashi, K-I. (1972) Efficiencies of solar energy conversion in rice varieties. *Bulletin of the National Institute of Agricultural Sciences (Japan)* Series D, No. 23.

He, L., Li, L. and Li, Z. (1988) A study on the differentiation and development stages of wheat and barley. *Journal of Beijing Agricultural College* 3, 18–24.

Heide, O.M. (1980) Studies of flowering in *Poa pratensis* L. ecotypes and cultivars. *Meldinger. fra Norges Landbrukshogskole* 59(14), 1–27.

Heide, O.M. (1984) Flowering requirements in *Bromus inermis*, a short-long-day plant. *Physiologia Plantarum* 62, 59–64.

Heide, O.M., Bush, M.G. and Evans, L.T. (1987) Inhibitory and promotive effects of gibberellic acid on floral initiation and development in *Poa pratensis* and *Bromus inermis*. *Physiologia Plantarum* 69, 342–350

Helenius, J. and Jokinen, K. (1994) Yield advantage and competition in intercropped oats (*Avena sativa* L.) and faba bean (*Vicia faba* L.): Application of the hyperbolic yield-density model. *Field Crops Research* 37, 85–94.

Hodges, T. (1991) *Predicting Crop Phenology*. CRC Press, Boca Raton, FL.

Hoogendoorn, J. (1984) A comparison of different vernalization techniques in wheat. *Journal of Plant Physiology* 116, 11–20.

Hoogendoorn, J. (1985) The physiology of variation in the time of ear emergence among wheat varieties from different regions of the world. *Euphytica* 34, 559–571.

Horie, T. and Nagagewa, H. (1990) Modelling and prediction of the developmental process in rice. II. Structure and method of parameter estimation of model for simulating developmental process towards heading. *Japanese Journal of Crop Science* 59, 687–695.

Horie, T., Nagagawa, H. and Kira, T. (1986) Modelling and prediction of developmental processes of rice. I. Dynamic model for predicting the flowering data. *Japanese Journal of Crop Science* 55, 214–215.

Hu, C. and Luo, C. (1988) A botanical criterion of completed vernalization and combination effect of temperature and light in wheat. *Journal of Beijing Agricultural College* 3, 1–7.

Huang, J., Zhou, L., Hai, W. and Jian, Z. (1988) The vernalization stage of wheat and its inflorescence development. *Journal of Beijing Agricultural College* 3, 25–31.

Huffman, W.E. and Evenson, R.E. (1993) *Science for Agriculture: A Long Term Perspective*. Iowa State University Press, Ames, IA.

Hunt, L.A. (1979) Photoperiodic responses of winter wheats from different climatic regions. *Journal of Plant Physiology (Zeitschrift feur Pflanzenphysiologie)* 82, 70–80.

Hyne, V., Kearsey, M.J., Martinez, O., Gang, W. and Snape, J.W. (1995) A partial genome assay for quantitative trait loci in wheat (*Triticum aestivum*) using different analytical techniques. *Theoretical and Applied Genetics* 89, 735–741.

Izhar, S. and Wallace, D.H. (1967) Studies of the physiological basis for yield differences. III. Genetic variation in photosynthetic efficiency of *Phaseolus vulgaris* L. *Crop Science* 7, 457–460.

Jackson, P., Robertson, R., Cooper, M., and Hammer, G. (1996) The role of physiological understanding in plant breeding; from a breeding perspective. *Field Crops Research* 49, 11–37.

Jain, S.M., Sopory, S.K. and Veilleux, R.E. (eds) (1996) In vitro *Haploid Production in Higher Plants. Volume I. Fundamental Aspects and Methods*. Kluwer Academic Publishers, New York, 368 pp.

Jamieson, P.D., Brooking, I.R., Porter, J.R. and Wilson, D.R. (1995) Prediction of leaf appearance in wheat: a question of temperature. *Field Crops Research* 41, 35–44.

Jansen, R.C., Van Ooijen, J.W., Stam, P., Lister, C. and Dean, C. (1995) Genotype-by-environment interaction in genetic mapping of multiple quantitative trait loci. *Theoretical and Applied Genetics* 91, 33- 37.

Jedel, P.E., Evans, L.E. and Scarth, R. (1986) Vernalization responses of a selected group of spring wheat cultivars. *Canadian Journal of Plant Science* 66, 1–9.

Jensen, N.F. (1994) Historical perspectives on plant breeding methodology. In: Frey, K.F. (ed.) *Historical Perspectives in Plant Science*. Iowa State University Press, Ames, IA, pp. 179–194.

Jing, Shangbao (1991) *Wheat Ecology of China*. Academic Press, Beijing, China, 560 pp.

Jing, Shangbao (1993) *Theoretical and Applied Wheat Ecology of China*. Academic Press, Beijing, China.

Jokinen, K. (1991) Assessment of competition and yield advantage in addition series of barley variety mixtures. *Journal of Agricultural Science in Finland* 63, 307–320.

Jordan, B.R. (ed.) (1993) *The Molecular Biology of Flowering.* CAB International. Wallingford, Oxon, UK, 226 pp.

Kacser, H., Burns, J.A. and Fell, D.A. (1995) The control of flux. *Biochemical Society Transactions* 23, 341–366.

Kang, M.S. (ed.) (1990) *Genotype-by-environment Interaction and Plant Breeding.* Louisiana University, Baton Rouge, LA, pp. 141–140.

Kang, M.S. and H.G. Gauch, Jr (1996) *Genotype-by-Environment Interaction.* CRC Press, Boca Raton, FL, 416 pp.

Karlsson, B.H., Sills, G.R. and Nienhuis, J. (1993) Effects of photoperiod and vernalization on the number of leaves at flowering in 32 *Arabidopsis thaliana (Brassicaceae)* ecotypes. *American Journal of Botany* 80(6), 646–648.

Kato, K. and Yamagata, H. (1988) Method for evaluation of chilling requirement and narrow sense earliness of wheat cultivars. *Japanese Journal of Breeding* 38, 172–186.

Kato, K. and Yamashita, S. (1991) Varietal variation in photoperiodic response, chilling requirement and narrow-sense earliness and their relation to heading time in wheat. *Japanese Journal of Breeding* 41, 475–484.

Katsuta, M. (1975) Effects of chilling and photoperiod treatments on cone development in *Pinus thunbergii* Parl. and *P. densiflora* Sieb. et Zuaa. *Bulletin of the Tokyo University Forests* 67, 67–104.

Keiler, D and Smith, H. (1989) Control of carbon partitioning by light quality mediated by phytochrome. *Plant Science* 63, 25–29.

Kenworthy, W.J. and Brim, C.A. (1979) Recurrent selection in soybeans. I. Seed yield. *Crop Science* 19, 315–318.

King, W.M. and Murfet, I.C. (1985) Flowering in *Pisum*: A sixth locus, *Dne. Annals of Botany* 56, 835–846.

Kirby, E.J.M. (1990) Co-ordination of leaf emergence and leaf and spikelet primordium initiation in wheat. *Field Crops Research* 25, 253–264.

Kirby, E.J.M. (1992) A field study of the number of main shoot leaves in wheat in relation to vernalization and photoperiod. *Journal of Agricultural Science* 118, 271–278.

Kirby, E.J.M. and Perry, M.W. (1987) Leaf emergence rate of wheat in a Mediterranean environment. *Australian Journal of Agricultural Research* 38, 455–464.

Kirby, E.J.M., Appleyard, M. and Fellowes, G. (1985) Effect of sowing date and variety on the main shoot leaf emergence and number of leaves of barley and wheat. *Agronomie* 5, 117–126.

Kirby, E.J.M., Porter, J.R., Day, W., Adam, J.S., Appleyard, M., Ayling, S., Baker, C.K., Beale, P., Belford, R.K., Biscoe, P.V., Chapman, A., Fuller, M.P., Hampson, J., Hay, R.K.M., Hough, M.N., Matthews, S., Thompson, W.J. Weir, A.H., Willington, V.B.A. and Wood. D.W. (1987) An analysis of primordia initiation in Avalon winter wheat crops with different sowing dates at nine sites in England and Scotland. *Journal of Agricultural Science* 109, 123–134.

Knott, P.R. (1972). Effects of selection for F2 plant yield on subsequent generations in wheat. *Canadian Journal of Plant Science* 52, 721–726.

de Koning, G.J. and Weller, J.I. (1994) Efficiency of direct selection on quantitative trait loci for a two-trait breeding objective. *Theoretical and Applied Genetics* 88, 669–677.

Kooman, P.L. and Rabbinage, R. (1996) An analysis of the relation between dry matter allocation to the tuber and earliness of a potato crop. *Annals of Botany* 77, 235–242.

Koornneef, M. and Kendrick, R.E. (1994) Photomorphogenic mutants of higher plants. In: Kendrick, R.E. and Kronenberg, G.H.M. (eds) *Photomorphogenesis in Plants.* 2nd edn. Kluwer, Norwell, MA, pp. 601–628.

Koornneef, M., Hanhart, C.J. and van der Veen, J.H. (1991) A genetic and physiological analysis of the late flowering mutants in *Arabidopsis thaliana. Molecular and General Genetics* 229, 57–66.

Kueneman, E.A. (1978) Evaluation of the yield potential of growth habits of dry bean (*Phaseolus vulgaris* L.) and determination of plant types for high density plantings. PhD thesis, Cornell University, Ithaca, NY.

Kueneman, E.A., Sandsted, R.F., Wallace, D.H., Bravo, A. and Wien, H.C. (1979) Effect of plant arrangements and densities on yields of dry beans. *Agronomy Journal* 71, 419–424.

Kulkarni, R.N. (1990) Honeycomb and simple mass selection for herb yield and inflorescence to leaf plus stem ratio in palmarosa grass. *Euphytica* 47, 147–152.

Lang, A. (1965) Physiology of flower initiation. In: Ruhland, W. (ed.) *Encyclopedia of Plant Physiology* 15, Part 1. Springer-Verlag, New York, pp. 1380–1536.

Law, C.N. (1995) Genetic manipulation in plant breeding – prospects and limitations. *Euphytica* 85, 1–12.

Lawn, R.J., Summerfield, R.J., Ellis, R.H., Qi, A., Roberts, E.H., Chay, P.M., Brouwer, J.B., Rose, J.L. and Yeates, S.J. (1995) Towards the reliable prediction of time to flowering in six annual crops. VI. Applications in crop improvement. *Experimental Agriculture* 31, 89–108.

Levy, J. and Peterson, M.L. (1972) Response of spring wheat to vernalization and photoperiod. *Crop Science* 12, 487–490.

Linnemann, A.R. and Craufurd, P.Q. (1994) Effect of temperature and photoperiod on phenological development in three genotypes of Bambara groundnut (*Vigna subterranea*). *Annals of Botany* 74, 675–681.

Linneman, A.R, Westphal, E., Wessel, M. (1995) Photoperiod regulation of development and growth in bambara groundnut (*Vigna subterranea*). *Field Crops Research* 40, 39–47.

Littlejohn, G.M., Pienaar, R.D.V. and Roux, H.S. (1994) Yield response to modified honeycomb and recurrent selection in an open pollinated rye population. *Applied Plant Science* 8, 27–30.

Loomis, R.S. (1993) Optimization theory and crop improvement. In: Buxton, D.R., Shibles, R., Forsberg, R.A., Blad, B.L., Asay, K.H., Paulsen, G.M. and Wilson, R.F. (eds). *International Crop Science I.* Crop Science Society of America. Madison, WI, pp. 583–588.

Loomis, R.S., and Connor, D.J. (1992) *Crop Ecology: Productivity and Management in Agricultural Systems.* Cambridge University Press, Cambridge, UK, 538 pp.

Lucas, D. (1971) Effects of environment on morphogenesis of the shoot apex in wheat. PhD thesis, University of Adelaide, Adelaide, Australia.

Lulsdorf, M.M. and McVetty, P.B.E. (1986) Comparison of honeycomb and independent culling of spaced plants for yield potential in faba bean *(Vicia faba)*. *Canadian Journal of Plant Science* 66, 483–490.

Major, D.J. (1980) Photoperiod response characteristics controlling flowering of nine crop species. *Canadian Journal of Plant Science* 60, 777–784.

Major, D.J., Johnson, D.R., Tanner, J.W. and Anderson, I.C. (1975) Effect of daylength and temperature on soybean development. *Crop Science* 15, 174–179.

Major, D.J., Rood, S.B. and Miller, F.R. (1990) Temperature and photoperiod effects mediated by the sorghum maturity genes. *Crop Science* 30, 305–310.

Malse, J., Doussinault, G. and Sun, B. (1989) Response of wheat to temperature and photoperiod in natural conditions. *Crop Science* 19, 712–721.

Mamni, A., Erskine, W. and Gates, P. (1991) Relationships among economic characters in Lentil. *Euphytica* 57, 109–116.

Manupeerapan, T., Davidson, J.L., Pearson, C.J. and Christian, K.R. (1992) Differences in flowering responses of wheat to temperature and photoperiod. *Australian Journal of Agricultural Research* 43, 575–584.

Marcellos, H. and Single, W.V. (1971) Quantitative responses of wheat to photoperiod and temperature in the field. *Australian Journal of Agricultural Research* 22, 343–357.

Marshall, D.R. (1991) Alternative approaches and perspectives in breeding for higher yields. *Field Crops Research* 26, 171–190.

Martinez-Zapater, J.M. and Somerville, C.R. (1990) Effect of light quality and vernalization on late flowering mutants of *Arabidopsis thaliana*. *Plant Physiology* 92, 770–776.

Masaya, P.N. (1978) Genetic and environmental control of flowering in *Phaseolus vulgaris* L. PhD thesis, Cornell University, Ithaca, NY.

McKenney, H.H. and Sando, W.J. (1935) Earliness of sexual reproduction in wheat as influenced by temperature and light in relation to growth phases. *Journal of Agricultural Research* 51, 621–641.

McMaster, G.S. (1997) Phenology, development and growth of the wheat (*Triticum aestivum* L.) shoot apex: A review. *Advances in Agronomy* 59, 63–118.

McMaster, G.S., Klepper, B., Rickman, R.W., Wilhelm, W.W. and Willis, W.O. (1991) Simulation of shoot vegetative development and growth of unstressed winter wheat. *Ecological Modelling* 53, 189–204.

McMaster, G.S., Morgan, J.A. and Wilhelm, W.W. (1992) Simulating winter wheat spike development and growth. *Agricultural and Forest Meteorology* 60, 193–220.

McPherson, H.G., Warrington, I.J. and Turnbull, H L. (1985) The effects of temperature and daylength on the rate of development of pigeonpea. *Annals of Botany* 56, 597–611.

Medrano, H. and Vadell, J. (1993) Photosynthesis improvement as a way to increase crop yield. In: Abrol, Y.P., Mohanty, P. and Govindjee (eds.) *Photosynthesis: Photoreactions to Plant Productivity*. Kluwer Academic Publishers. Boston, pp. 561–582.

Miao, G.Y., Zhang, Y.T., Hou, Y.S., Yin, J. Wang, S.Y. and Li, H.Z. (1988) Studies on the photo-thermal responses of wheat (I). *Journal of Beijing Agricultural College* 3, 8–17.

Miao, G.Y., Zhang, Y.T., Hou, Y.S., Yin, J. and Wang, S.Y. (1990) A simulated study of temperature and daylength effects on wheat development. *Acta Agriculturae Boreali-Sinica* 5, 15–23.

Miao, G.Y., Zhang, Y.T., Hou, Y.S., Yin, J. and Wang, S.Y. (1991) Studies on the development of wheat under controlled environments. In: Jing, S. (ed.) *Wheat Ecology of China*. Academic Press, Beijing, pp. 479–497.

Miao, G.Y., Zhang, Y.T., Hou, Y.S., Yin, J. and Wang, S.Y. (1992) Relationship between effect of thermo-photoperiod and leaf number of main stem in wheat. *Acta Agronomica Sinica* 18, 321–330.

Miglietta, F. (1989) Effect of photoperiod and temperature on leaf initiation rates in wheat. *Field Crops Research* 21, 121–130.

Miglietta, F. (1991a) Simulation of wheat ontogenesis: I. Appearance of wheat main stem leaves in the field. *Climate.Research* 1, 145–150.

Miglietta, F. (1991b) Simulation of wheat ontogenesis: II. Predicting dates of ear emergence and main stem final leaf number. *Climate Research* 1, 151–160.

Miller, M.H. and Kay, B.D. (1993) New approaches needed for research in sustainable cropping systems. In: Buxton, D.R., Shibles, R., Forsberg, R.A., Blad, B.L., Asay, K.H., Paulsen, G.M. and Wilson, R.F. (eds) *International Crop Science I*. Crop Science Society of America, Madison, WI, pp. 15–18.

Misa, A.L., Isoda, A., Nojima, H., Takasaki, Y., Yoshimura, T. and Ishikawa, T. (1994) Plant type and dry matter production in peanut (*Arachis hypogaea* L.) cultivars. *Japanese Journal of Crop Science* 63, 289–297.

Mitchell, J.W., Baker, R.J. and Knott, D.R. (1982) Evaluation of honeycomb selection for single plant yield in durum wheat. *Crop Science* 22, 840–843.

Mmopi, S., Wallace, D.H., Masaya, P.N., Zobel, R.W. (1995) Control of days to flowering of bean (*Phaseolus vulgaris* L.) by interaction of a photoperiod and a nonphotoperiod gene. In: Pessarakli, M. (ed.) *Handbook of Plant and Crop Physiology*. Marcel Dekker, New York, pp. 917–939.

Morgan, D.G., Morgan, C.B. (1984) Photoperiod and the abscission of flower buds in *Phaseolus vulgaris*. In: Vince-Prue, D., Thomas, B., Cockshull K.E. (eds) *Light and the Flowering Process*. Academic Press, New York, pp. 227–240.

Moss, D.N. (1976) Studies on increasing photosynthesis in crop plants. In: Burris, R.H. and Black, C.C. (eds) *CO_2 Metabolism and Plant Productivity*. University Press, Baltimore, pp. 31–41.

Mozley, D. and Thomas, B. (1995) Developmental and photobiological factors affecting photoperiodic induction in *Arabidopsis thaliana* Heynh. Landsberg erecta. *Journal of Experimental Botany* 46, 173–179.

Muhammad, A.F. (1983) The effects of temperature on days to flower and maturity in dry beans *Phaseolus vulgaris* L. PhD thesis, Cornell University, Ithaca, NY.

Murfet, I.C. (1977a) Environmental interaction and the genetics of flowering. *Annual Review of Plant Physiology* 28, 253–78.

Murfet, IC. (1977b) The physiological genetics of flowering. In: Sutcliffe, J.F. and Pate, J.S. (eds) *The Physiology of the Garden Pea*. Academic Press, New York, pp. 385–430.

Murfet, I.C. (1985) *Pisum sativum*. In: Halevy, A.H. (ed.) *CRC Handbook of Flowering, Vol. IV*. CRC Press, Boca Raton, Florida, pp. 97–126.

Murfet, I.C. and Reid, J.B. (1974) Flowering in *Pisum*, the influence of photoperiod and vernalization temperatures on the expression of genes *Lf* and *Sn*. *Zeitschrift für Pflanzenphysiologie* 71, 323–331.

Murneek, A.E. (1948) History of research in photoperiodism. In: Murneek, A.E. and Whyte, R.O. (eds) *Vernalization and Photoperiodism*. Chronica Notanica Company, Waltham, MA, pp. 39–61.

Napp-Zinn, K. (1984) Light and vernalization. In: Vince-Prue, D., Thomas, B., and Cockshull, K.E. (eds) *Light and the Flowering Process*. Academic Press, London, pp. 75–88.

Napp-Zinn, K. (1985) *Arabidopsis thaliana*. In: Halevy, A.H. (ed.) *CRC Handbook of Flowering, Volume I*. CRC Press, Boca Raton, LA, pp. 492–503.

Nichiporovich, A. (1967) Aims of research on the photosynthesis of plants as a factor of productivity. In: Nichiporovich, A. (ed.) *Photosynthesis of Productive Systems*. Russian to English translation by the Israel Program for Scientific Translations, Jerusalem, pp. 3–36.

Ogren, W.L. (1976) Search for higher plants with modifications of the reductive pentose phosphate pathway of CO_2 assimilation. In: Burris, R.H. and Black, C.C. (eds) *CO_2 Metabolism and Plant Productivity*. University Press, Baltimore, pp. 19–29.

Olsen, J.K., McMahon, C.R. and Hammer, G.L. (1993) Prediction of sweetcorn phenology in subtropical environments. *Agronomy Journal* 85, 410–415.

Omanga, P.A., Summerfield, R.J. and Qi, A. (1995) Flowering of pigeonpea in Kenya: Responses of early-maturing genotypes to location and date of sowing. *Field Crops Research* 41, 25–34.

Oosterom, E.J.V. and Acevedo, E. (1992) Adaptation of barley to harsh Mediterranean environments. III. Plant ideotype and grain yield. *Euphytica* 62, 29–38.

Owen, H.R., Veilieux, R.E. and Haynes, K.G. (1988) Photoperiod effects on 2N pollen production, response to anther culture and net photosynthesis of a diplandrous clone of *Solanum phureja*. *American Potato Journal* 63, 131–139.

Padda, D.S. and Munger, H.M. (1969) Photoperiod, temperature, and genotype interactions affecting time of flowering in beans (*Phaseolus vulgaris L.*). *Journal of the American Society of Horticultural Science* 94, 157–160.

Pan, A., Hayes, P.M., Chen, F., Chen, T.H.H., Blake, T., Wright, S., Karsal, I. and Bedo, Z.

(1994) Genetic analysis of the components of winter hardiness in barley (*Hordeum vulgare* L.). *Theoretical and Applied Genetics* 89, 900–910.

Pantone, D.J. and Baker, J.B. (1991) Reciprocal yield analysis of red rice (*Oryza sativa*) competition in cultivated rice. *Weed Science* 39, 42–47.

Paton, D.M. (1968) Photoperiodic and temperature control of flower initiation in the late pea cultivar Greenfeast. *Australian Journal of Biological Science* 21, 609–617.

Penrose, L.D.J., Martin, R.H. and Landers, C.F. (1991) Measurement and response to vernalization in Australian wheats with winter habit. *Euphytica* 57, 9–18.

Perry, M.W., Siddique, K.H.M. and Wallace, J.F. (1987) Predicting phenological development for Australian wheats. *Australian Journal of Agricultural Research* 38, 809–820.

Pfeiffer, T.W. (1996) Choosing soybean varieties from yield trials: Multiple maturity groups and yield variability. *Journal of Production Agriculture* 9, 371–376.

Pinthus, M.J. (1985) *Triticum*. In: Halevy, A.H. (ed.) *CRC Handbook of Flowering, Volume VI.* CRC Press, Boca Raton, FL, pp. 418–443.

Porter, J.R., Kirby, E.J.M., Day, W., Adam, J.S., Appleyard, M., Ayling, S., Baker, P., Beale, R.K., Belford, P.V., Biscoe, A., Chapman, M.P., Fuller, J., Hampson, R.K.M., Hay, C.K., Hough, M.N., Matthews, S., Thompson, W.J., Weir, A.H., Willington, V.B.A., and Wood, D.W. (1987) An analysis of morphological development stages in Avalon winter wheat crops with different sowing dates and at 10 sites in England and Scotland. *Journal of Agricultural Science* 109, 107–121.

Prasad, K.L.N. (1996) Analysis of genotype × environment interaction in pigeonpea (*Cajanus cajan* (L.) Millsp.) subjected to waterlogging. MS. Thesis. Andra Pradesh Agricultural University. Rajendranagar, Hyderabad 500 030, India.

Pugsley, A.T. (1966) The photoperiodic sensitivity of some spring wheats with special reference to the variety Thatcher. *Australian Journal of Agricultural Research* 17, 591–599.

Pugsley, A.T. (1971) A genetic analysis of the spring-winter habit of growth in wheat. *Australian Journal of Agricultural Research* 22, 21–31.

Purohit, A.N. (1970) The qualitative and quantitative photoperiodic response of Indian potato varieties. *New Phytologist* 69, 521–27.

Purvis, O.N. and Gregory, F.G. (1937) Studies in vernalization of cereals. I. A comparative study of vernalization of winter rye by low temperature and by short days. *Annals of Botany* 4, 569–591.

Quinby, J.R., Hesketh, J.D. and Voigt, A.L. (1973) Influence of temperature and photoperiod on floral initiation and leaf number in sorghum. *Crop Science* 13, 243–246.

Rahman, M.S. (1980) Effect of photoperiod and vernalization on the rate of development and spikelet number per ear in 30 varieties of wheat. *Journal of the Australian Institute of Agricultural Science* 46, 68–70.

Rahman, M.S. and Wilson, J.H. (1977) Determination of spikelet number in wheat. I. Effect of varying photoperiod on ear development. *Australian Journal of Agricultural Research* 28, 565–574.

Rahman, M.S. and Wilson, J.H. (1978) Determination of spikelet number in wheat. III. Effect of varying temperature on ear development. *Australian Journal of Agricultural Research* 29, 459–467.

Rahman, M.S., Wilson, J.H. and Aitken, Y. (1977) Determination of spikelet number in wheat. II. Effect of varying light level on ear development. *Australian Journal of Agricultural Research* 28, 575–581.

Rasmusson, D.C. (1987) An evaluation of ideotype breeding. *Crop Science* 27, 1140–1146.

Rasmusson, D.C. (1991) A plant breeder's experience with ideotype breeding. *Field Crops Research* 26, 191–200.

Ratinam, M and Thurling, N. (1989) Early generation selection for grain yield in narrow-leaf lupine *Lupinus angustifolius* L. I. Studies with a simulated F2 population. *Plant Breeding* 102, 237–247.

Rawson, H.M. (1971) An upper limit for spikelet number per ear in wheat, as controlled by photoperiod. *Australian Journal of Agricultural Research* 22, 537–546.

Rawson, H.M. (1993) Radiation effects on the rate of development in wheat grown under different photoperiods and high and low temperatures. *Australian Journal of Plant Physiology* 20, 719–727.

Rawson, H.M. and Richards, R.A. (1993) Effects of high temperature and photoperiod on floral development in wheat isolines differing in vernalization and photoperiod genes. *Field Crops Research* 32, 181–192.

Rawson, H.M. and Zajac, M. (1993) Effects of higher temperatures, photoperiod and seed vernalization on development in two spring wheats. *Australian Journal of Plant Physiology* 20, 211–222.

Razumov, V.I. and Limar, R.S. (1974) The influence of short days on the development of some winter barley varieties. *Plant Breeding Abstracts* 44, 661.

Reynolds, M.P., Acevedo, E., Sayre, K.D. and Fischer, R.A. (1994) Yield potential in modern wheat varieties: its association with a less competitive ideotype. *Field Crops Research* 37, 149–160.

Rezende, G.D.S.P. and Ramalho, M.A.P. (1994) Competitive ability of maize and common bean (*Phaseolus vulgaris*) cultivars intercropped in different environments. *Journal of Agricultural Science* 123, 185–190.

Richards, R.A. (1991) Crop improvement for temperate Australia: Future opportunities. *Field Crops Research* 26, 141–169.

Riddle, J.A. and Gries, G.A. (1958) Development of spring wheat. III. Temperature of maturation and age of seeds as factors influencing their response to vernalization. *Agronomy Journal* 50, 743–746.

Roberts, E.H., Muehlbauer, F.J. and Short, R.W. (1986) Flowering in lentil (*Lens culinaris* Medic.): The duration of the photoperiod inductive phase as a function of accumulated daylength above the critical photoperiod. *Annals of Botany* 58, 235–248.

Roberts, E.H., Summerfield, R.J., Cooper, J.P. and Ellis, R.H. (1988a) Environmental control of flowering in barley. I. Photoperiod limits to long-day response, photoperiod insensitive phases and effects of low-temperature and short-day vernalization. *Annals of Botany* 62, 127–144.

Roberts, E.H., Summerfield, R.J., Ellis, R.H. and Stewart, K.A. (1988b) Photothermal time for flowering in lentils and the analysis of potential vernalization response. *Annals of Botany* 61, 29–39.

Roush, M.L., Radosevich, S.R., Wagner, R.G., Maxwell, B.D. and Petersen, T.D. (1989) A comparison of methods for measuring effects of density and proportion in plant competition experiments. *Weed Science* 37, 268–275.

Russell, W.A. (1986) Contribution of breeding to maize improvement in the United States, 1920s–1980s. *Iowa State Journal of Research* 61, 5–34.

Salisbury, F.B. (1963). *The Flowering Process*. Pergamon Press, New York. 234 pp.

Salisbury, F.B. and Ross, C.W. (1981) *Plant Physiology*. Wadsworth Publishing, Belmont, CA.

Salisbury, P.A., Berry, G.J. and Halloran, G.M. (1979) Expression of vernalization genes in near-isogenic wheat lines: Methods of vernalization. *Canadian Journal of Genetics and Cytology* 21, 429–434.

Sampson, D.R. and Burrows., V.D. (1972) Influence of photoperiod, short-day vernalization, and cold vernalization on days to heading in *Avena* species and cultivars. *Canadian Journal of Plant Science* 52, 471–482.

Scarth, R.E, Kirby, J.M. and Law, C.N. (1985) Effect of the photoperiod genes Ppd1 and

Ppd2 on growth and development of the shoot apex in wheat. *Annals of Botany* 55, 351–359.

Schrader, L.E. (1976) CO_2 metabolism and productivity in C3 plants: An assessment. In: Burris, R.H. and Black, C.C. (eds) *CO_2 Metabolism and Plant Productivity.* University Press, Baltimore, pp. 384–403.

Scully, B.T., Wallace, D.H. and Viands, D.R. (1991) Heritability and correlation of biomass, growth rates, harvest index, and phenology to the yield of common beans. *Journal of the American Society for Horticultural Science* 116, 127–130.

Sedgley, R.H. (1991) An appraisal of the Donald ideotype after 21 years. *Field Crops Research* 26, 221–226.

Settr, T.L., Conocono, E.A., Egdane, J.A. and Kropff, M.J. (1995) Possibility of increasing yield potential of rice by reducing panicle height in the canopy. I. Effects of panicles on light interception and canopy photosynthesis. *Australian Journal of Plant Physiology* 22, 441–451.

Sharma, H.C. and Mascia, P.N. (1987) Vernalization of immature embryos of winter wheat genotypes. *Euphytica* 36, 161–165.

Sharma, R.C. and Smith, E.L. (1986) Selection for high and low harvest index in three winter wheat populations. *Crop Science* 26, 1147–1150.

Shaykewich, C.G. (1995) An appraisal of cereal crop phenology modeling. *Canadian Journal of Plant Science* 75, 329–341.

Shroyer, J.P. and Cox, T.S. (1993) Productivity and adaptive capacity of winter wheat landraces and modern cultivars grown under low fertility conditions. *Euphytica* 70, 27–33.

Simmonds, N.W. (1989) How frequent are superior genotypes in plant breeding populations. *Biological Reviews* 64, 341–365.

Sinclair, T.R., Kitani, S., Hinson, K., Bruniard, J., and Horie, T. (1991) Soybean flowering date: Linear and logistic models based on temperature and photoperiod. *Crop Science* 31, 786–790.

Slafer, G.A. and Rawson, H.M. (1994a) Sensitivity of wheat phasic development to major environmental factors: a re-examination of some assumptions made by physiologists and modelers. *Australian Journal of Plant Physiology* 21, 393–426.

Slafer, G.A. and Rawson, H.M. (1994b) Does temperature affect final numbers of primordia in wheat? *Field Crops Research* 39, 111–117.

Slafer, G.A. and Rawson, H.M. (1995a) Intrinsic earliness and basic development rate assessed for their response to temperature in wheat. *Euphytica* 83, 175–183.

Slafer, G.A. and Rawson, H.M. (1995b) Base and optimum temperatures vary with genotype and stage of development in wheat. *Plant, Cell and Environment* 18, 671–679.

Slafer, G.A., Halloran, G.M. and Connor, D.J. (1994a) Development rate in wheat as affected by duration and rate of change of photoperiod. *Annals of Botany* 73, 671–677.

Slafer, G.S., Connor, D.J. and Halloran, G.M. (1994b) Rate of leaf appearance and final number of leaves in wheat: effects of duration and rate of change of photoperiod. *Annals of Botany* 74, 427–436.

Squire, G.R. 1990. *The Physiology of Tropical Crop Production.* Bookcraft, Bath, UK, 236 pp.

Steer, B.T. (1980) The bulbing response to daylength and temperature of some Australian cultivars of onion (*Allium cepa* L.). *Australian Journal of Agricultural Research* 31, 511–518.

Stelmak, A.F. (1993) Genetic effects of *Vrn* genes on heading date and agronomic traits in bread wheat. *Euphytica* 65, 53–60.

Stern, W.R. and Kirby, E.J.M. (1979) Primordium initiation at the shoot apex in four contrasting varieties of spring wheat in response to sowing date. *Journal of Agricultural Science* 109, 107–121.

Stoskopf, N.C, Tomes, D.W. and Christies, B.R. (1993) *Plant Breeding: Theory and Practice.* Westview, Boulder, CO, 531 pp.

Stuber, C.W. (1995) Mapping and manipulating quantitative traits in maize. *Trends in Genetics* 11, 477–481.

Summerfield, R.J. and Roberts, E.H. (1987) Effects of illuminance on flowering in long- and short-day grain legumes: a reappraisal and unifying model. In: Atherton, J.G. (ed.) *Manipulation of Flowering.* Butterworths, London, pp. 203–223.

Summerfield, R.J. and Roberts, E.H. (1993) The Plant Environment Laboratory. In: Harris, P.M. (ed.) *The Silent Fields: One Hundred Years of Agricultural Research at Reading.* Reading University-Holdsworth Associates, Reading, UK, pp. 144–150.

Summerfield, R.J., Roberts, E.H., Ellis, R.H. and Lawn, R J. (1991) Towards the reliable prediction of time to flowering in six annual crops. I. The development of simple models for fluctuating field environments. *Experimental Agriculture* 27, 11–31.

Summerfield, R.J., Collinson, S.T., Ellis, R.H., Roberts, E.H. and Penning de Vries, F.W.T. (1992) Photothermal responses of flowering in rice (*Oryza sativa*). *Annals of Botany* 69, 101–112.

Summerfield, R.J., Lawn, R.J., Qi, A. Ellis, R.H., Roberts, E.H., Chay, P.M., Brouwer, J.B., Rose, J.L., Shanmugasunderam, S., Yeates S.J. and Sandover, S. (1993) Towards the reliable prediction of time to flowering in six annual crops. II. Soyabean (*Glycine max*). *Experimental Agriculture* 29, 253–289.

Syme, J.R. (1968) Ear emergence of Australian, Mexican and European wheats in relation to time of sowing and their response to vernalization and daylength. *Australian Journal of Experimental Agriculture and Animal Husbandry* 8, 578–581.

Syme, J.R. (1973) Quantitative control of flowering time in wheat cultivars by vernalization and photoperiod sensitivities. *Australian Journal of Agricultural Research* 24, 657–665.

Takeda, K and Frey, K.J. (1987) Improving grain yield in backcross populations from *Avena-sativa* × *Avena-sterilis* matings by using independent culling for harvest index and vegetative growth index or unit straw weight. *Theoretical and Applied Genetics* 74, 659–665.

Tanaka, A., Kawano, K., and Yamaguchi, J. (1966) Photosynthesis, respiration and plant type of the tropical rice plant. *Technical Bulletin* 7, 3–46. International Rice Research Institute, Los Banos, Philippines.

Tanksley, S.D. and McCouch, S.R. (1997) Seed banks and molecular maps: Unlocking genetic potential from the wild. *Science* 277, 1063–1066.

Tanksley, S.D. and Nelson, J.C. (1996) Advanced backcross QTL analysis: a method for the simultaneous discovery and transfer of valuable QTLs from unadapted germplasm into elite breeding lines. *Theoretical and Applied Genetics* 92, 191–203.

Thomas, J.B., Schaalje, G.B. and Grant, M.N. (1994) Height, competition and yield potential in winter wheat. *Euphytica* 74, 9–17.

Toenniessen, G. (1993) Progress and prospects for using biotechnology in plant breeding. In: You, C.B., Chen, Z. and Ding, Y. (eds) *Biotechnology in Agriculture.* Kluwer Academic Publishers, New York, pp. 125–129.

Tollenaar, M. (1989) Genetic improvement in grain yield of commercial maize hybrids grown in Ontario from 1959 to 1988. *Crop Science* 29, 1365–1371.

Tollenaar, M., McCullough D.E., and Dwyer, L.M. (1994) Physiological basis of genetic improvement of corn. In: Slafer, G.A. *Genetic Improvement of Field Crops.* Marcel Dekker, New York, pp. 183–286.

Trewavas, A. (1986) Understanding the control of plant development and the role of growth substances. *Australian Journal of Plant Physiology* 13, 447–457.

Tribe, D. (1994) *Feeding and Greening the World.* CAB International, Wallingford, UK, 274 pp.

Trione, E.J. and Metzger, R.J. (1970) Wheat and barley vernalization in a precise temperature gradient. *Crop Science* 10, 390–392.

Ud-Din, N., Carver, B.F. and Clutter, A.C. (1992) Genetic analysis and selection for wheat yield in drought-stressed and irrigated environments. *Euphytica* 62, 89–96.

Upadhyay, A.P., Ellis, R.H., Summerfield, R.J., Roberts, E.H. and Qi, A. (1994) Characterization of photothermal flowering responses in maturity isolines of soybean [*Glycine max* (L.) Merrill] cv. Clark. *Annals of Botany* 74, 87–96.

Valentine, J. (1979) The effect of competition and method of sowing on efficiency of single plant selection for grain yield, yield components and other characters in spring barley. *Plant Breeding* 83, 193–204.

Veldboom, L.R. and Lee, M. (1995) Molecular-marker-facilitated studies of morphological traits in maize. II. Determination of QTLs for grain yield and yield components. *Theoretical and Applied Genetics* 89, 451–458.

Vince-Prue, D. (1975) *Photoperiodism in Plants.* McGraw Hill, New York, 444 pp.

Vince-Prue, D. (1983) Photoperiodic control of plant reproduction. In: Meudt, W.J. (ed.) *Strategies of Plant Reproduction*. Allenheld, Osmun Publishers, Granada, pp. 73–97.

Wall, P.C. and Cartwright, P.M. (1974) Effects of photoperiod, temperature and vernalization on the phenology and spikelet numbers of spring wheat. *Annals of Applied Biology* 76, 299–309.

Wallace, D.H. (1973) Commentary upon: Plant architecture and physiological efficiency in the field bean. In: *Seminar on: Potentials of Field Beans and other Food Legumes in Latin America.* Series Seminar No. 2E. Centro Internacional de Agricultura Tropical, Cali, Colombia.

Wallace, D.H. (1976) Efficiency of photosynthate accumulation and partitioning. In: Ozbun, J.L. (ed.) *Researchable Areas Which Have Potential for Increasing Crop Production. Report to the National Science Foundation.* Cornell University, Agricultural Experiment Station, Ithaca, NY, pp. 5–14.

Wallace, D.H. (1980) The genetics of photosynthesis and crop productivity – With emphasis on beans. In: *Well-being of Mankind and Genetics.* Proceedings of the XIV International Congress of Genetics. Volume 1 (Book 2). MIR Publishers, Moscow, pp. 306–324.

Wallace, D.H. (1985) Physiological genetics of plant maturity, adaptation and yield. *Plant Breeding Reviews* 3, 21–168.

Wallace, D.H. (1991) System analysis of yield trials can raise efficiency of breeding for yield. In: Prakash, J. and Peirik, R.L.M. (eds) *Horticulture – New Technologies and Applications.* Kluwer Academic Publishers, Dordrecht, pp. 1–11.

Wallace, D.H. and Enriquez, G.A. (1980) Daylength and temperature effects on flowering time of photoperiod sensitive and insensitive bean, *Phaseolus vulgaris* L. *Journal of the American Society of Horticultural Science* 105, 583–591.

Wallace, D.H. and Munger, H.M. (1965) Studies of the physiological basis for yield differences. I. Growth analysis of six dry bean varieties. *Crop Science* 5, 343–348.

Wallace, D.H. and Munger, H.M. (1966) Studies of the physiological basis for yield differences. II. Variations in dry matter distribution among aerial organs for several dry bean varieties. *Crop Science* 6, 503–507.

Wallace, D.H. and Zobel, R.W. (1982) The biology of crop yield. In: Rechcigl, M. (ed.) *CRC Handbook of Agricultural Productivity, Volume I Plant Productivity.* CRC Press, Boca Raton, FL, pp. 137–142.

Wallace, D.H. and Zobel, R.W. (1995) Whole system research complements reductive research. In: Pessarakli, M. (ed.) *Handbook of Plant and Crop Physiology.* Marcel Dekker, New York, pp. 833–847.

Wallace, D.H., Ozbun, J.L. and Munger, H.M. (1972) Physiological genetics of crop yield. *Advances in Agronomy* 24, 97–146.

Wallace, D.H., Peet, M.M. and Ozbun, J.L. (1976) Studies of CO_2 metabolism in *Phaseolus vulgaris* L. and applications in breeding. In: Burris, R.H. and Black, C.C. (eds) *CO_2 Metabolism and Plant Productivity*. University Press, Baltimore, pp. 43–58.

Wallace, D.H., Gniffke, P.A., Masaya, P.N. and Zobel, R.W. (1991) Photoperiod, temperature and genotype interaction effects on days and node to flowering of bean (*Phaseolus vulgaris* L.). *Journal of the American Society of Horticultural Science* 116, 534–543.

Wallace, D.H., Baudoin, J.P., Beaver, J., Coyne, D.P., Halseth, D.E., Masaya, P.N., Munger, H.M., Myers, J.R., Silbernagel, M., Yourstone, K.S. and Zobel, R.W. (1993a) Improving efficiency of breeding for higher crop yield. *Theoretical and Applied Genetics* 86, 27–40.

Wallace, D.H., Yourstone, K.S., Masaya, P.N. and Zobel, R.W. (1993b) Photoperiod gene control over partitioning between reproductive vs. vegetative growth. *Theoretical and Applied Genetics* 86, 6–16.

Wallace, D.H., Zobel, R.W. and Yourstone, K.S. (1993c) Whole-system reconsideration of paradigms about photoperiod and temperature control of crop yield. *Theoretical and Applied Genetics* 86, 17–26.

Wallace, D.H., Masaya, P.N., Rodriguez, R. and Zobel, R.W. (1995a) Genotype, temperature and genotype × temperature interaction effects on yield of bean (*Phaseolus vulgaris* L.). In: Pessarakli, M. (ed.) *Handbook of Plant and Crop Physiology*. Marcel Dekker, New York, pp. 893–915.

Wallace, D.H., Yourstone, K.S., Baudoin, J.P., Beaver, J., Coyne, D.P., White, J.W. and Zobel, R.W. (1995b) Photoperiod × temperature interaction effects on the days to flowering of bean (*Phaseolus vulgaris* L.). In: Pessarakli, M. (ed.) *Handbook of Plant and Crop Physiology*. Marcel Dekker, New York, pp. 863–891.

Wang, S-Y., Ward, R.W., Ritchie, J.T., Fischer, R.A. and Schulthess, U. (1995) Vernalization in wheat. I. A model based on the interchangeability of plant age and vernalization duration. *Field Crops Research* 41, 91–100.

Watson, D.J. (1952) The physiological basis of variation in yield. *Advances in Agronomy* 5, 101–145.

Watkinson, A.R., Lawn, R.J., Ellis, R.H., Qi, A. and Summerfield, R.J. (1994) *RoDMod: A Computer Program for Characterizing Genotypic Variation in Flowering Responses to Photoperiod and Temperature*. CSIRO Division of Crops & Pastures, Plant Environment Laboratory, Reading University, RG2 9AD, England, 30 pp.

Watt, K.E.F. (1966) *Systems Analysis in Ecology*. Academic Press, New York, 276 pp.

Way, R.D., Sanford, J.C. and Lakso, A.N. (1983) Fruitfulness and productivity. In: Moore, J.N. and Janick, J. (eds) *Methods in Fruit Breeding*. Purdue University Press, West Lafayette, IN, pp. 353–367.

Weaver. R.D. (ed.) (1993) *U.S. Agricultural Research: Strategic Challenges and Options*. Agricultural Research Institute, Bethesda, MD, 380 pp.

Whan, B.R., Carton, G.P., Siddique, K.H.M.J., Regan, K.L., Turner N.C. and Anderson, W.K. (1993) Integration of breeding and physiology: Lessons from a water-limited environment. In: Buxton, D.R., Shibles, R., Forsberg, R.A., Blad, B.L., Asay, K.H., Paulsen, G.M. and Wilson, R.F. *International Crop Science I*. Crop Science Society of America. Madison, WI, pp. 607–613.

Whelan, E.D.P. and Schaalje, G.B. (1992) Vernalization of embryogenic callus from immature embryos of winter wheat. *Crop Science* 32, 78–80.

Whyte, R.O. (1948) History of research in vernalization. In: Murneek, A.E. and Whyte, R.O. (eds) *Vernalization and Photoperiodism*. Chronica Notanica Company, Waltham, MA, pp. 1–38.

Williams, J.H. (1992) Concepts for the application of crop physiological models to crop breeding. In: Nigam, S.N. (ed.) *Groundnut – A Global Perspective*. International Crops

Research Institute for the Semi-Arid Tropics, Patancheru, Andhra Pradesh 502 324, India, pp. 345–352.

Williams, J.H. and Saxena, N.P. (1991) The use of non-destructive measurement and physiological models of yield determination to investigate factors determining differences in seed yield between genotypes of 'desi' chickpeas (*Cicer arietum*). *Annals of Applied Biology* 119, 105–112.

Witzenberger, A., Williams, J.H. and Lenz, F. (1985) Yield, components of yield and quality responses of groundnut cultivars (*Arachis hypogaea*) as influenced by photoperiod and a growth regulator. *Field Crops Research* 12, 347–361.

Witzenberger, A., Williams, J.H. and Lenz, F. (1988) Influence of daylength on yield-determining processes in six groundnut cultivars (*Arachis hypogaea*). *Field Crops Research* 18, 89–100.

Wolokolie, A.Z. (1984) A study of field growth characteristics and photoperiod and temperature effects on flowering and pod set in lima beans (*Phaseolus lunatus* L.) in controlled environments. MS thesis, Cornell University, Ithaca, NY.

Worland, A.J., Appendino, M.L. and Sayers, E.J. (1994) The distribution in European winter wheats of genes that influence ecoclimatic adaptability whilst determining photoperiodic insensitivity and plant height. *Euphytica* 80, 219–228.

Wynne, J.C., Emery, D.A. and Downs, R.J. (1973) Photoperiodic responses of peanuts. *Crop Science* 13, 511–514.

Xiao, J., Li, J., Yuan, L., Tanksley, S.D. (1995) Dominance is the major genetic basis of heterosis in rice as revealed by QTL analysis using molecular markers. *Genetics* 140, 745–754.

Xiao, J., Li, J., Yuan, L. and McCouch, S.R. (1996) Genetic diversity and its relationship to hybrid performance and heterosis in rice as revealed by PCR-based markers. *Theoretical and Applied Genetics* 92, 637–643.

Yan, W-K. (1989) Preliminary investigation into the productivity of wheat genotypes differing in plant height. *Shaanxi Journal of Agricultural Science* 1, 17–20.

Yan, W-K. (1992) The tendency towards spring habit of wheat cultivars in China. *Shaanxi Journal of Agricultural Science* 1, 37–38.

Yan, W-K. (1993) A hypothesis on the interrelations among the traits of wheat and its implication on breeding for higher yield. *Agronomy Abroad – Wheat And Its Relatives* 1, 43–45.

Yan, W-K. (1994) On selection of parents for hybrid wheat breeding. *Seed* 1, 34–35 (in Chinese).

Yan, W-K. and Wallace, D.H. (1995a) Breeding for negatively associated traits. *Plant Breeding Reviews* 17, 141–147.

Yan, W-K. and Wallace, D.H. (1995b) A physiological-genetic model of photoperiod-temperature interactions in photoperiodism, vernalization and male sterility of plants. *Horticultural Reviews* 17, 73–122.

Yan, W-K. and Wallace, D.H. (1996) A model of photoperiod × temperature interaction effects on plant development. *Critical Reviews in Plant Sciences* 15, 63–96.

Yan, W-K. and Wang, C-H. (1992) Two determinants of wheat crop yield potential: single plant productivity and tolerance to density. *Shaanxi Journal of Agricultural Sciences* 4, 30–31.

Yan, W-K., He, P-R. and Liu, S-D. (1991a) Discussions on the methodology of wheat breeding. II. Progeny selections. *Agronomy Abroad – Wheat And Its Relatives* 5, 41–44.

Yan, W., Wang, X. and Wang, H. (1991b) Analysis of leaf area distribution within wheat canopies. *Shaanxi Journal of Agricultural Sciences* 3, 36–37.

Yan, W-K, Zhao, X-K., Wang, K-G. and Liu, Y-B. (1993) A study on wheat developmental phases and developmental patterns of different wheat genotypes. *Acta Universitis Agriculturae Boreali-Occidentalis* 2 (2), 50–56.

Yasuda, S. and Shimoyama, H. (1965) Analysis of internal factors influencing the heading time of wheat varieties. *Berichte des Ohara Instituts Für Landwirtschaffliche Biologie*, Okayama University, Japan 13, 23–38.

Yau, S.K. (1995) Regression and AMMI analysis of genotype × environment interactions: An empirical comparison. *Agronomy Journal* 87, 121–126.

Zobel, R.W. (1990) A powerful statistical model for understanding genotype-by-environment interaction. In: M.S. Kang. (ed.) *Genotype-by-environment Interaction and Plant Breeding*. Louisiana University, Baton Rouge, LA, pp. 126–140.

Zobel, R.W. and Wallace, D.H. (1995) The AMMI statistical model and interaction Analysis. In: Pessarakli, M. (ed.) *Handbook of Plant and Crop Physiology*. Marcel Dekker, Inc. New York, pp. 349–862.

Index

This index is arranged to illustrate the system-established interconnections within and between the biological systems focused on within this book: the physiological genetics of the process of accumulating crop yield, and the physiological genetics of expression of the days to flowering and its carry through effect on the days to harvest maturity, adaptation and yield. The whole-system approach used involves comparing yield and the days to flowering (the system outputs) with the smallest conceivable number of physiological processes that account for each. There are three such components for both days to flowering and for the accumulation of the crop yield. Days to flowering is the short to intermediate to large span of time the genotype uses (requires) to develop to flowering, which usage of time strongly controls the days required to develop to harvest maturity, and thereby strongly establishes each genotype's adaptation and yield for the environment it is grown in.

Most concepts in the text are listed within this index, followed by some pages on which that concept is discussed. For many, there follows the suggestion – *see also* some other indexed concept. These suggestions lead to system-established interconnections to other parts of the overall plant, developmental and yield systems. New concepts and those most emphasized within the text are in bold.

Abbreviations used within this index are: AMMI = additive main effects and multiplicative interaction analysis; DTF = days to flowering; G × E = genotype × environment interaction; GPTmodel = genotype × photoperiod × temperature interaction model; IPCA = interaction principal component axis; P = photoperiod; T = temperature.